ADVANCES IN PHARMACEUTICAL BIOTECHNOLOGY

Volume 1

Exploring Cutting-Edge Innovations and Applications

Advances in Pharmaceutical Biotechnology, 2-volume set:
ISBN: 978-1-774919-76-7 (hbk)

Volume 1: Exploring Cutting-Edge Innovations and Applications
ISBN: 978-1-774919-78-1 (hbk)
ISBN: 978-1-774919-79-8 (ebk)

Volume 2: Current Developments in Biopharmaceuticals
ISBN: 978-1-774919-80-4 (hbk)
ISBN: 978-1-774919-81-1 (ebk)

ADVANCES IN PHARMACEUTICAL BIOTECHNOLOGY

Volume 1

Exploring Cutting-Edge Innovations and Applications

Vivek P. Chavda, MPharm

Zhe-Sheng (Jason) Chen, PhD

Editors

APPLE ACADEMIC PRESS

First edition published 2026

Apple Academic Press Inc.
1265 Goldenrod Circle, NE,
Palm Bay, FL 32905 USA

760 Laurentian Drive, Unit 19,
Burlington, ON L7N 0A4, Canada

CRC Press
2385 NW Executive Center Drive,
Suite 320, Boca Raton FL 33431

4 Park Square, Milton Park,
Abingdon, Oxon, OX14 4RN, UK

Library and Archives Canada Cataloguing in Publication

Title: Advances in pharmaceutical biotechnology / Vivek P. Chavda, MPharm, Zhe-Sheng (Jason) Chen, PhD, editors.
Names: Chavda, Vivek P., editor. | Chen, Zhe-Sheng, editor
Description: First edition. | Includes bibliographical references and indexes. | Contents: Volume 1. Exploring cutting-edge innovations and applications.
Identifiers: Canadiana (print) 20250193361 | Canadiana (ebook) 2025019337X | ISBN 9781774919781 (v. 1 ; hardcover) | ISBN 9781774919798 (v. 1 ; ebook)
Subjects: LCSH: Pharmaceutical biotechnology—Technological innovations.
Classification: LCC RS380 .A38 2025 | DDC 615.1/9—dc23

Library of Congress Cataloging-in-Publication Data

CIP data on file with US Library of Congress

ISBN: 978-1-774919-78-1 (hbk)
ISBN: 978-1-774919-79-8 (ebk)

DOI: 10.1201/9781774919798

Dedication

Dedicated to the 75th year anniversary of the LM College of Pharmacy, Ahmedabad, Gujarat, India

Vivek P. Chavda also wants to dedicate this book to his father and mother. Also, both the editors want to dedicate this book to their students.

About the Editors

Vivek P. Chavda, MPharm
Assistant Professor, Department of Pharmaceutics and Pharmaceutical Technology, L. M. College of Pharmacy, Ahmedabad, Gujarat, India

Vivek P. Chavda, MPharm, is an Assistant Professor in the Department of Pharmaceutics and Pharmaceutical Technology at the L. M. College of Pharmacy, Ahmedabad, Gujarat, India. Before joining academia, he served in the biologics industry for almost eight years in research and development. He has more than 250 national and international publications, nearly 20 book chapters, several books, seven patents in progress as well as numerous newsletter articles to his credit. His research interests include development of biologics processes and formulations, medical device development, nano-diagnostics, long-acting parenteral formulations, and nano-vaccines. He earned his BPharm and MPharm degrees (with gold medals) at Gujarat Technological University, Gujarat, India. He is also listed among the world's top 2% of scientists in 2023 and 2024 by Stanford University and Elsevier.

Zhe-Sheng (Jason) Chen, MD, PhD
Professor, College of Pharmacy and Health Sciences at St. John's University, Queens, New York, USA;
Editor-in-Chief, Journal of Cancer Research Updates

Zhe-Sheng (Jason) Chen, MD, PhD, is currently a Professor at the College of Pharmacy and Health Sciences at St. John's University, Queens, New York, USA. Dr. Chen is an editor-in-chief of *Drug Resistance Updates, Recent Patents on Anticancer Drugs Discovery, Journal of Cancer Research Updates, and New Developments on Chemistry* as well as a guest editor for *Current Pharmaceutical Biotechnology* and an editorial board member of more than 20 journals. He is also a reviewer for more than 300 peer-reviewed journals and an ad hoc reviewer of grants for the National Institutes of Health (USA) and NCI Special Emphasis Review Panel, OmniBus SEP-10, as well as a state grant reviewer for Louisiana and a reviewer for grants for New Zealand, Poland, China, Hungary, Netherlands, etc. He has published around 580 articles in journals such as *Chemical Society*

Reviews, Nature Communications, Cell Research, Advanced Materials, Advanced Science, Molecular Cancer, Drug Resistance Updates, Cancer Research, Journal of Biological Chemistry, etc. His H-index is 97 (Google Scholar list). His publications have been cited more than 39,000 times. He was listed on Stanford/Elsevier's World's Top 2% Scientists List in 2024. As a consultant to pharmaceutical, biotechnology, and health industry organizations, Dr. Chen, commands invitations for presentations, keynote addresses, and conference chairman.

Contents

Contributors

Nasima Ahmed
Department of Pharmaceutical Sciences, Faculty of Science and Engineering, Dibrugarh University, Dibrugarh, India

Farak Ali
Department of Pharmaceutical Sciences, Faculty of Science and Engineering, Dibrugarh University, Dibrugarh, India
Department of Pharmaceutical Chemistry, Girijananda Chowdhury Institute of Pharmaceutical Science-Tezpur, Sonitpur, Assam, India

Shahnaz Alom
Department of Pharmaceutical Sciences, Faculty of Science and Engineering, Dibrugarh University, Dibrugarh, India
Department of Pharmacology, Girijananda Chowdhury Institute of Pharmaceutical Science, Tezpur, Sonitpur, Assam, India

Pankti C. Balar
Pharmacy Section, L.M. College of Pharmacy, Ahmedabad, Gujarat, India

Vishwa Jyoti Baruah
Centre for Biotechnology and Bioinformatics, Faculty of Biological Sciences, Dibrugarh University, Dibrugarh, Assam, India

Rajashri Bezbaruah
Department of Pharmaceutical Sciences, Faculty of Science and Engineering, Dibrugarh University, Dibrugarh, India; Institute of Pharmacy, Assam Medical College and Hospital, Dibrugarh, Assam, India

Xiang Chen
Institute for Biotechnology, St. John's University, New York, USA

Vivek P. Chavda
Department of Pharmaceutics and Pharmaceutical Technology, L.M. College of Pharmacy, Ahmedabad, Gujarat, India

Binay Dalai
Institute of Science, Nirma University, Ahmedabad, Gujarat, India

Sarat K. Dalai
Institute of Science, Nirma University, Ahmedabad, Gujarat, India

Sathvika Kamaraj
Department of Biotechnology, School of Bio Sciences and Technology (SBST), Vellore Institute of Technology, Vellore, Tamil Nadu, India

Rupshikha Malakar
Department of Pharmacology, Girijananda Chowdhury Institute of Pharmaceutical Science, Tezpur, Sonitpur, Assam, India

Tejal Mehta
Department of Pharmaceutics, Institute of Pharmacy, Nirma University, Ahmedabad, Gujarat, India

Toshika Mishra
Department of Biotechnology, School of Bio Sciences and Technology (SBST), Vellore Institute of Technology, Vellore, Tamil Nadu, India

Radhika Pandya
Department of Pharmaceutics and Pharmaceutical Technology, L. M. College of Pharmacy, Ahmedabad, Gujarat, India

Digna Patel
Institute of Science, Nirma University, Ahmedabad, Gujarat, India

Himali Patel
Institute of Science, Nirma University, Ahmedabad, Gujarat, India

Jahanvi Patel
Department of Pharmaceutics, Institute of Pharmacy, Nirma University, Ahmedabad, Gujarat, India

Swati Punetha
Department of Biotechnology, School of Bio Sciences and Technology (SBST), Vellore Institute of Technology, Vellore, Tamil Nadu, India

Bhaswati Sarmah,
Department of Plant Breeding and Genetics, Assam Agricultural University, Jorhat, Assam, India
Plant Biology Section, School of Integrative Plant Science, Cornell University, Ithaca, NY, USA

Jagrut Shah
Institute of Science, Nirma University, Ahmedabad, Gujarat, India

Yesha Shah
PharmD Section, L.M. College of Pharmacy, Ahmedabad, Gujarat, India

Nikita Sharma
Department of Biotechnology, School of BioSciences and Technology (SBST), Vellore Institute of Technology, Vellore, Tamil Nadu, India

Priya Sharma
Institute of Science, Nirma University, Ahmedabad, Gujarat, India

Qiuxu Teng
College of Pharmacy and Health Sciences, St. John's University, New York, USA

Dixa Vaghela
Department of Pharmaceutics and Pharmaceutical Technology, L. M. College of Pharmacy, Ahmedabad, Gujarat, India

Suneetha Vuppu
Department of Biotechnology, School of Bio Sciences and Technology (SBST), Vellore Institute of Technology, Vellore, Tamil Nadu, India

Yuhao Xie
Department of Pharmaceutical Sciences, College of Pharmacy and Health Sciences, St. John's University, Queens, New York, USA

Dong-Hua Yang
New York College of Traditional Chinese Medicine, Mineola, NY, USA

Abbreviations

6-APA	6-amino-penicillanic acid
AAV	adeno-associated virus
ABC	ATP-binding cassette
aBLA	Abbreviated Biologic License Application
AC	affinity chromatography
ACDP	advisory committee on dangerous pathogens
ADA	adenosine deaminase
ADCs	antibody-drug conjugates
ADD	androst-1, 4-dien-3-17-dione
ADEPT	antibody-directed enzyme prodrug therapy
ADP	adenosine di-phosphate
AIDS	acquired immune deficiency syndrome
AMP	adenosine mono phosphate
AMR	antimicrobial resistance
ANDA	abbreviated new drug application
Anti-VEGF	vascular endothelial growth factor
APCs	activation of antigen-presenting cell
API	active pharmaceutical ingredients
ATP	adenosine triphosphate
ATR	Acori Tatarinowii Rhizoma
BC	buffy coat
BCG	Bacillus Calmette-Guerin
BDEPT	bacterial-directed enzyme prodrug therapy
BEHP	bis(2-ethylhexyl) phthalate
BLA	Biologic License Application
BSA	bovine serum albumin
Cas9	CRISPR-associated protein 9
CBD	cell-binding domain
CBER	Center for Biologics Evaluation and Research
CC	column chromatography
CDER	Center for Drug Evaluation and Research

cDNA	complementary DNA
CDSCO	Central Drugs Standard Control Organisation
CE	capillary electrophoresis
CFD	computational fluid dynamics
CGD	chronic granulomatous disease
cGMP	current good manufacturing practices
CHO	Chinese hamster ovary
CNS	central nervous system
CPDA	citrate–phosphate–dextrose–adenine
CRISPR	clustered regularly interspaced short palindromic repeats
CRISPR-Cas9	clustered regularly interspaced short palindromic repeats-CRISPR-associated protein 9
crRNA	CRISPR RNA
CSF	cerebro spinal fluid
CTB	cholera toxin b
CTD	common technical document
CZE	capillary zone electrophoresis
DC	differential centrifugation
DCGI	Drugs Controller General of India
DCs	dendritic cells
DDS	drug delivery system
DDT	dichloro-diphenyl-trichloroethane
DEAE	diethyl amino ethyl
DEPT	directed enzyme prodrug therapy
DFPP	double filtration plasmapheresis
DGC	density gradient centrifugation
DMSO	dimethyl sulfoxide
DNA	deoxyribose nucleic acid
dNTPs	deoxyribonucleotide triphosphates
dsRNA	double-stranded RNA
DTA	diphtheria toxin
EI	electron ionization
EMA	European medicines agency
EORD	early onset retinal disease
EPC	European patent convention
EPO	erythropoietin
EpPCR	error-prone PCR

EPR	enhanced permeability reaction
ETEC	enterotoxigenic *Escherichia coli*
EU	European Union
Fab	fragment antigen binding
FAV	favipiravir
FCC	flash column chromatography
FcRn	Fc receptors
FDA	Food and Drug Administration
FDCA	Federal Food, Drug, and Cosmetic Act
FFP	fresh frozen plasma
FIAs	fluor immunoassays
FISH	fluorescence in situ hybridization
FLU	flumequine
FRET	Forster resonance energy transfer
FSH	follicle-stimulating hormone
FTIR	Fourier Transforms Infrared Spectroscopy
FUS	focused ultrasound system
FZE	free zone electrophoresis
GC	gel chromatography
GC-MS	gas chromatography-mass spectrometry
GCP	good clinical practices
G-CSF	growth factors colony stimulatory factor
GDEPT	gene-directed enzyme prodrug therapy
GFC	gel filtration chromatography
GLC	gas–liquid chromatography
GLP	good laboratory practices
GMOs	genetically modified organisms
GMP	good manufacturing practices
GMP	guanosine monophosphate
GPC	gel permeation chromatography
GPSP	good post-marketing study practices
gRNA	guide RNA
GSC	gas–solid chromatography
HACL	hydroxyacyl-CoA-lyase
HBF	fetal hemoglobin
HBsAg	hepatitis B surface antigen
HBV	hepatitis B virus

HCV	hepatitis C virus
HCV	hepatitis C virus
HDR	homology-directed repair
HES	hetastarch
hGH	human growth hormone
HIV	human immunodeficiency virus
HPLC	high-performance liquid chromatography
HPTLC	high-performance thin-layer chromatography
HSC	hematopoietic stem cells
HSCT	hematopoietic stem cell transplantation
HSV	herpes simplex virus
IEC	ion exchange chromatography
IgA	immunoglobulin A
IgD	immunoglobulin D
IgE	immunoglobulin E
IgG	immunoglobulin G
IgM	immunoglobulin M
IL-1	interleukin-1
IL-2	interleukin-2
IR	infrared spectroscopy
ISCOMs	immunostimulating complexes
IUPAC	International Union of Pure and Applied Chemistry
IZC	isopycnic-zonal centrifugation
LC	liquid chromatography
LC-MS	liquid chromatography- mass spectrometry
LIF	laser-induced fluorescence
LMW-HES	pentastarch
LRS	lactated Ringer's solution
mAbs	monoclonal antibodies
MALT	mucosa-associated lymphoid tissue
MC	Monte Carlo
MCPA	2-methyl-4-chlorophenoxyacetic acid
MDM	maridomycin
MedDRA	medical dictionary for regulatory activities
MHC	major histocompatibility complex
MIBK	methylisobutyl ketone
miRNA	micro RNA

MMHs	mesoporous magnetic hollow nanoparticles
MMPs	matrix metalloproteinases
MNP	magnetically responsive nanoparticles
MPCC	medium-pressure column chromatography
mRNA	messenger RNA
MS	mass spectroscopy
MSP	merozoite surface protein
MYXV	myxoma virus
NAAR	non-native amino acid replacement
NaCl	sodium chloride
nCas	nickase Cas
NCC	normal column chromatography
NCE	new chemical entity
NCI	new clinical investigation exclusivity
NHEJ	non-homologous end joining
NIH	national institution of health
NIR	near-infrared
NK	natural killer cells
NMA	normal mode analysis
NMR	nuclear magnetic resonance
NP-HPLC	normal-phase-high-performance liquid chromatography
NRA	national regulatory authority
OBRP	office of biologics research and review
ODE	orphan drug exclusivity
PAM	protospacer adjacent motif
PAT	process analytical technique
PC	paper chromatography
PC	platelet concentrate
PCBs	polychlorinated biphenyls
PCC	prothrombin complex concentrate
PCR	polymerase chain reaction
PDE	pediatric exclusivity
PDGF	platelet-derived growth factor
PEG	polyethylene glycol
PHS	Public Health Service Act
PID	primary immunodeficiency
PINE-CONE	prime-induced nucleotide engineering creator of new edits

PIP	paediatric investigation plan
PPF	plasma protein fractions
PPP	platelet poor plasma
pRBCs	packed red blood cells
PRP	platelet-rich plasma
PS	polystyrene
PTGS	post-transcriptional gene silencing
PVC	polyvinyl chloride
QBC	quantitative buffy coat
RBC	red blood cells
RCGM	review committee of genetic manipulation
rDNA	recombinant DNA
REM	remdesivir
Rf	retardation factor
RLD	reference listed drug
RNA	ribose nucleic acid
RNAi	ribonucleic acid interference
RP	reference product
RP-HPLC	reverse-phase–high-performance chromatography
RTA	ricin toxin
RZC	rate-zonal centrifugation
SAGM	saline, adenine, glucose, mannitol
SCD	sickle cell disease
SCID	severe combined immunodeficiency
SEC	size exclusion chromatography
SFC	supercritical fluid chromatography
SGE	slab gel electrophoresis
sgRNA	small guide RNA
SID	strong ion difference
siRNA	small interference RNA
SLC	solute carrier transporters
SMA	spinal motor atrophy
SMN	survival motor neuron
SNPs	single-nucleotide polymorphism
SpCas-9	*Streptococcus pyrogens* Cas9
SPR	surface plasmon resonance
TACO	transfusion-associated circulatory overload

TALE	transcription activator-like effector
TCRs	t-cell receptors
TDE	tetrachlorodiphenylethane
TGF-β	transforming growth factor β
TH	t-helper cells
TIL	tumor-infiltrating lymphocytes
TLC	thin-layer chromatography
TLRs	toll-like receptors
tracrRNA	trans-activating crispr RNA
TRALI	transfusion-related acute lung injury
TTI	transfusion-transmitted infections
USFDA	United State Food and Drug Administration
USPTO	United States Patent and Trademark Office
UV	ultraviolet
UV–vis	ultraviolet visible
VACV	vaccinia virus
VEGF	vascular endothelial growth factor
VOC	volatile organic compound
WAS	Wiscott–Aldrich syndrome
WASP	Wiscott–Aldrich syndrome protein
WHO	World Health Organization
ZFN	zinc finger nuclease
ZFP	zinc finger proteins

Foreword

Biotechnology pushes humanity forward on many fronts.

Advancements in technology have allowed scientists to explore new frontiers in their research. Some of us have witnessed groundbreaking methods such as growing proteins in corn or tobacco plants, creating transgenic animals for organ transplants, and attempting to create test-tube embryos. While some of these methods yielded successful results, others have resulted in failure. Through trial and error, researchers have been able to develop groundbreaking medical treatments such as human insulin, growth factors, and human monoclonal antibodies. These developments have revolutionized the medical field and are now considered staples in modern medicine.

As someone with more than 30 years of experience in the biotechnology industry, I can say with confidence that the cell and gene therapy are among the most exciting and promising areas of research today. These innovative treatment methods have the potential to revolutionize healthcare, offering newfound hope to patients with conditions that were once thought to be incurable.

Cell therapy utilizes living cells to treat disease, while gene therapy involves introducing new genetic material into a patient's cells to correct or modify a genetic defect. Both approaches hold immense promise for treating a wide range of diseases, from genetic disorders to cancer and beyond.

I heartily congratulate the authors for their creative and informative compilation of this book. Each chapter presents essential knowledge and key information. This book is a must read for individuals of all levels of expertise, from novices to established scholars. Its pages are brimming with knowledge to be discovered and ideas waiting to be explored.

—Dr. Basant Sharma
Former Global Head of Biotherapeutics Development
Janssen Pharmaceutical, Johnson & Johnson

Preface

Advances in pharmaceutical biotechnology continue to revolutionize drug discovery, development, and therapeutic interventions, integrating cutting-edge innovations with practical applications. This volume aims to provide a comprehensive overview of key technological strides shaping the future of biopharmaceutical sciences.

The opening chapter, *Introduction to Advances in Pharmaceutical Biotechnology*, lays the foundation by highlighting emerging trends and overarching themes driving the field forward. *Advancement in Genetic Engineering* explores the novel genetic manipulation techniques enabling precise control over biological systems for improved therapeutic outcomes. The third chapter, *Advancement in Protein Engineering*, delves into engineering proteins with enhanced functionalities critical for drug design and delivery. *Recombinant DNA Technology* offers insights into the methodologies and applications of recombinant techniques in producing biopharmaceuticals. Finally, *Microbial Biotransformation* presents evolving technologies transforming microbial processes to develop novel bioactive compounds. Together, these chapters provide readers with a multi-faceted understanding of pharmaceutical biotechnology innovations, fostering knowledge to accelerate translational research and clinical applications.

—Vivek P. Chavda
Zhe-Sheng (Jason) Chen

CHAPTER 1

Introduction to Advances in Pharmaceutical Biotechnology

RADHIKA PANDYA[1*], JAHANVI PATEL[2], TEJAL MEHTA[2], and DIXA VAGHELA[1]

[1]*Department of Pharmaceutics and Pharmaceutical Technology, L. M. College of Pharmacy, Ahmedabad, Gujarat, India*

[2]*Department of Pharmaceutics, Institute of Pharmacy, Nirma University, Ahmedabad, Gujarat, India*

**Corresponding author.*

ABSTRACT

The branch of biotechnology has gained increased recognition over the past 20 years. Advances happening in this area have been depicted in the present chapter. Enzymes are biocatalysts, and they have widespread use in various aspects like the textile industry, leather industry, food industry, metabolic engineering, pharmaceutical industry, etc. Various modifications can be made to enzymes, which are now widely used for the treatment of diseases like cancer, tissue damage, and antibacterial and anti-HIV agents. Cell cultures are now used to detect a novel drug or chemical, dosage form's efficacy, and toxicity by using a particular cell type. It is also used to check the selectivity of drugs like anticancer drugs for specific cell types. Recombinant DNA technology is used in producing transgenic plants, animals, vaccines, antibiotics, monoclonal antibodies, etc. OMICS technology has gained popularity in drug discovery because of its toxicity and efficacy studies. Biosensors are devices that convert the biological response into an electrical signal. There are various types of biosensors that are used for innumerable applications. Bioprocess technology, commonly known as fermentation technology, is widely used to produce biopharmaceuticals. Various biotherapeutics are now

Advances in Pharmaceutical Biotechnology, Volume 1: Exploring Cutting-Edge Innovations and Applications. Vivek P. Chavda and Zhe-Sheng (Jason) Chen (Eds.)

DOI: 10.1201/9781774919798-1

commercially available for the treatment of autoimmune diseases and cancer types. Gene therapies are gaining popularity, and various products like Luxturna and Zolgensma are available for the treatment of genetic disorders like inherited retinal disease, spinal muscular dystrophy, and many more. Thus, the amalgamation of bioengineering with science has created a special opportunity to tackle the main problems associated with pharmacology, biology, and human well-being. The increasing quest for research and development, technological advancements, and demand for novel therapies are considered drivers for global advancement in the field of biotechnology.

1.1 INTRODUCTION

Biotechnology is the integration of natural science to achieve the application of cells, microorganisms, plants, and molecular analogues for products and services. The first biotechnology word was used by Karoly Ereky in 1919.[1] According to the European Federation of Biotechnology, "Biotechnology is a set of techniques by which human beings modify living things or use them as tools". The American Chemical Society defines biotechnology as "a technology that utilizes biological systems, living organisms, or parts of them to develop or create different products". It has a huge range of applications such as in the biofuels, food, agricultural, brewing, paper and pulp, and pharmaceutical industries.

1.1.1 HISTORICAL ASPECTS OF BIOTECHNOLOGY

In the last 20 years, the term biotechnology has gained extensive attention. Biotechnology affects and has many applications in fields like food, health, agriculture, biofuel, aquaculture, etc.

If we consider the development of biotechnology, then it can be divided into three phases:

a. **Ancient biotechnology:** All the discoveries before 1800 are called ancient biotechnology.
b. **Classical biotechnology:** Development after 1800 to the 20th century in biotechnology is called classical biotechnology.
c. **Modern biotechnology:** After 1945, important discoveries were reported. So, since then, it has been considered modern biotechnology.

Following Table 1.1 shows the development of modern biotechnology from ancient biotechnology.

TABLE 1.1 Evolution of Biotechnology.

Scientist name	Year	Discovery
Robert Hook	1665	Observed little boxes or cells in a thin slice of cork.
Antoni von Leeuwenhoek	1673	Observed live microorganisms through magnifying lenses.
John Needham	1745	Observed the container full of microorganisms, even though it was heated with nutrient fluids, before filling it into the covered container.
Koelreuter	1761	Successful crossbreeding of crop plants in different species.
Jenner	1798	The first vaccine was discovered (smallpox).
Schleiden and Schwann	1835–1855	"Every cell arises from a cell."
Louis Pasteur (Father of microbiology)	1857	Fermentation process.
	1861	Disproved spontaneous generation theory.
	1864	Invented pasteurization process.
N/A	1857–1914	The golden age of microbiology.
Joseph Lister	1860	Surgical wound infection caused by microorganisms.
Gregor Mendel	1865	Studied garden peas and discovered genetic traits passed from parents to offspring.
Robert Koch	1876	Germ theory of disease.
	1877	A technique for staining and identifying bacteria.
Fleming	1879	Discovered chromatin, the rod-like structure inside the nucleus.
-	1902	The term 'immunology' first appeared.
-	1906	The term 'genetics' first appeared.
Paul Ehrlich	1910	Found salvarsan effective against syphilis.
Evans and Long	1920	Discovered the human growth hormone.
Morgan	1926	Published "Theory of Genes".
Alexander Fleming	1928	Found the antibiotic penicillin.
-	1938	The term 'molecular biology' was confirmed.
Waksman	1944	Isolated the antibiotic streptomycin.
McClintock	1947	Discovered transposable elements or 'jumping genes' in corn.
James D. Watson, Frances H.C. Crick, and Maurice A.F. Wilkins	1953	The structure of DNA was identified.
George Otto Gey	1953	Developed a human HeLa cell line. It was the first immortal human cell.
-	1955	An enzyme involved in the synthesis of nucleic acid was isolated for the first time.
USFDA	1961	The first registered biopesticide was Bacillus thuringiensis.

TABLE 1.1 *(Continued)*

Scientist name	Year	Discovery
Robert Holley Har Gobind Khorana Marshall W. Nirenberg	1968	Genetic code for amino acids.
Paul Berg	1980	Worked on gene splicing (recombinant DNA technology).
FDA supported	1982	The first recombinant protein.
Kary Mullis	1983	Discovered polymerase chain reaction.
Peter G Schultz	1986	Described how to conjugate antibodies and enzymes to produce therapeutics.
Susumu Tonegawa	1987	Explained the genetics of antibody production.
Maynard Olson and colleagues (Washington University)	1987	For large proteins, the expression vectors are the yeast artificial proteins.
Joseph E. Murray and E. Donald Thomas	1990	First successful organ transplants by utilizing immunosuppressive agents.
Richard J. Roberts and Phillips A. Sharp	1993	Found out that genes can be split into DNA fragments.
Michael Smith	1993	Discovered the procedure for the modification of DNA to make a new protein.
Researcher at the Roslin Institute situated in Scotland	1997	Developed a clone of sheep from the cell of an adult ewe named Dolly.
Dr. Lap-Chee Tsui	1999	Discovery of the gene responsible for developing cystic fibrosis.
Craig Venter	2000	Completion of a "rough draft" of the human genome in the Human Genome Project.
Sanger	2002	Rice becomes the first crop to have its genome decoded.
Craig Venter and Team	2003	The Human Genome Project was completed, which gave information on the locations and sequence of human genes on all 46 chromosomes.
Melina Fan, Kenneth Fan, and Benjie Chen	2004	Addgene is a global, nonprofit repository that was created to help scientists share plasmids.
Cedars-Sinai Heart Institute	2009	Cedars-Sinai Heart Institute uses modified SAN heart genes to create the first viral pacemaker in guinea pigs, now known as iSANs.
NA	2012	Thirty-one-year-old Zac Vawter successfully uses a nervous system-controlled bionic leg to climb the Chicago Willis Tower.
Lu You	2019	Scientists report, for the first time, the use of CRISPR technology to edit human genes to treat cancer patients with whom standard treatments were not successful.

TABLE 1.1 *(Continued)*

Scientist name	Year	Discovery
Bryan Smith and team	2020	Scientists demonstrate a "Trojan horse" designer nanoparticle that makes blood cells eat away – from the inside out – portions of atherosclerotic plaque that cause heart attacks.
Rebecca Devine	2021	Scientists report the use of CRISPR/Cas9 genome editing to produce a tenfold increase in superbug-targeting formicamycin antibiotics.
University of Oxford	2021	A malaria vaccine with 77% efficacy after 1 year—and the first to meet the WHO's goal of 75% efficacy—is reported by the University of Oxford.
NA	2022	• The first successful xenogeneic heart transplant, from a genetically modified pig to a human patient. • Researchers report the development of a deep learning software that can design proteins that contain prespecified functional sites.

1.1.2 *DIFFERENT FIELDS OF BIOTECHNOLOGY*

Biotechnology has rapidly emerged as an area having an impact on all aspects of human welfare, ranging from food processing to high-yielding varieties of crops to developing harmless biofertilizers to producing biofuels, bioinformatics and computer science, biological welfare, marine biotechnology and philosophy, and law biotechnology.[2] Fields of biotechnology are explained in Figure 1.1.

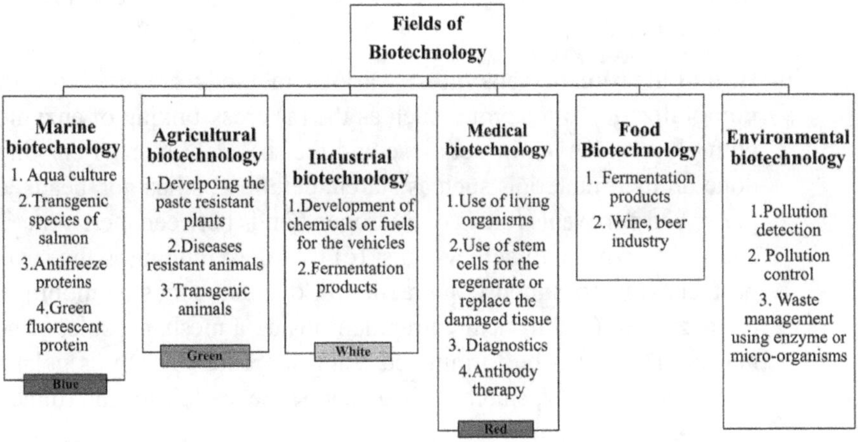

FIGURE 1.1 Various fields of biotechnology.

1.1.3 ENZYME TECHNOLOGY

Enzymes have great selectivity and activity under atmospheric conditions. It led to a chemical reaction without being used in the reaction. Enzymes are used in many bioprocesses. Nowadays, enzymes are used in many industries like food, pharmaceutical, chemical, biofuel industries, etc. Enzyme technology is associated with the isolation, purification, and production of enzymes.[3] Using recombinant DNA technology, the company is now able to produce the enzymes, which are more effective and used in microbiology, biotechnology, medicinal chemistry, diagnostics, biocatalysts, and structural biology to fulfill the requirements of a fast-growing population. Oxidoreductases, isomerases, transferases, ligases, and hydrolases are the classes of enzymes. There are various organisms, which can be used to produce various enzymes like fungi, bacteria, higher animals and plants, yeast, etc.

Following is the procedure for the production of enzymes:

1. **Selection of organism:** The organism selection is done based on the amount of enzyme produced in the minimum time, and it will produce as many as a smaller number of secondary metabolites. Inoculum is added to the sterile nutrient medium and allowed to ferment. The microbial inoculum is provided with suitable growth parameters like pH, temperature, nutrients, oxygen, etc.[4]
2. **Isolation and purification:** The enzyme may be produced within the cell or secreted into the nutrient medium. It is recovered by the cell-rupturing method and purified. During the isolation and purification, it will be taken care that the activity of the enzyme will remain intact.[5]
3. **Immobilization of enzymes:** Various methods are used for the immobilization of enzymes such as the (a) cross-linking of enzymes with glyceraldehyde or cellulose and the (b) adsorption of enzymes done on inert materials such as starch, cellulose, silica gel, beads, or glass. Van der Waals forces or hydrogen bonds between inert material and the enzyme will achieve this. (c) Covalent binding is the strongest enzyme to support interaction. Covalent bonds immobilized the enzymes. (d) Physical entrapment inside a mesh, capsule, or gel matrix. These can be formulated using ingredients such as gelatin, polyacrylamide gel, starch, collagen, silicone, cellulose, and rubber.[6]

Enzymes play an important role in metabolic engineering. Enzyme 2-hydroxyacyl-CoA-lyase (HACL) successfully catalyzes the reaction between formaldehyde, 2-hydroxyisobutyrate, and acetone for the production of glycolate in E.coli. Enzymes can be used for the production of a highly important natural substance, namely, isoprenoids. Prenyl transferase is used for the reaction between the isoprenoid precursors isopentenyl pyrophosphate and dimethylallyl pyrophosphate.[7] Enzyme technology can be used in the following areas:

a. Textile industry, e.g., cellulase is used in the finishing of fabric and for dye uptake in cotton.
b. Enzymes more efficiently remove the stains on the cloth, so they can be used in the detergent industry. Without enzymes, it will take more energy to remove stains. Lipase and amylase are used in the detergent industry.
c. In the leather industry, proteolytic enzymes are used for the dehairing of skin.
d. In the food industry, proteolytic enzymes are used to extend the shelf-life of meat and seafood. Some enzymes are used as preservatives in food like glucose oxidase and catalase. Fructose and fructose-rich syrups are manufactured using glucose isomerase.
e. In the pharmaceutical industry, enzymes like pancreatic enzymes are utilized against the fat malabsorption in HIV patients; chondrotinases treat spinal injuries; hyaluronidase aids in the rebuilding of damaged nerve tissues; lysozyme is used as an antibacterial and anti-HIV agent; and PEGylated arginine deaminase inhibits human melanoma and hepatocellular carcinomas.[7,8]

1.1.4 CELL, TISSUE, AND ORGAN CULTURE

Cell, tissue, and organ culture are laboratory methods that aid the growth of eukaryotic or prokaryotic cells, tissues, and organs in controlled physiological conditions. It is generally done outside the natural environment. While working with cell culture techniques, one must take care of the potential hazards like pathogenic organisms and reagents, which are toxic, corrosive, and mutagenic in nature. This can be ingested through inhalation, ingestion, or absorption. Health and safety executives manage the Advisory Committee on Dangerous Pathogens (ACDP). It publishes guidelines for cellular lab safety. For successful cell culture work, asepsis is the primary or important requirement. So, a distinct or separate room should be dedicated to this work, and no other work should be done in this area.[9] The necessary conditions should be maintained after isolation such as being kept under controlled conditions in an incubator at 37°C. These conditions vary for each cell type, but they should be kept in a nutrient-rich medium.[10] There are three types of cell lines:

a. **Primary cell line:** It is derived from tissue slices. After a few generations, they tend to die out. Examples: fibroblasts obtained from skin biopsies and hepatocytes isolated from liver explants. They are good representatives of the original tissue; they are used in biomedical and translational research.[11]

b. **Diploid cell line:** It is derived from a human embryo. It can be alive for 100 generations. It is used for growing viruses, which require a human host. Example: The rabies virus for the rabies vaccine is called the human diploid culture vaccine.[12]

c. **Continuous cell line:** These are cancerous or transformed cells, which can be maintained for an indefinite number of generations. It is also called an immortal cell line. Example: HeLa cell lines were isolated from a woman (Henrietta Lacks) who suffered from cancer. This is the oldest continuous cell line and was first used to culture and identify poliovirus. Chinese hamster ovary (CHO) and human umbilical endothelial cells are other examples of continuous cell lines. The basic requirement for culturing the cell lines is the selection of acceptable growth conditions. Parameters that are considered for the optimum growth of cell lines are a suitable nutrient medium, oxygen, temperature, pH, and osmotic pressure. Cell lines can be grown either in suspension form or in plated form. Recombinant proteins can more suitably be grown in the suspension, while epithelial cells, in which polarity is an important parameter for the cell functionality, are grown in plated form.[13] Cells grown in suspension are spherical in shape, while cells grown in plates are spiked or polygonal in shape. After growing the cells, quantification and analysis are done. Cell quantification can be done using the trypan blue method. After subculturing, the surplus cells are going to be preserved by freezing in the presence of cryoprotective agents like dimethyl sulfoxide or glycerol. Cell cultures can be used to detect the novel drug or chemical, dosage form's efficacy, and toxicity in particular cell types. It is also used to check the selectivity of drugs like anticancer drugs for specific cell types. Cell cultures are also used for the production of genetically engineered proteins, antibodies, hormones, and biopharmaceuticals. Cell cultures can be used in tissue regeneration. For example, embryonic stem cells and adult stem cells can differentiate into specific cell types that can be utilized in tissue or organ replacement. The impact of inserting the gene into a cultured cell has been examined. So, cell cultures can be used in genetic engineering.[14,15]

1.1.5 GENETICS

Genetics is "the science of heredity." It is the study of what genes are, how they carry information, and how they are replicated and passed to subsequent generations of cells. Modern genetics focuses on the chemical substance in which genes are made of DNA and the way it affects the chemical reactions. The genetic information in a cell is called the genome. The genome is organized into chromosomes. Chromosomes are structures containing DNA that physically carry hereditary information. Chromosomes contain genes. Genes are regions of DNA that code for functional products.[16]

Gene expression is the process by which information from a gene is used in the synthesis of a functional gene product. Products are generally proteins. Gene expression includes the following process:

- Transcription: It is a process in which mRNA is synthesized from DNA. Transcription occurs in the eukaryotic nucleus. The enzyme RNA polymerase attaches to the DNA template and initiates transcription for the production of complementary RNA.[17]
- Translation: From the genetic code on the mRNA, a polypeptide chain is synthesized. During translation, the tRNA reads according to the genetic code and attaches complementary amino acids for the synthesis of proteins.[18]

The language of mRNA is in the form of codons. Codons are a group of three nucleotides such as AUG and AAA. The sequence of codons on mRNA governs the sequence of amino acids. Each codon codes for a specific amino acid. This is called genetic code. There are 64 codons and 20 amino acids. One amino acid has more than one codon, which is known as the degeneracy of the code.[19] The difference between genotype and phenotype is described below in Table 1.2. Genetic mutations can lead to phenotypic changes. The genotype–phenotype mapping can represent the interdependency between the genotype and phenotype.

TABLE 1.2 Differences between Genotype and Phenotype.

Genotype	Phenotype
Genetic makeup	Actual expressed properties
Genetic makeup means it will code the characteristics of an organism	An organism's ability to perform a particular chemical reaction
Collection of genes	Collection of proteins
Represents the potential properties but not the properties themselves	They are the physical properties of a particular organism like the color of the eyes

*Data derived from Ref. [20].

1.1.6 MUTATION

Mutation is any chemical change occurring in a cell's DNA that can alter the sequence of a regular base pair in DNA. Various factors can lead to mutation, including radiation (X-rays), viruses, toxic compounds, heavy metals (mercury, arsenic, and lead), and biological and chemical substances. These factors that cause mutation are called mutagens or mutagenic agents. The detailed mutagens and their examples are described in Table 1.3. It may occur in two types of cells, either in somatic cells (which are not passed on to offspring) or in gametes (eggs and sperm), and may be passed on to the offspring. Almost all mutations are silent, which means they will lead to no change in product activity.[21]

There are two types of mutations as follows.

1.1.6.1 SPONTANEOUS MUTATION

Spontaneous mutations occur naturally during the replication of DNA or during the growing chain of DNA. It occurs in the absence of mutagens due to errors made by DNA polymerase during replication or repair, during recombination, and during the movement of genetic elements that spontaneously damage the actual base pair of DNA structure. Spontaneous mutations play an important role in forward genetics, where base pairs responsible for specific physical characteristics are determined.[22]

Examples: Lactose fermentation in *Escherichia coli*, sickle cell anemia, radiation resistance traits in E. coli, sugary traits in maize are due to spontaneous mutation, etc.

1.1.6.2 INDUCED MUTATION

The induced mutation arises from exposure to any mutagenic agents. Mutagens can cause direct or indirect mutations. Induced mutations can lead to either minor changes or more extensive mutations. Small mutations are more common and involve minor changes in the DNA base pair, while large mutations are less common and result in massive changes in nucleotides.

Examples: UV radiation, gamma radiation, and chemicals are some examples of mutagens.[23]

Physical agents: It involves UV radiation; gamma radiation induces mutations and causes breaks in chromosome structure.

Chemical agents: It involves alkylating agents, nitrous acid, acridine, and base analogues; their chemical properties and reactions cause mutations in DNA.[24]

a. Point mutation:

A point mutation occurs when a single nucleotide base is changed, inserted, or removed from the sequence of the genome. Point mutation happens during DNA replication, modification, recombination, or due to any mutagens. It can be induced by exposure to physical and chemical agents. There are two types of point mutations: transversion mutations and transition mutations.[25]

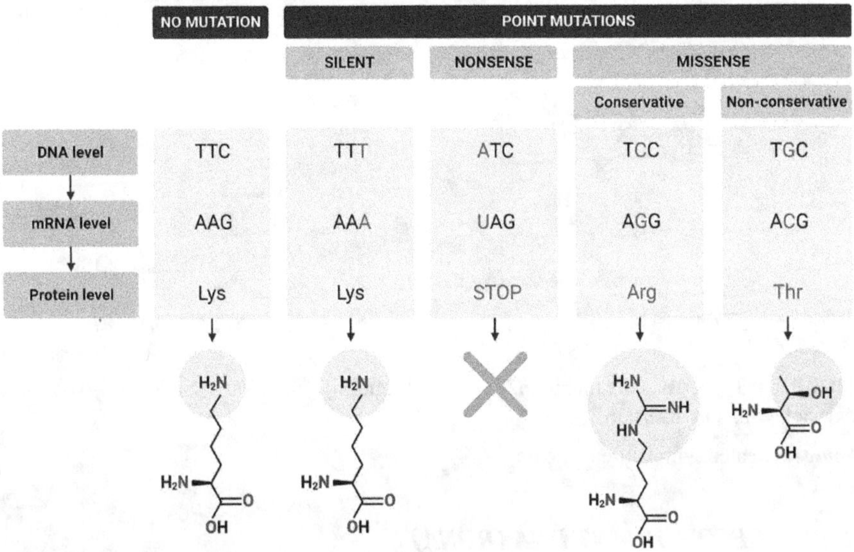

FIGURE 1.2 Point mutations. The silent mutation takes place due to the change of a single DNA nucleotide; it does not affect the sequence of amino acids. Then a nonsense mutation is a pointing mutation in DNA that results in the formation of a stop codon or a nonsense codon. A missense mutation is also a genetic alteration in which a single base pair substitutes with another base. It is of two types: one is conservative, in which the resulted protein is functional, and another is nonconservative, in which the resulted protein is nonfunctional.

Source: Created with Biorender.com

b. Frameshift mutation:

Frameshift mutation in a gene refers to the addition or deletion of a base pair in the DNA. When the normal sequence of codons is disrupted by the addition or deletion of the nucleotides, it can change the reading frame, resulting in a completely different translation from the original. It leads to a disturbed reading frame by the incorporation of many incorrect amino acids or deletions, resulting in the stoppage of protein synthesis from mRNA.[26] Frameshift mutation is illustrated in Figure 1.3 with suitable examples. The mutation may be lethal or beneficial. Enzymes repair many mutations.[27,28]

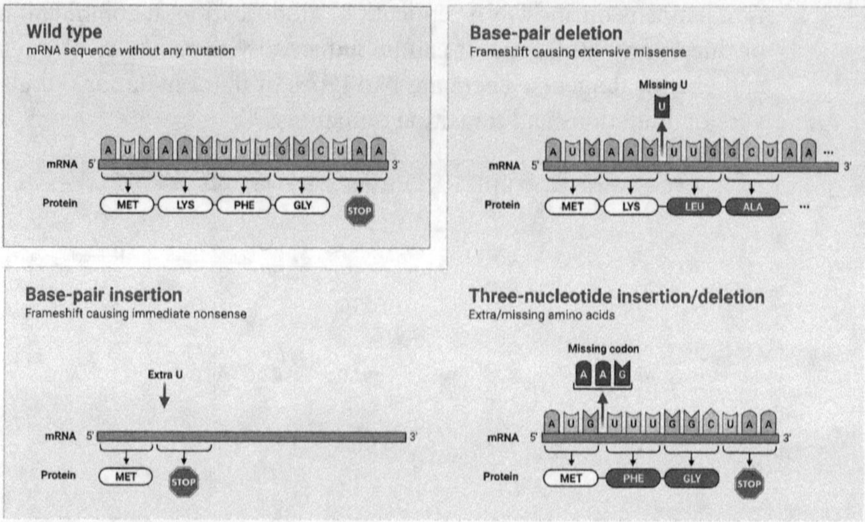

FIGURE 1.3 Frameshift mutation. Types of frameshift mutation include wild-type, base-pair deletion, and insertion.

Source: Created with Biorender.com

1.1.7 RECOMBINANT DNA (RDNA)

Recombination of DNA is a process by which two DNA pieces are broken and recombined to form a new combination of genes. The artificial techniques used to make the rDNA are called rDNA technology or genetic engineering. rDNA can be created by three processes: (a) transformation; (b) conjugation; and (c) transduction.[35]

TABLE 1.3 Examples of Mutagens and Mechanisms Causing Mutation.

Mutagen	Mechanism by which it causes mutation
Nitrous acid[29]	Causes adenine (A) to bind with cytosine (C) instead of thymine (T).
2 Aminopurines (nucleoside analogue of adenine nucleoside)[30]	2-aminopurine is structurally similar to adenine. When it is incorporated into the DNA chain, it will bind to cytosine instead of thiamine, which will form a CG pair instead of an AT pair in the next DNA replication.
Bromouracil (a nucleoside analogue of thymine nucleoside) (an anticancer drug)[31]	Bromouracil is structurally similar to thymine. In the next DNA replication, it will form a GC pair instead of an AT pair as it will bind to cytosine instead of thiamine.
Aflatoxin and benzopyrene[32]	They are examples of frameshift mutagens. They will cause the DNA base pair to be added or removed, which will lead to shifting sequence reading by one base pair.
Gamma radiation and X-rays[33]	It will ionize the molecules, which will lead to the generation of free radicals. Free radicals will combine with the bases of DNA, which will lead to defects in DNA replication.
Ultraviolet radiation[34]	UV light causes dimer formation. UV light will form the covalent bond between two adjacent thymine bases and form a thymine dimer. This formation of dimers will lead to the death of cells as dimers cannot further replicate properly.

1.1.7.1 GENERAL PROCEDURE FOR MAKING RDNA

Plasmid, which is called a vector, is isolated from the bacteria. By using a specific enzyme, the DNA containing the gene of interest is cut. This plasmid now comprises recombinant DNA. The other bacteria take up this plasmid, which is now called transformed bacteria. Transformed bacteria are allowed to multiply. The final aim will either be to collect copies of the gene of interest or to make protein products (enzymes, hormones) from the gene of interest. By using rDNA technology, we can produce many therapeutic products and new vaccines that can improve the health conditions of humans.[36] By rDNA technology, we can prepare the following:

a. Transgenic plants: They are resistant to herbicides and insects.
b. Transgenic animals: They have a greater breeding rate, which gives them commercial benefits.
c. Hormones: Examples: insulin, somatostatin, somatotropin, etc.
d. Vaccines: Examples: polio, malaria, hepatitis, rabies, etc.
e. Interferon: They have antiviral and anticancer activities. Interferon is produced from the gene of the human fibroblast.

f. Antibiotics: A greater amount of antibiotics can be produced by improving the microorganism strains. Examples: penicillin, bacitracin, streptomycin, etc.
g. Industrially important chemicals: Examples: alcohol, citric acid, acetic acid, etc.
h. Monoclonal antibodies: They are produced by the hybridoma technique and utilized for diagnosis.

rDNA technology also helps in gene therapy. To cure the specific disease, the gene is delivered to humans. By rDNA technology, the inserted gene will be expressed in the cell, restoring the normal cell activity.

1.1.8 OMICS TECHNOLOGY

'OMICS' is defined as a set of molecules that analyze a specific large amount of data and constitute the structure and function of a biological system at a particular level.[37] It is also called high-dimensional biology. It will include genomics, transcriptomics, proteomics, and metabolomics. Integration of these technologies will be called system biology. It can be studied not only in the healthy subjects but also in the diseased conditions, where we can study the etiology of diseases. In drug discovery and its toxicity and efficacy studies, OMICS technology has gained popularity.[38,39]

Table 1.4 will give an idea of the different technologies included in the OMICS technology, and Figure 1.4 provides data regarding factors affecting this and the connection of phenotypes with different omics technologies.

1.1.9 BIOSENSORS AND OTHER ADVANCED ANALYTICAL TECHNIQUES

"A biosensor is a device that consists of a biological sensing system that converts the biological response into an electrical signal." It has applications in drug discovery, biomedicine, diagnosis, the food industry, defense, etc. Clark and Lyons have invented the first biosensor for glucose detection in biological samples. It contains an immobilized glucose oxidase electrode, which will detect oxygen or hydrogen peroxide. After that, more sensitive biosensors are invented using electrochemistry and nanotechnology in bioelectronics.[48] Table 1.5 contains types of biosensors along with their applications.

TABLE 1.4 Technologies Included in OMICS Technology.

Technology	Description
Genomics[40]	The branch of molecular biology that deals with the structure, function, evolution, and mapping of genomes is called genomics. It is related to the study of genes. The entire DNA is called the genome of an organism. It contains 3.2 billion bases and 30 to 40 thousand protein-coding genes. Nowadays, DNA microarray technology has gained popularity for DNA sequencing as well as for the expression of thousands of genes. By using this technique, one can reveal the mutations or abnormalities within the chromosomes.
Proteomics[41]	It involves the detection of proteins in the sample. It includes the detection of quantity, modification, and interaction of peptides. Mass spectroscopy is used to detect the quantity of proteins. Interaction between proteins can be detected by two methods, i.e., phage display and a yeast two-hybrid assay. By affinity purification methods, a specific molecule is isolated using an antibody. Then mass spectroscopy is utilized to detect any associated proteins. Then post-translational modification is done for the functions of a large fraction of proteins. Modifications play key roles in intracellular signaling, control of enzyme activity, protein turnover and transport, and maintaining overall cell structure. Comparison between the modified protein and the parent protein can be done by the MS method.
Transcriptomics[42]	The study of mRNA within cells is called transcriptomics. It includes the study of RNA qualitatively as well as quantitatively. Qualitatively means the type of transcript and editing site of RNA. Quantitatively means in how much quantity each transcript is expressed. RNA is considered as the link between DNA and proteins. Packaged mRNA is measured by the gene expression microarray technique.
Metabolomics[43]	Metabolomes are a group of small molecular-weight compounds such as amino acids, fatty acids, and carbohydrates (metabolites) that participate in metabolic reactions. The normal body functioning shows the normal metabolite levels and relative ratios. A diseased condition reflects the abnormal value of the metabolite level and ratios. The detailed study of metabolites in organisms is called metabolomics. The study of metabolites also needs fractionation. Fractionation can be done by gas or liquid chromatography or capillary electrophoresis. After fractionation, mass spectroscopy is generally utilized for the detection of analytes.
Ayurgenomics[44]	According to the Ayurvedic system, every individual is born with his or her basic constitution or Prakriti (vata, pita, and kapha). Depending on the Prakriti, there are differences in individuals in health parameters, vulnerability to infections, and response to external environments, together with food, medicine, and standard of living. Ayurgenomics is exploring whether any genetic bases are there for Prakriti. It combines genetics with ayurveda. Genomic studies show that there is a vast difference in the genomics of different Prakriti individuals. Research reveals that in vata and kapha prakriti individuals, there is lower hemoglobin gene expression in comparison to pita individuals. Metabolic syndrome and chronic inflammation markers occur more frequently in kapha males than in vata males due to overexpression of inflammatory response genes.

TABLE 1.4 *(Continued)*

Technology	Description
	Nowadays, diseases like asthma, cardiovascular diseases, and diabetes are growing very fast. These types of diseases require lifelong treatment and have a variable response to modern drug therapy. For that, one needs personalized and predictive medicine. For that, we need to understand the genetics of these diseases. By knowing the genetics of the diseases, we can predict their progression. Ayurgenomics has been gaining attention in personalized medicine.
Phenomics[45]	Phenomics is the systematic study of phenotypes. It will include biochemical, clinical, and imaging methodologies for the detection of phenotypes. Phenomics needs detailed knowledge of genetics, molecular biology, cell biology, and higher levels of phenotypic expression. It also requires mathematical modeling, statistics, and the information sciences, so phenomics is also called "transdisciplinary". Phenomics has applications in the pharmaceutical and agricultural industries, in phylogenetics, and in metabolic engineering.
Ionomics[46]	Ionomes are the inorganic components of living organisms. It includes minerals and trace elements. Ionomics is defined as the quantitative determination of the ionomes. It needs very sensitive analytical technologies that detect elements precisely. Inductively coupled plasma mass spectroscopy, coupled optical emission spectroscopy, x-ray fluorescence, and neutron activation analysis are some techniques used for elemental analysis. Along with ionomics, there are several genetic tools, and bioinformatics is also required. By ionomics, one can detect the functional state of an organism.

Apart from the above-mentioned applications, biosensors can also be used in plant biology and in the fermentation process to monitor the products, biomass, enzymes, antibodies, or byproducts in the fermentation process. Factors that influence biosensor development are sensitivity, nontoxicity, specificity toward a particular molecule, and the cost of raw materials. Using a definite combination of biosensing materials and biosensing principles will lead to the development of more sensitive and specific biosensors for the modern world.

1.1.10 BIOPROCESS TECHNOLOGY

Bioprocess technology is a processing technology that utilizes complete living organisms or their components (e.g., bacteria, enzymes, and chloroplasts) to get the required products. This process is also commonly referred to as fermentation[57, 58]. Figure 1.5 describes the stages involved in bioprocess technology. Stage 1 is upstream processing, which involves a liquid medium, separation of chemicals, and contains mechanical parts that provide proper

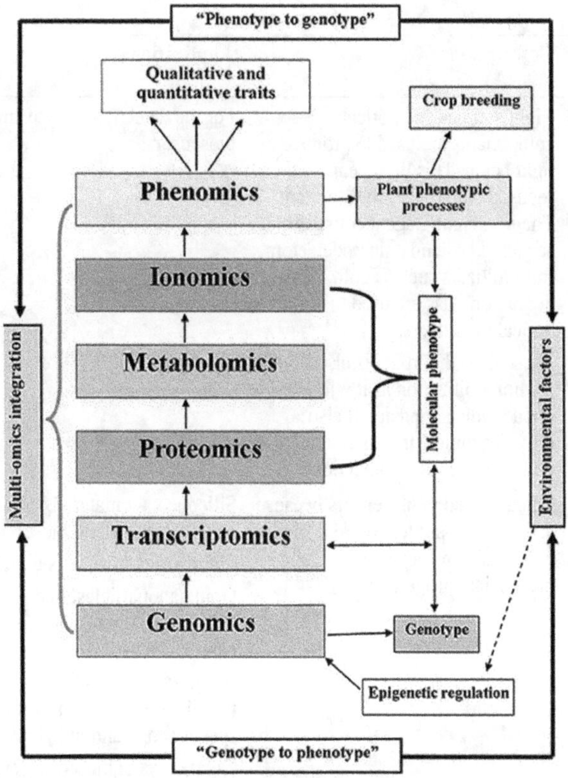

FIGURE 1.4 Types of OMICS technology and linkage with phenotypes.

Source: Reprinted from Ref. [47]. https://creativecommons.org/licenses/by/4.0/

TABLE 1.5 Biosensors and Their Applications.[49]

Types of biosensors	Description	Application
Electrochemical biosensors[50]	Prepared by modification of the surface of metal and carbon electrodes by enzymes, antibodies, or DNA. DNA-based biosensors are more sensitive because they are more gene-specific and do not require the amplification step.	Measurement of glucose, uric acid, and hormones. Detection of miRNA. Detection of pesticidal residue, which will help in environmental monitoring. Piezoelectric biosensor for the detection of organophosphate and carbamate pesticides. Detection of human papillomavirus DNA that is responsible for the invasive cervical cancer by a novel leaky surface acoustic wave peptide nucleic acid biosensor with double two-port resonators.

TABLE 1.5 *(Continued)*

Types of biosensors	Description	Application
Optical or visual biosensors[51]	Light sources and optical components are used to produce a light beam. This light beam is then focused on the modulating agent. The modulating agent consists of a sensing head and a photodetector. Immobilizers such as gold, silica, quartz, or glass are used to make the optical biosensors. These days, hydrogel is also used for immobilization as it will provide good loading capacity. It also provides optical transparency, which will ease the visual detection.	For the detection of cytochrome P450 enzyme biosensors.
Silica, quartz, crystal, or glass biosensors[52,53]	Silica is used in biosensors because it is biocompatible, nontoxic, and has suitable electric, optical, and mechanical properties.	Silicone nanomaterials are used in bioimaging, cancer therapy, and biosensing. Fluorescent silicone is used in bioimaging. Gold nanoparticles, in combination with the silicone nanowires, are used as reagents in cancer therapy. A thiol-modified DNA oligomer attached to the silica forms a film, which is used in UV spectroscopy and in hybridization techniques. Quartz crystal biosensors are used for the detection of proteins in liquids.
Nanomaterial-based biosensors[54]	Gold, silver, silicon, platinum, and copper nanoparticles; graphite; grapheme; and carbon nanotubes. Nanoparticle-based biosensors have great sensitivity.	To target the tumor antigens, semiconductor quantum dots and iron oxide nanocrystals can be used. It can effectively target the tumor antigen because it has optical as well as magnetic properties that are linked with tumor-targeting ligands, such as monoclonal antibodies, peptides, or small molecules.
Synthetic fluorescent or genetically encoded biosensors[55]	Biosensors tagged with genetic codes or synthetic fluorescence can be used to recognize the molecular pathways of cells. Optical-based biosensors with fluorescence and nanomaterials have better sensitivity and a broad range of applications.	To analyze the motor proteins, fluorescent biosensors were developed using a single molecule with a specific analyte concentration. Genetically encoded biosensors are used to understand the physiology of mitochondria. Forster resonance energy transfer (FRET)-based biosensors have been developed for visualizing cGMP, cAMP, and Ca^{2+} in cells. Detection of Kinase, which contains fluorescence resonance energy transfer probes. A FRET-based biosensor provides diagnostic procedures for evaluating the efficacy of imatinib treatment in chronic myeloid leukemia.

TABLE 1.5 *(Continued)*

Types of biosensors	Description	Application
Microbial biosensors[56]	Use microorganisms for specific responses.	Bioremediation To detect the biochemical oxygen demand and toxicity in the atmosphere, a microbial fuel-based biosensor has been used. Pesticidal and heavy metal detection by biosensors are prepared from eukaryotic microorganisms as they have more sensitivity to different toxic molecules and have significance for higher animals. Environmental monitoring and energy production.

conditions inside the chamber, along with the selection of strains that were fermented and the media for their growth included in the upstream stage. Stage 2 fermentation involves the conversion of substrate into the desired product by using microorganisms, and stage 3 is downstream processing involving the separation of cells from the fermentation broth, the purification of the desired product, and waste disposal.[59]

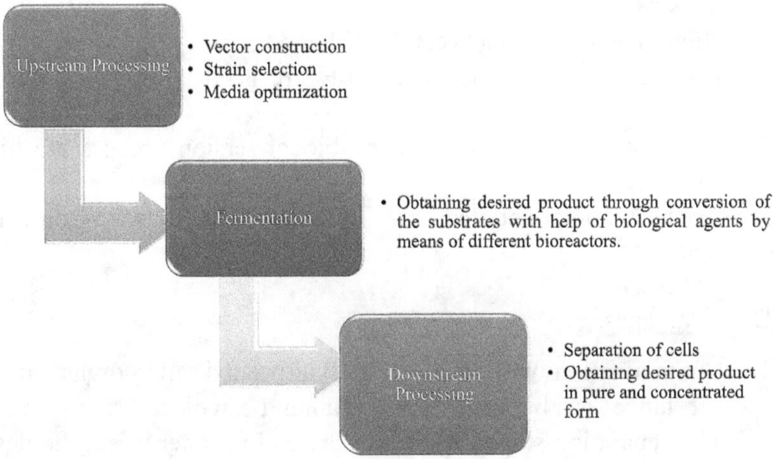

FIGURE 1.5 Flow chart depicting stages of bioprocess technology.

Source: Created with Biorender.com

The main purpose of bioprocess technology is to produce a product of biological origin by utilizing the necessary resources and promoting the growth of microorganisms in a controlled environment. Each step of bioprocessing technology is unique, and different kinds of separation technologies are involved depending on the type of product. The applications of the fermentation process are categorized into five groups. They are as follows:

1. Microbial cell production (biomass) – Examples are Baker's yeast and single-cell protein.
2. Microbial enzyme production – Examples are amylases, proteases, pectinases, and lactases.
3. Microbial metabolites – Examples like ethanol, citric acid, lysine, and vitamins.
4. Recombinant products – Examples include insulin, interferon, and epidermal growth factor.
5. Transformed products – Examples are steroids, antibiotics, vinegar, and prostaglandins.[60]

Advantages and Disadvantages of Bioprocessing in Comparison to a Chemical Process of Manufacturing[61]

Advantages:

1. Complex products like proteins and antibodies can be made through bioprocess engineering, which are difficult to obtain through chemical processes.
2. High yields of products can be obtained.
3. Low temperatures and neutral pH are needed for the systems to operate.
4. The reactions happening in the bioconversion process are highly specific in nature.
5. Isomeric compounds can also be produced exclusively on a large scale.

Disadvantages:

1. Contamination with foreign and unwanted microorganisms and substances is a big problem that cannot be avoided.
2. The separation step is needed at the end in order to get the desired specified substance as the final product obtained from the bioprocessing step is a complex compound.
3. Large volumes of water are needed for the process, so its proper availability, handling, and disposal have to be managed.
4. It is a lengthier process in terms of time consumption when compared with conventional chemical processes.

1.1.11 BIOTHERAPEUTICS

According to the US National Library of Medicine and the National Institutes of Health, biotherapeutics are "antibody–drug cell therapy products where the active substance is extracted or produced from a biological source." The WHO states that "biotherapeutic technology describes biological processes that have been engineered." Growth factors, stem cell and gene therapies, vaccines, monoclonal antibodies, proteins, hormones, etc. are products that fall under the category of biotherapeutics.[62] In recent times, demand and access to such products have increased as they are very effective in treating chronic and life-threatening diseases. This is achieved only due to advancements and improvements in the diagnosis of the diseases. In the last 30 years, there has been a significant advancement in biotherapeutic medicines focused on chronic diseases like hepatitis C, cancer, diabetes, and chronic renal failure, as well as less common ones like hemophilia, Fabry's disease, growth deficiency, multiple sclerosis, and Crohn's disease. The first approved biotherapeutic product was human insulin, known as Humulin, developed through the use of rDNA technology using a culture of E. coli in 1982[63]. The list of other available biotherapeutic products is shown in Table 1.6 below. The delivery of a drug by binding it with an antibody has a four-step mechanism described in Figure 1.6.

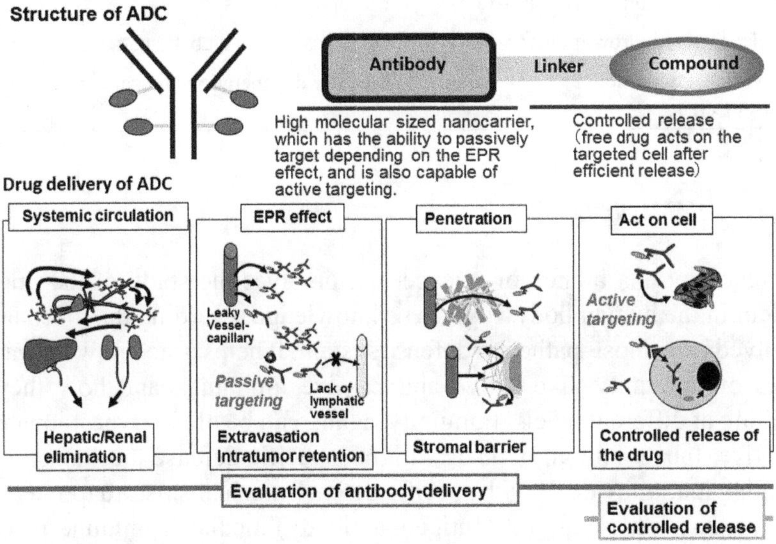

FIGURE 1.6 Structure and drug delivery of an antibody–drug conjugate (ADC). ADC has action in four steps: it reaches systemic circulation, enhanced permeability and retention (EPR), penetration, and action on cells.

Source: Reprinted from Ref. [64]. https://creativecommons.org/licenses/by/4.0/

TABLE 1.6 Available Biotherapeutic Products.[65,66]

Biotherapeutic class	Used in treatment for
Anti-CD20	Rheumatoid arthritis and cancer
Anti-HER2	Cancer
Anti-TNFs	Rheumatoid arthritis, psoriasis, Crohn's disease, and ulcerative colitis
Anti-Interleukin-6	Rheumatoid arthritis
Consensus Interferon	Hepatitis C
Glucagon	Hypoglycemia
Granulocyte Colony-Stimulating Factor	Cancer, neutropenia
Granulocyte-Macrophage Colony-Stimulating Factor	Cancer, bone marrow transplantation
Erythropoietin (EPO)	Chronic anemia
Follicle-Stimulating Hormone (FSH)	Infertility
Human Chorionic Gonadotropin	Infertility
Human Insulin	Diabetes mellitus
Interleukin-2	Cancer
Interferon α-2a and 2b	Cancer, hepatitis
Interferon β-1b and Interferon β-1a	Multiple sclerosis
Platelet-Derived Growth Factor (PDGF)	Ulcers caused in diabetic patients
Antivascular endothelial growth factor (Anti-VEGF)	Macular degeneration and cancer

1.1.12 IMMUNOLOGY

Immunology is a branch of science that involves the study of the immune system in the human body. It gives us knowledge regarding the mechanisms involved in the host–pathogen defense system. It helps us to know the various types of immunity, like active and passive immunity, and how they are working at different levels. Immunity, again, can be of two types: innate and adaptive. Innate immunity deals with nonspecific defense mechanisms, and adaptive immunity deals with specific defense mechanisms. Adaptive immunity consists of a humoral (antigen–antibody)-mediated immune response and a cell-mediated immune response. Figure 1.7 shows the mechanism of action of humoral and cellular immunity to fight against pathogens or foreign invasions.[67]

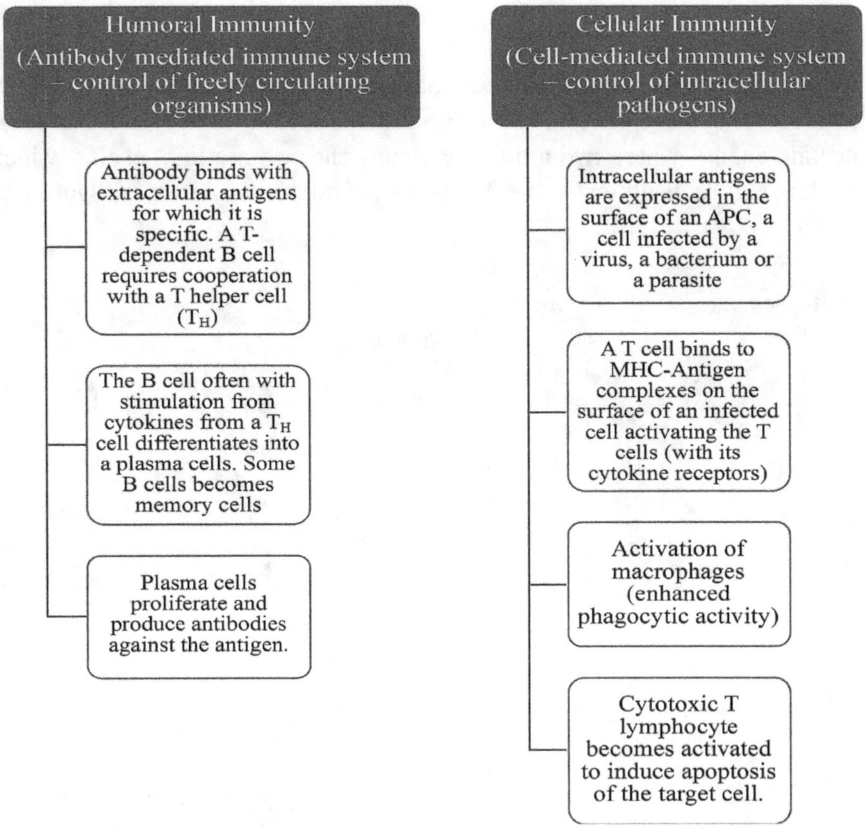

FIGURE 1.7 Diagram representing types of immunity.

Source: Created with Biorender.com

1.1.12.1 HUMORAL IMMUNITY

Antigen–antibody-based immune reactions are called humoral immunity. Here, antigen and antibody interactions play a major role in evading the pathogens. Antigen is a macromolecule that elicits the formation of an antibody that can subsequently react with that antibody. Examples are all proteins, most polysaccharides, nucleoproteins, and lipoproteins. Two essential properties of an antigen are immunogenicity (the ability to stimulate antibody formation) and specific reactivity with antibody molecules. The epitope, or antigenic determinant, is a portion that reacts biochemically with the antibody molecule.[68,69]

Immunoglobulins (Igs) are globular glycoproteins found in the serum fraction of blood tissue fluids and some secretions. Plasma cells derived from activated B-lymphocytes synthesize antibodies in response to the detection of foreign antigens. They are made in response to specific antigens and react with those antigens. Immunoglobulins are of five types such as IgG, IgA, IgM, IgD, and IgE. They have the same basic molecular structure comprising four polypeptide chains, where two are heavy chains and two are light chains, which are linked by a disulfide bond.[70] The detailed structure is shown in Figure 1.8.

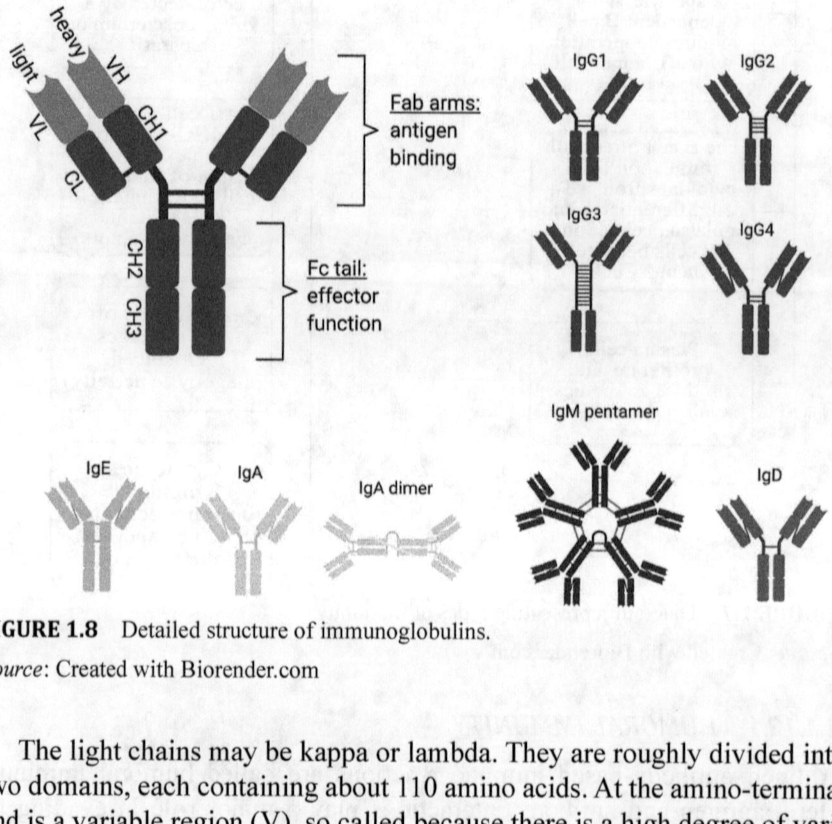

FIGURE 1.8 Detailed structure of immunoglobulins.

Source: Created with Biorender.com

The light chains may be kappa or lambda. They are roughly divided into two domains, each containing about 110 amino acids. At the amino-terminal end is a variable region (V), so called because there is a high degree of variability in amino acid sequence from one lambda or kappa chain to another. Within each V region are three domains that have even greater amino acid variability than the rest of the V region. These three domains are hypervariable regions, and together with the heavy chain hypervariable regions, they form the antigen recognition and combining sites.[71] The carboxy-terminal of a molecule is a constant region (C). The amino acid sequences are always in conservation with both the light chains. Heavy chains are denoted as alpha, gamma, delta, epsilon, and mu and are different from each other. Differences in heavy chains are responsible for five major classes or isotypes. The heavy chains are roughly divided into several domains, each containing approx. 110 amino acids. Gamma, delta, and alpha chains form three domains, while mu and epsilon chains form four domains. The N-terminal domain contains the variable region, within which are three hypervariable regions. The remaining two or three domains of the heavy chain form the C or constant region.[72] The specific characteristics of different classes of Ig are shown in Table 1.7.

1.1.12.2 CELL-BASED IMMUNITY

Cellular immunity consists of different types of T-lymphocytes, or T-cells. They have on their surface the presence of T-cell receptors (TCRs) that identify antigen-presenting cells (APCs) like macrophages or dendritic cells, which have the presence of antigenic peptide bonds. T-cells are activated, and some of them destroy cells that have a particular antigenic peptide (Table 1.8). Others can multiply and increase the secretion of chemical messengers like cytokines. This type of immunity plays an important role in acting against an antigen that has entered a cell.[73] Therefore, in viral infections and some intracellular bacterial infections caused by organisms like Listeria monocytogenes or Mycobacterium leprae, cellular immunity plays an important role in providing protection and fighting them off. Different types of T-cells have different types of glycoproteins on their cell surface known as clusters of differentiation, or CD. Among them, CD4 and CD8 are of greatest interest. If CD4 and CD8 are present on their surface, they are designated as CD4+ and CD8+ cells, respectively. T-helper cells have CD4+, and it binds to MHC class II molecules on B cells and other antigen-presenting cells. Cytotoxic T-cells have CD8+, which binds to MHC class I molecules. Cytotoxic T-lymphocytes, or CTLs, emerge from the thymus gland as CTLp cells and are inert in nature. So, they are not able to destroy any target cells. The activation of the CTLp is required, which is a sequential and complex process whereby an antigen-presenting cell presents an antigen to it along with the aid of a T-helper cell. The processed CTL is an activated cell that recognizes and kills target cells. It can also destroy our body's cells that have been infected by pathogens, especially viruses. The tumor cells and transplanted foreign tissue are also considered to be the target cells. The CTL recognizes endogenous antigens on the target cell's surface that are in combination with an MHC class I molecule. A CTL attaches to the target cell and releases a pore-forming protein, perforin. Pore formation contributes to the subsequent death of the cell. Granzymes and proteases that induce apoptosis are then able to enter through the pore. It is also called programed cell death.[74]

TABLE 1.7 Classes of Immunoglobulins and Their Characteristics.

Properties	**IgG**	**IgM**	**IgA**	**IgD**	**IgE**
Types of Heavy Chains	γ	μ	α	δ	ε
Number of epitopes	2	10	4	2	2
Molecular weight (Daltons)	150,000	900,000	385,000	180,000	200,000
Amount of antibody found in serum	80%	6%	13%	<1%	<1%
Permeable to the placental barrier	Yes	No	No	No	No
Complement fixation	Yes	Yes	No	No	No
Mechanism of action	Neutralization, Agglutination, Complement activation, Opsonization, and antibody-dependent cell-mediated toxicity	Neutralization, Agglutination, and Complement activation	Neutralization and trapping of pathogens in mucus	B-cell receptor	Activation of basophils and mast cells
Major characteristics	Most abundant and combats infecting bacteria and toxin.	An effective agglutinator is produced early in the immune response.	Major in mucosal secretions, protects external body surfaces.	Mostly present on the lymphocyte surface.	Protects external body surfaces and is responsible for atopic allergies.
Location	Blood and lymph	Blood and lymph	Tears, saliva, and mucus	B-cell surface	Mast cells and basophils
Half life	25 days	6 days	5 days	3 days	2 days

TABLE 1.8 T-Cells and Their Function.[75,76]

Cell	Function
T-Helper Cell	T_H1 – Helps in macrophage activation and stimulation of CTLs and natural killer cells. T_H2 – Increases the production of eosinophils, IgM, and IgE. T_H17 – Helps in recruitment of neutrophils and stimulation of antimicrobial protein production.
Cytotoxic T-lymphocyte	Responsible for causing apoptosis in the targeted cells on coming in contact.
T-regulatory cell (T_{reg})	Regulation of the immune response and helps in maintaining self-tolerance.
Macrophages	Enhancement in phagocytic activity and attacking of cancer cells.
Natural Killer Cells (NKCs)	Attacking and destroying target cells, participating in antibody-dependent cell-mediated cytotoxicity.

1.1.12.3 HYPERSENSITIVITY REACTIONS

Hypersensitivity refers to undesirable reactions produced by the normal immune system, including allergy and autoimmunity. Hypersensitivity reactions occur in those people who have been sensitized by previous exposure to allergens. Allergens such as food and environmental changes, parasites, early exposure to microorganisms, and pollen grains cause allergic reactions, and sometimes they may cause asthma-like diseases.[67]

There are four types of hypersensitivity reactions as described below.

1.1.12.3.1 Type 1 Anaphylactic Reactions

Type 1 hypersensitivity reaction, also called immediate hypersensitivity, occurs 15 to 30 min after a sensitized person re-exposes to an antigen, when antigens combined with IgE antibodies, which were located on mast cells. Upon reaction, mast cells and basophiles release granules and chemical mediators. The released mediator histamine is released in blood flow and causes allergic reactions, swelling, redness, and increased mucus secretion, which lead to the contraction of the bronchi and result in breathing difficulty. The allergens that cause hypersensitivity reactions may be harmless or more hazardous such as insect venoms. There are certain risk factors that increase the risk of allergic diseases, including geographical distribution, environmental risk factors, and genetic predisposition. Depending upon exposure and its response, these reactions are further divided into two types: systemic anaphylaxis reactions, which can be fatal sometimes, and localized anaphylactic reactions, which are immediate, temporary, and less severe than systemic anaphylaxis reactions.[77]

Systemic anaphylactic reaction: when mediators are released, it causes dilation of blood vessels and a resulting drop in blood pressure. Other symptoms include narrowing airway passages, skin rashes, and tingling sensations. A self-administration of epinephrin injections is used in treatment, which causes a rise in blood pressure.[78]

Localized anaphylactic reaction: this reaction is associated with antigens that are ingested, and symptoms depend on the route by which antigens enter the body. Allergic reactions involve the upper respiratory tract, in which antigen binds to the IgE antibody, which is located on the mast cells. Upon reaction, mast cells release histamine and cause allergy, while asthma mainly affects the lower respiratory tract. Symptoms such as wheezing and shortness of breath are caused by the contraction of the bronchial tubes. Omalizumab is a monoclonal antibody used to treat allergic asthma or hives.[77]

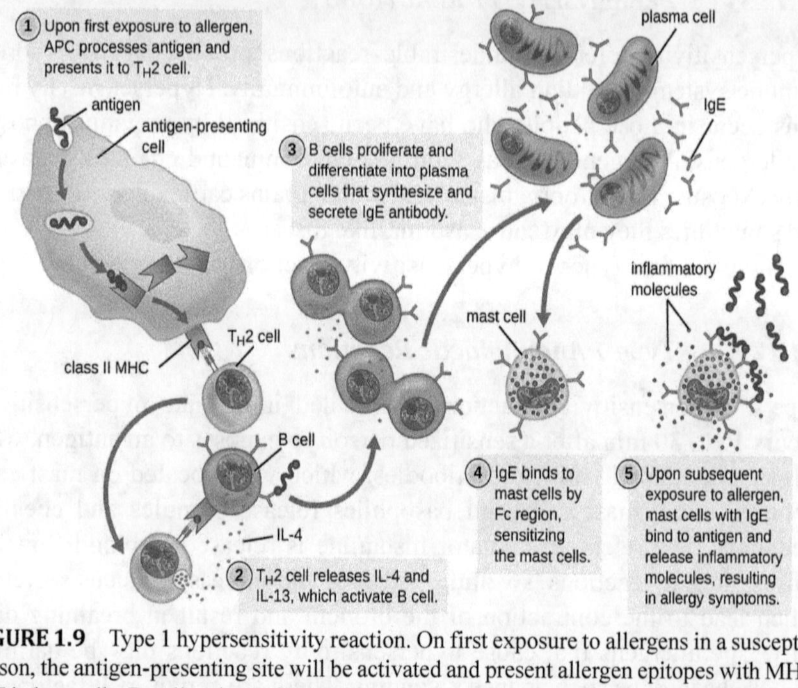

FIGURE 1.9 Type 1 hypersensitivity reaction. On first exposure to allergens in a susceptible person, the antigen-presenting site will be activated and present allergen epitopes with MHC 2 to T-helper cells. B-cells release the cytokines IL-4 and IL-13 to stimulate the proliferation of secreting plasma cells to release IgE antibodies. IgE binds to mast cells, activating them with subsequent exposure to the allergen. With the activation of mast cells, it releases chemical mediators by the degranulation process and causes allergy.

Source: Reprinted from Ref. [79].

1.1.12.3.2 Type 2 Cytotoxic Reactions

Type 2 reactions generally involve the activation of complement by IgG or IgM antibodies. This activation stimulates the complement system and causes cellular damage within 5–8 hours by the action of macrophages and other cells attacking antibody-containing cells (Figure 1.10). The most common cytotoxic hypersensitivity reactions are transfusion reactions, which involve blood group systems and Rh antigens.[80] ABO blood group incompatibility is a common example of a cytotoxicity reaction. ABO blood types are inherited as alleles, and they have been associated with dominant and codominant inheritance. The alleles for A and B blood types are codominant to each other, and both are dominant over blood type O. Therefore, type O blood group individuals can donate blood to type A, type B, and AB blood group individuals, and type AB blood group individuals can take blood from any healthy person. The whole blood is not used in transfusion because antigen–antibody reactions cause agglutination and activate the complement system, which is the cause of the lysis of the donor's RBCs. For example, when a type A blood group person's blood is transfused with type B blood, the antigen on type A blood cells will react with anti-A antibodies in the patient's serum. A detailed explanation is given in Figure 1.11.[81] The Rh incompatibility factor between mother and fetus can also cause a type 2 hypersensitivity hemolytic reaction, referred to as hemolytic disease of the newborn (HDN), as is shown in Figure 1.12. When the father is Rh+ and the mother is Rh-, the baby in the fetus carries Rh+ antigens, which enter the mother's blood during delivery, so the first child has no severe effect, and entering Rh+ antigen into the mother's blood causes the body to develop anti-Rh+ antibodies to fight against those antigens. If the mother becomes pregnant with another Rh+ fetus, her anti-Rh antibodies cross the placenta and damage red blood cells of the fetus by activating complement-mediated hemolysis.[82]

1.1.12.3.3 Type 3 Immune Complex Reactions

Type 3 reactions involve IgG-mediated antigen–antibody complex reactions that are deposited in organs and cause inflammatory damage. This reaction may be general or involve individual organs including the skin, kidneys, joints, or other organs. It is illustrated in Figure 1.13. The reaction may take 3–10 hours after exposure to the antigen. Immune complexes are deposited in the wall of the blood vessel; they activate the complement system and attract inflammatory cells. Enzymes will release from these inflammatory cells and cause damage to the basement membrane. Type 3 reactions can occur in glomeruli

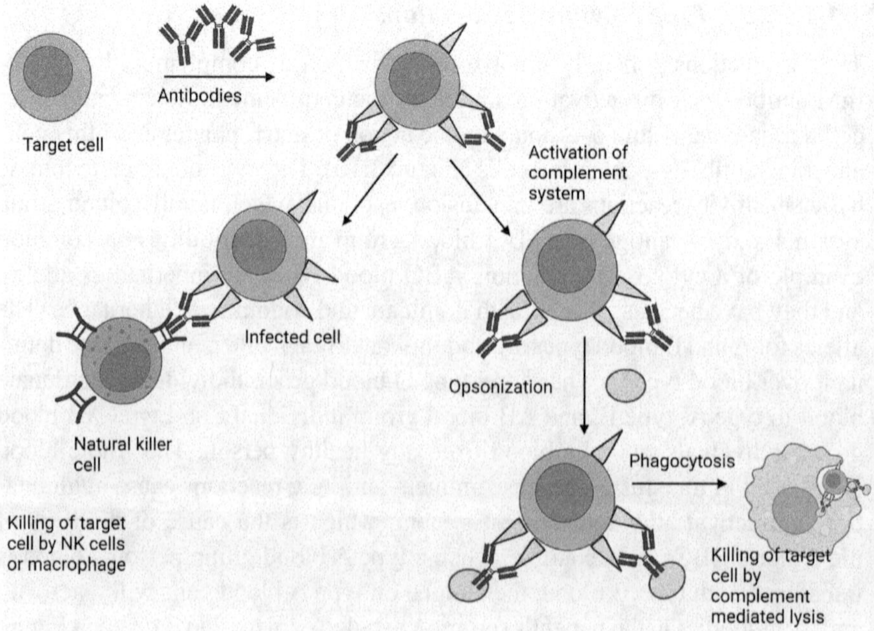

FIGURE 1.10 Type 2 hypersensitivity reaction. It involves the activation of IgG and IgM antibodies. Antibodies attached with antigens activate the complement system, which causes cell lysis, which leads to tissue damage.

Source: Reprinted from Ref. [79]. https://creativecommons.org/licenses/by/4.0/

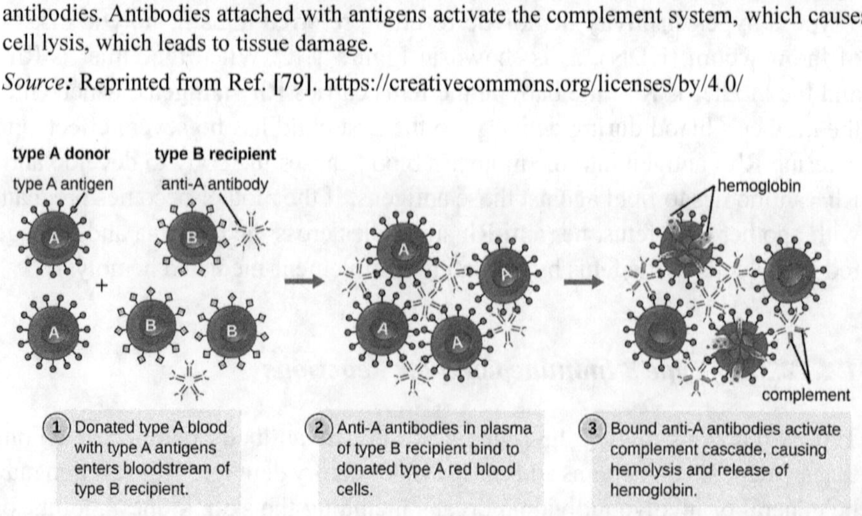

FIGURE 1.11 Type A blood is administered to a patient having blood of type B. An anti-A antibody is activated, which binds to type A blood cells and activates the complement system, which leads to the destruction or lysis of the donor's RBCs.

Source: Reprinted from Ref. [79]. https://creativecommons.org/licenses/by/4.0/

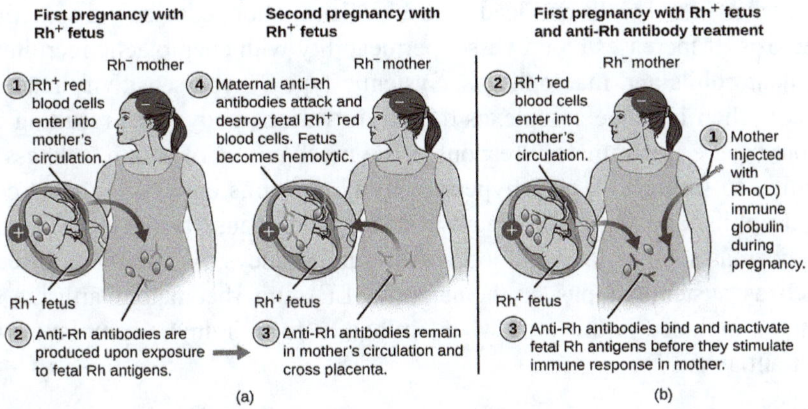

FIGURE 1.12 (A) When a Rh- mother has a Rh+ fetus, before or during birth, fetal erythrocytes enter the mother's circulatory system, leading to the production of anti-Rh IgG antibodies. If a second child has a Rh+, they can cross the placenta and attach to fetal Rh+ erythrocytes. The activation of the complement system causes hemolysis of fetal erythrocytes. (B) Mother is injected with Rho(D) immune globulin during pregnancy to prevent hemolysis.

Source: Reprinted from from Ref. [79]. https://creativecommons.org/licenses/by/4.0/

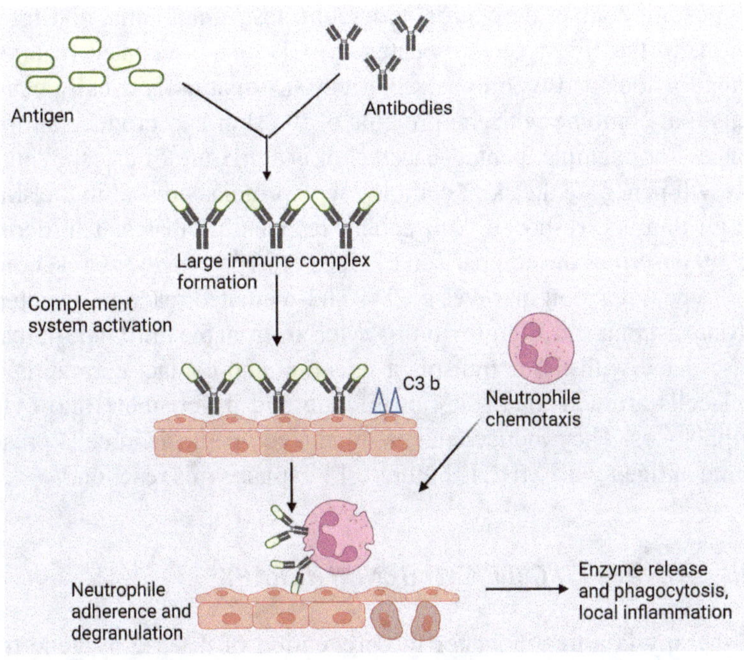

FIGURE 1.13 Type 3 hypersensitivity reaction. It is facilitated by the formation of antigen–antibody complexes, which leads to local inflammation and phagocytosis.

Source: Created with Biorender.com

and other vessel walls and lead to local inflammation, edema, and sometimes necrosis. It increases blood vessel permeability with chemotactic recruitment of neutrophils and macrophages. Systemic type 3 hypersensitivity reactions occur when immune complexes deposit in various body sites, resulting in a more severe inflammatory response. The mechanisms of serum sickness are similar to localized type 3 hypersensitivity reactions but involve mast cells, neutrophils, and the complement system, which cause tissue destruction in body areas such as joints, blood vessels, and kidneys. Autoimmune diseases such as systemic lupus erythematous (SLE) and rheumatoid arthritis can also be included in this type when antibodies form immune complexes with self-antigens.[79,83]

1.1.12.3.4 Type 4 Delayed Cell-Mediated Reactions

Type 4 reactions involve a cell-mediated immune response and are mainly caused by T-cells. A factor in delay is that the participating T-cells and macrophages migrate and accumulate near foreign antigens; the reaction happened after 48–72 h, so it is called a 'delayed cell-mediated reaction.' These reaction types are contact dermatitis, tuberculin test, granuloma, and transplant rejection. In the first type 4 subcategory, CD4 Th1-mediated reactions are described as delayed-type hypersensitivity. Dermatitis is usually caused by haptens that combine with the proteins of the skin and produce an immune response. For example, pentadecacatechol is a mixture of catechol and plant oils. It will penetrate the skin, combine with proteins present in the skin, and cause an immune response. Subsequent responses will result in dermatitis. Table 1.9 describes the comparison of types of hypersensitivity reactions. The second type 4 reaction involves a CD4 Th2-mediated reaction, which results in chronic asthma and allergic rhinitis. Moreover, in the third type 4, reactions are associated with tissue transplant rejection and contact dermatitis. When CD8 T-cells are activated, they proliferate and differentiate into cytotoxic T-lymphocytes. They induce enzyme-mediated apoptosis in cells presenting the same antigen with MHC1. Figure 1.14 explains this reaction.[84]

1.1.13 ADVANCED CELL AND GENE THERAPY

Gene therapy is a treatment for the prevention of disease by gene transfer that involves the use of RNA or DNA (nucleic acids). The mode of therapy is either through the delivery of functional, therapeutic genes for the correction

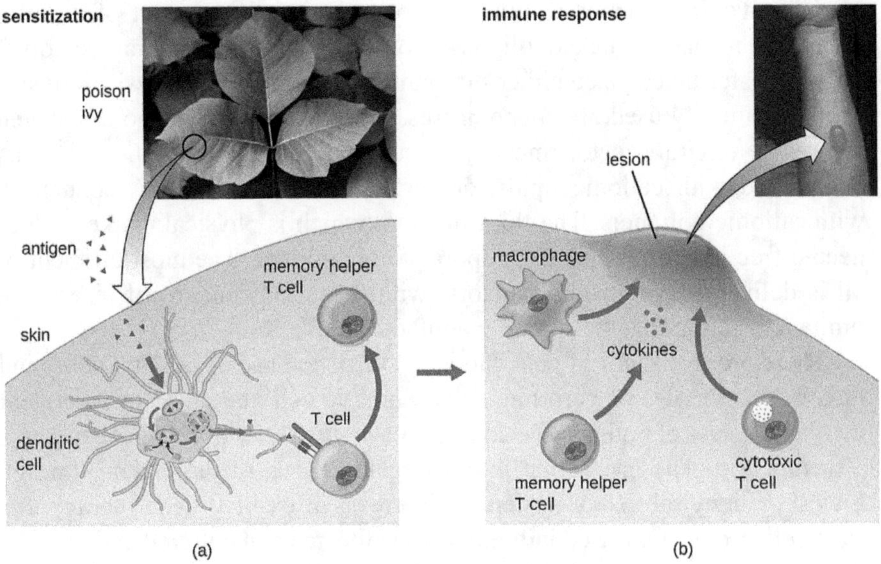

FIGURE 1.14 A poison ivy that comes in contact with the skin on exposure to antigens can cause contact dermatitis. (a) The first exposure does not lead to a hypersensitivity reaction; it only stimulates helper T-cells and leads to the production of memory T-cells. (b) On secondary exposure, memory T-cells become reactivated and will produce inflammatory mediators that induce inflammatory lesions at the exposure site.

Source: Reprinted from Ref. [79]. https://creativecommons.org/licenses/by/4.0/

TABLE 1.9 Comparison of Different Types of Hypersensitivity.

Characteristics	Type 1 anaphylactic	Type 2 cytotoxic	Type 3 immune complex	Type 4 delayed cell-mediated
Antibody	IgE	IgG, IgM	IgG, IgM	None
Antigen	Exogenous	Cell surface	Soluble	Tissue and organs
Appearance	Weal and flare	Lysis and necrosis	Erythema and edema	Erythema and induration
Response time	15–30 min	Minutes to hours	3–8 h	48–72 h
Transferred with	Antibody	Antibody	Antibody	T-cells
Histology	Basophiles and eosinophiles	Antibody and complement	Complements and neutrophiles	Monocytes and lymphocytes
Examples	Allergy, asthma, and hay fever.	Erythroblastosis fetalis	SLE, dermatitis.	Tuberculin test, granuloma.

Source: Ref. [85].

of defective endogenous counterparts or reducing the amount of harmful genes by means of naked oligonucleotides, viral, or nonviral vectors.[86] Gene transfer takes place either through viral, nonviral, or physical means. More commonly used are adenoviruses, retroviruses, and adeno-associated viruses. Nonviral vector means the utilization of naked DNA, DNA complexed with cationic lipids, or particles comprising DNA condensed with cationic polymers. The third method, which is physical, makes use of needle-free injections and electroporation techniques. The most efficient of all is delivery through viral vectors, with the disadvantages of insert size limitation, immunogenicity, and manufacturing.[87]

There are two types of gene therapy. One is the use of somatic cells, and the other is the use of germline cells. Somatic cell gene therapy treatment involves the use of both in vivo and ex vivo methods. In in-vivo gene therapy, a therapeutic gene is inserted into a targeted tissue of the patient through a viral or nonviral vector system.[88] Whereas, in ex-vivo gene therapy, the stem cells from an affected individual are collected and cultured and then the therapeutic genes are inserted in the cultured cells. Once grown in sufficient amounts, these cells with the therapeutic gene are infused or transplanted back into the patient. There is no adverse immunological response occurring in patients' bodies due to the use of a patient's own cells (autologous cells).[89]

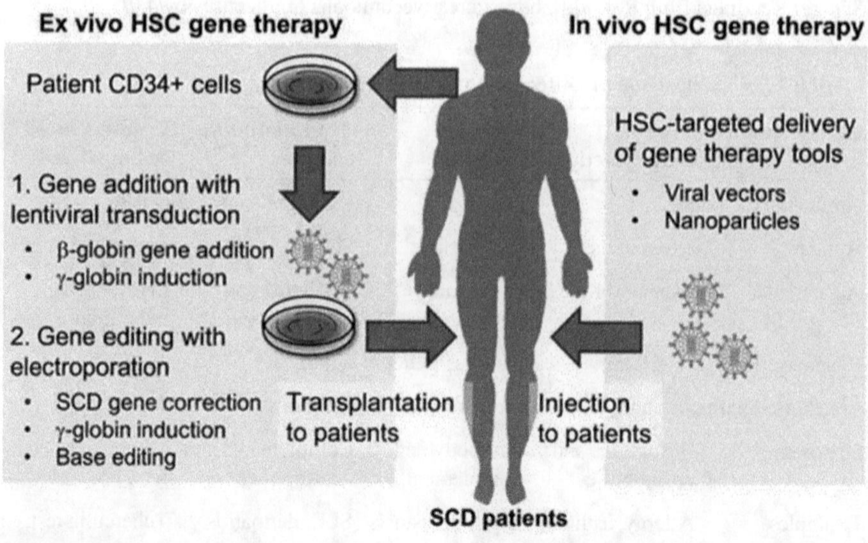

FIGURE 1.15 Differentiation between ex vivo and in vivo HSC-targeted gene therapies in sickle cell disease.

Source: Reprinted from Ref. [90]. https://creativecommons.org/licenses/by/4.0/

The therapy involving the use of germ line cells can result in permanent changes that can be transferred to future generations. It has the potential to bring permanent therapeutic effects to all those who inherit the targeted gene. During embryonic development, if the gene is inserted, it will affect all the cells of the developing embryo. Thus, it has the potential to eliminate particular familial hereditary diseases forever.[91] Exosome-based therapies are also a niche segment of research.[92–94] Gene therapies that are approved by the FDA are listed in Table 1.10.

TABLE 1.10 FDA-Approved Gene Therapies.[95]

Name of the product	Approval year	Disease targeted	Formulation
Kymriah	August 2017	Acute Lymphoblastic Leukemia	CD19-directed genetically modified autologous T-cell immunotherapy (IV infusion)
Yescarta	October 2017	Large B-cell lymphoma and follicular lymphoma	IV infusion
Luxturna	December 2017	Inherited retinal disease	Subretinal injection
Zolgensma	May 2019	Spinal muscular atrophy in children <2 years old	IV infusion
Tecartus	July 2020	Mantle cell lymphoma	IV infusion
Breyanzi	February 2021	Large B-cell lymphoma	IV infusion
Abecma	March 2021	Multiple myeloma	IV infusion

1.1.14 FUTURE OUTLOOK

The invention and exploration related to biotechnology were stated before 200 years, but they have gained more attention since 20 years. Nowadays, biotechnology has applications in almost every field including pharmaceuticals, textiles, food, farming, environmental, forensic, marine, diagnostics, etc. Biotechnology empowers the above-stated industries to produce superior products by utilizing less power and time. Genetics, mutation, cell tissue and organ culture, and recombinant DNA are the bases of biotechnology. We should have in-depth knowledge of these before applying biotechnology. During the application of biotechnology, we should understand the genes, mRNA, proteins, and metabolites in a specific sample. These can be achieved by OMICS technology.[96] The amalgamation of bioengineering with science has also created a special opportunity to tackle the main problems associated with pharmacology, biology, and human well-being. The development of pharmaceuticals using bioengineered tools is well-recognized by the food development authority. 3D cell culture, which uses model human tissue in vivo and not the actual animal test subjects, has attained tremendous growth and will continue to grow in the future for the purposes of drug discovery and safety testing. Integration of omics technology with 3D cell culture and analysis will be an exciting possibility. This new paradigm shift will help to understand millions of protein–protein interactions known as interactomes, along with the structure and function of translated products. The testing of stem cell-derived therapies, which play an important role in regenerative medicine, is also done by 3D cell culture technology.[97] One of the upcoming areas of application of bioprocess technology is the manufacturing of key oligosaccharides. As they have several health benefits, they are predicted to be the superfood of the future. The increasing quest for research and development, higher demand for novel vaccines, progress in the field of technology, and growth and development in the field of biopharmaceuticals are considered to be the most promising drivers in the global bioprocess technology market. In the case of gene therapy, the design of new experimental vectors with increased efficiency, specificity, and greater understanding of inflammatory response induction can result in safer and more clinically applicable therapies for a wide range of health conditions.[98]

KEYWORDS

- **cell culture**
- **OMICS technology**
- **biosensors**
- **biotherapeutics**
- **gene therapy**
- **immunology**

REFERENCES

1. Ravindran, R.; Jaiswal, A. K. Microbial Enzyme Production Using Lignocellulosic Food Industry Wastes as Feedstock: A Review. *Bioengineering* **2016,** *3* (4), 30. https://doi.org/10.3390/bioengineering3040030.

2. Biotechnology: Scope and Branches of Biotechnology. https://www.biologydiscussion.com/biotechnology/branches-biotechnology/biotechnology-scope-and-branches-of-biotechnology/15653 (accessed Feb 10, 2023).
3. Technology Prospecting on Enzymes: Application, Marketing and Engineering - PMC. https://www.ncbi.nlm.nih.gov/pmc/articles/PMC3962110/ (accessed Feb 9, 2023).
4. Barros, F. F. C.; Simiqueli, A. P. R.; Pastore, G. M.; de Andrade, C. J. Production of Enzymes from Agroindustrial Wastes by Biosurfactant-Producing Strains of *Bacillus Subtilis*. *Biotechnol. Res. Int.* **2013,** *2013,* e103960. https://doi.org/10.1155/2013/103960.
5. Overview of the Purification of Recombinant Proteins - PMC. https://www.ncbi.nlm.nih.gov/pmc/articles/PMC4410719/ (accessed Feb 9, 2023).
6. Enzyme Immobilization: An Overview on Techniques and Support Materials – PMC. https://www.ncbi.nlm.nih.gov/pmc/articles/PMC3563746/ (accessed Feb 9, 2023).
7. Katsimpouras, C.; Stephanopoulos, G. Enzymes in Biotechnology: Critical Platform Technologies for Bioprocess Development. *Curr. Opin. Biotechnol.* **2021,** *69,* 91–102. https://doi.org/10.1016/j.copbio.2020.12.003.
8. Schmid-Dannert, C.; López-Gallego, F. Advances and Opportunities for the Design of Self-Sufficient and Spatially Organized Cell-Free Biocatalytic Systems. *Curr. Opin. Chem. Biol.* **2019,** *49,* 97–104. https://doi.org/10.1016/j.cbpa.2018.11.021.
9. Carrel, A.; Burrows, M. T. Cultivation of Tissues In Vitro and its Technique. *J. Exp. Med.* **1911,** *13* (3), 387–396.
10. Segeritz, C.-P.; Vallier, L. Cell Culture. *Basic Sci. Methods Clin. Res.* **2017,** 151–172. https://doi.org/10.1016/B978-0-12-803077-6.00009-6.
11. Vangipuram, M.; Ting, D.; Kim, S.; Diaz, R.; Schüle, B. Skin Punch Biopsy Explant Culture for Derivation of Primary Human Fibroblasts. *J. Vis. Exp. JoVE* **2013,** *77,* 3779. https://doi.org/10.3791/3779.
12. Chen, T. R.; Hay, R. J.; Macy, M. L. Intercellular Karyotypic Similarity in Near-Diploid Cell Lines of Human Tumor Origins. *Cancer Genet. Cytogenet.* **1983,** *10* (4), 351–362. https://doi.org/10.1016/0165-4608(83)90092-4.
13. What is the Difference Between Finite and Continuous Cell Lines - Pediaa.Com https://pediaa.com/what-is-the-difference-between-finite-and-continuous-cell-lines/ (accessed Feb 11, 2023).
14. Kodack, D. P.; Farago, A. F.; Dastur, A.; Held, M. A.; Dardaei, L.; Friboulet, L.; von Flotow, F.; Damon, L. J.; Lee, D.; Parks, M.; et al. Primary Patient-Derived Cancer Cells and Their Potential for Personalized Cancer Patient Care. *Cell Rep.* **2017,** *21* (11), 3298–3309. https://doi.org/10.1016/j.celrep.2017.11.051.
15. Mirabelli, P.; Coppola, L.; Salvatore, M. Cancer Cell Lines Are Useful Model Systems for Medical Research. *Cancers* **2019,** *11* (8), 1098. https://doi.org/10.3390/cancers11081098.
16. Genetics | History, Biology, Timeline, & Facts | Britannica https://www.britannica.com/science/genetics (accessed Feb 10, 2023).
17. Girbig, M.; Misiaszek, A. D.; Müller, C. W. Structural Insights into Nuclear Transcription by Eukaryotic DNA-Dependent RNA Polymerases. *Nat. Rev. Mol. Cell Biol.* **2022,** *23* (9), 603–622. https://doi.org/10.1038/s41580-022-00476-9.
18. Translation: DNA to mRNA to Protein | Learn Science at Scitable https://www.nature.com/scitable/topicpage/translation-dna-to-mrna-to-protein-393/ (accessed Feb 10, 2023).

19. Ahnert, S. E. Structural Properties of Genotype–Phenotype Maps. *J. R. Soc. Interface* **2017,** *14* (132), 20170275. https://doi.org/10.1098/rsif.2017.0275.
20. Genotype vs Phenotype: Examples and Definitions http://www.technologynetworks.com/genomics/articles/genotype-vs-phenotype-examples-and-definitions-318446 (accessed Feb 12, 2023).
21. mutation | Learn Science at Scitable. https://www.nature.com/scitable/definition/mutation-8/ (accessed Feb 11, 2023).
22. Smith, K. C. Spontaneous Mutagenesis: Experimental, Genetic and Other Factors. *Mutat. Res.* **1992,** *277* (2), 139–162. https://doi.org/10.1016/0165-1110(92)90002-q.
23. Somatic Genome Editing - Human Genome Editing - NCBI Bookshelf. https://www.ncbi.nlm.nih.gov/books/NBK447271/ (accessed Feb 9, 2023).
24. Spontaneous vs. Induced Mutations Overview & Examples | What is Spontaneous Mutation? - Video & Lesson Transcript. https://study.com/academy/lesson/spontaneous-vs-induced-mutations.html (accessed Feb 10, 2023).
25. Point Mutation | Genetics | Britannica. https://www.britannica.com/science/point-mutation (accessed Feb 10, 2023).
26. Frameshift Mutations and the Genetic Code. http://symposium.cshlp.org/content/31/77.extract (accessed Feb 10, 2023).
27. Kodym, A.; Afza, R. Physical and Chemical Mutagenesis. *Methods Mol. Biol. Clifton NJ* **2003,** *236,* 189–204. https://doi.org/10.1385/1-59259-413-1:189.
28. Mutagen Types & Examples | What is a Mutagen? - Video & Lesson Transcript https://study.com/academy/lesson/mutagens-how-the-environment-affects-mutation-rates.html (accessed Feb 10, 2023).
29. Weiss, B. Evidence for Mutagenesis by Nitric Oxide during Nitrate Metabolism in Escherichia Coli. *J. Bacteriol.* **2006,** *188* (3), 829–833. https://doi.org/10.1128/JB.188.3.829-833.2006.
30. de Serres, F. J.; Brockman, H. E.; Hung, C. Y.; Overton, L. K. Mutagenicity of 2-Aminopurine, 6-N-Hydroxylaminopurine, and 2-Amino-N6-Hydroxyadenine in Neurospora Crassa. *Basic Life Sci.* **1985,** *31,* 381–389. https://doi.org/10.1007/978-1-4613-2449-2_23.
31. Lasken, R. S.; Goodman, M. F. The Biochemical Basis of 5-Bromouracil-Induced Mutagenesis. Heteroduplex Base Mispairs Involving Bromouracil in G x C----A x T and A x T----G x C Mutational Pathways. *J. Biol. Chem.* **1984,** *259* (18), 11491–11495. https://doi.org/10.1016/S0021-9258(18)90888-4.
32. Preisler, V.; Caspary, W. J.; Hoppe, F.; Hagen, R.; Stopper, H. Aflatoxin B1-Induced Mitotic Recombination in L5178Y Mouse Lymphoma Cells. *Mutagenesis* **2000,** *15* (1), 91–97. https://doi.org/10.1093/mutage/15.1.91.
33. Comparison and Characterization of Mutations Induced by Gamma-Ray and Carbon-Ion Irradiation in Rice (Oryza sativa L.) Using Whole-Genome Resequencing - PMC https://www.ncbi.nlm.nih.gov/pmc/articles/PMC6829151/ (accessed Feb 12, 2023).
34. Mutation Accumulation Under UV Radiation in Escherichia Coli | Scientific Reports. https://www.nature.com/articles/s41598-017-15008-1 (accessed Feb 12, 2023).
35. Khan, S.; Ullah, M. W.; Siddique, R.; Nabi, G.; Manan, S.; Yousaf, M.; Hou, H. Role of Recombinant DNA Technology to Improve Life. *Int. J. Genomics* **2016,** *2016,* 2405954. https://doi.org/10.1155/2016/2405954.
36. Alberts, B.; Johnson, A.; Lewis, J.; Raff, M.; Roberts, K.; Walter, P. Isolating, Cloning, and Sequencing DNA. *Mol. Biol. Cell 4th Ed.* **2002**.

37. Dai, X.; Shen, L. Advances and Trends in Omics Technology Development. *Front. Med.* **2022,** *9*.
38. Micheel, C. M.; Nass, S. J.; Omenn, G. S.; Committee on the Review of Omics-Based Tests for Predicting Patient Outcomes in Clinical Trials; Board on Health Care Services; Board on Health Sciences Policy; Institute of Medicine. *Omics-Based Clinical Discovery: Science, Technology, and Applications*, National Academies Press (US), 2012.
39. Romero, R.; Espinoza, J.; Gotsch, F.; Kusanovic, J. P.; Friel, L.; Erez, O.; Mazaki-Tovi, S.; Than, G.; Hassan, S.; Tromp, G. The Use of High-Dimensional Biology (Genomics, Transcriptomics, Proteomics, and Metabolomics) to Understand the Preterm Parturition Syndrome. *BJOG Int. J. Obstet. Gynaecol.* **2006,** *113* (Suppl 3), 118. https://doi.org/10.1111/j.1471-0528.2006.01150.x.
40. Goldman, A. D.; Landweber, L. F. What Is a Genome? *PLoS Genet.* **2016,** *12* (7), e1006181. https://doi.org/10.1371/journal.pgen.1006181.
41. Meyer, K.; Selbach, M. Peptide-Based Interaction Proteomics. *Mol. Cell. Proteomics MCP* **2020,** *19* (7), 1070–1075. https://doi.org/10.1074/mcp.R120.002034.
42. Kukurba, K. R.; Montgomery, S. B. RNA Sequencing and Analysis. *Cold Spring Harb. Protoc.* **2015,** *2015* (11), 951–969. https://doi.org/10.1101/pdb.top084970.
43. Metabolomics for Investigating Physiological and Pathophysiological Processes | Physiological Reviews https://journals.physiology.org/doi/full/10.1152/physrev.00035.2018 (accessed Feb 9, 2023).
44. Huang, Z.; Chavda, V. P.; Bezbaruah, R.; Uversky, V. N.; P., S.; Patel, A. B.; Chen, Z.-S. An Ayurgenomics Approach: Prakriti-Based Drug Discovery and Development for Personalized Care. *Front. Pharmacol.* **2022,** *13*.
45. Bilder, R. M.; Sabb, F. W.; Cannon, T. D.; London, E. D.; Jentsch, J. D.; Parker, D. S.; Poldrack, R. A.; Evans, C.; Freimer, N. B. Phenomics: The Systematic Study of Phenotypes on a Genome-Wide Scale. *Neuroscience* **2009,** *164* (1), 30–42. https://doi.org/10.1016/j.neuroscience.2009.01.027.
46. Salt, D. E.; Baxter, I.; Lahner, B. Ionomics and the Study of the Plant Ionome. *Annu. Rev. Plant Biol.* **2008,** *59* (1), 709–733. https://doi.org/10.1146/annurev.arplant.59.032607.092942.
47. Yang, Y.; Saand, M. A.; Huang, L.; Abdelaal, W. B.; Zhang, J.; Wu, Y.; Li, J.; Sirohi, M. H.; Wang, F. Applications of Multi-Omics Technologies for Crop Improvement. *Front. Plant Sci.* **2021,** *12*.
48. Biosensors: Design, Development and Applications | IntechOpen https://www.intechopen.com/chapters/76543 (accessed Feb 9, 2023).
49. Mehrotra, P. Biosensors and Their Applications – A Review. *J. Oral Biol. Craniofacial Res.* **2016,** *6* (2), 153–159. https://doi.org/10.1016/j.jobcr.2015.12.002.
50. Cho, I.-H.; Kim, D. H.; Park, S. Electrochemical Biosensors: Perspective on Functional Nanomaterials for on-Site Analysis. *Biomater. Res.* **2020,** *24* (1), 6. https://doi.org/10.1186/s40824-019-0181-y.
51. Damborský, P.; Švitel, J.; Katrlík, J. Optical Biosensors. *Essays Biochem.* **2016,** *60* (1), 91–100. https://doi.org/10.1042/EBC20150010.
52. Heydari, S.; Haghayegh, G. H. Application of Nanoparticles in Quartz Crystal Microbalance Biosensors. *J. Sens. Technol.* **2014,** *04* (02), 81–100. https://doi.org/10.4236/jst.2014.42009.

53. Akgönüllü, S.; Özgür, E.; Denizli, A. Recent Advances in Quartz Crystal Microbalance Biosensors Based on the Molecular Imprinting Technique for Disease-Related Biomarkers. *Chemosensors* **2022,** *10* (3), 106. https://doi.org/10.3390/chemosensors10030106.
54. Su, H.; Li, S.; Jin, Y.; Xian, Z.; Yang, D.; Zhou, W.; Mangaran, F.; Leung, F.; Sithamparanathan, G.; Kerman, K. Nanomaterial-Based Biosensors for Biological Detections. *Adv. Health Care Technol.* **2017,** *3,* 19–29. https://doi.org/10.2147/AHCT.S94025.
55. Wang, M.; Da, Y.; Tian, Y. Fluorescent Proteins and Genetically Encoded Biosensors. *Chem. Soc. Rev.* **2023**. https://doi.org/10.1039/D2CS00419D.
56. Lim, J. W.; Ha, D.; Lee, J.; Lee, S. K.; Kim, T. Review of Micro/Nanotechnologies for Microbial Biosensors. *Front. Bioeng. Biotechnol.* **2015,** *3.*
57. Engineering, N. R. C. (US) C. on B. *Current Bioprocess Technology, Products, and Opportunities*. National Academies Press (US), 1992.
58. Bioprocess Technology Market Latest Trends and Future Opportunities to 2025 | BioSpace https://www.biospace.com/article/bioprocess-technology-market-latest-trends-and-future-opportunities-to-2025/ (accessed Feb 9, 2023).
59. Bioprocess Technology http://bioprocessing.weebly.com/bioprocess-technology.html (accessed Feb 10, 2023).
60. Fermentation - an Overview | ScienceDirect Topics https://www.sciencedirect.com/topics/agricultural-and-biological-sciences/fermentation (accessed Feb 9, 2023).
61. What is the Advantages and Disadvantages of Biological Process(Bioprocess) - Brainly.in https://brainly.in/question/8557494 (accessed Feb 9, 2023).
62. Johnson, D. E. Biotherapeutics: Challenges and Opportunities for Predictive Toxicology of Monoclonal Antibodies. *Int. J. Mol. Sci.* **2018,** *19* (11), 3685. https://doi.org/10.3390/ijms19113685.
63. Sandow, J.; Landgraf, W.; Becker, R.; Seipke, G. Equivalent Recombinant Human Insulin Preparations and Their Place in Therapy. *Eur. Endocrinol.* **2015,** *11* (1), 10–16. https://doi.org/10.17925/EE.2015.11.01.10.
64. Yasunaga, M.; Manabe, S.; Tsuji, A.; Furuta, M.; Ogata, K.; Koga, Y.; Saga, T.; Matsumura, Y. Development of Antibody–Drug Conjugates Using DDS and Molecular Imaging. *Bioengineering* **2017,** *4* (3), 78. https://doi.org/10.3390/bioengineering4030078.
65. Charbonneau, M. R.; Isabella, V. M.; Li, N.; Kurtz, C. B. Developing a New Class of Engineered Live Bacterial Therapeutics to Treat Human Diseases. *Nat. Commun.* **2020,** *11,* 1738. https://doi.org/10.1038/s41467-020-15508-1.
66. Live Biotherapeutics – A Novel Way To Treat Disease | Technology Networks https://www.technologynetworks.com/drug-discovery/blog/live-biotherapeutics-a-novel-way-to-treat-disease-360566 (accessed Feb 12, 2023).
67. Marshall, J. S.; Warrington, R.; Watson, W.; Kim, H. L. An Introduction to Immunology and Immunopathology. *Allergy Asthma Clin. Immunol.* **2018,** *14* (S2), 49. https://doi.org/10.1186/s13223-018-0278-1.
68. Humoral vs Cell-Mediated Immunity | Technology Networks https://www.technologynetworks.com/immunology/articles/humoral-vs-cell-mediated-immunity-344829 (accessed Feb 10, 2023).
69. Charles A Janeway, J.; Travers, P.; Walport, M.; Shlomchik, M. J. The Humoral Immune Response. *Immunobiol. Immune Syst. Health Dis. 5th Ed.* **2001**.
70. Justiz Vaillant, A. A.; Jamal, Z.; Patel, P.; Ramphul, K. Immunoglobulin. In *StatPearls*, StatPearls Publishing: Treasure Island (FL), 2022.

71. Schroeder, H. W.; Cavacini, L. Structure and Function of Immunoglobulins. *J. Allergy Clin. Immunol.* **2010,** *125* (2 Suppl 2), S41–S52. https://doi.org/10.1016/j.jaci.2009.09.046.
72. Introduction to Immunoglobulins - IN https://www.thermofisher.com/in/en/home/life-science/antibodies/antibodies-learning-center/antibodies-resource-library/antibody-methods/introduction-immunoglobulins.html (accessed Feb 10, 2023).
73. Good, R. A.; Lorenz, E. Nutrition and Cellular Immunity. *Int. J. Immunopharmacol.* **1992,** *14* (3), 361–366. https://doi.org/10.1016/0192-0561(92)90165-H.
74. Schori, H.; Yoles, E.; Schwartz, M. T-Cell-Based Immunity Counteracts the Potential Toxicity of Glutamate in the Central Nervous System. *J. Neuroimmunol.* **2001,** *119* (2), 199–204. https://doi.org/10.1016/S0165-5728(01)00358-7.
75. B Cell and T Cell Structure and Function | Immunology. https://irepertoire.com/t-cell-and-b-cell-overview/ (accessed Feb 12, 2023).
76. Human T cell development, localization, and function throughout life - PMC https://www.ncbi.nlm.nih.gov/pmc/articles/PMC5826622/ (accessed Feb 12, 2023).
77. Justiz Vaillant, A. A.; Vashisht, R.; Zito, P. M. Immediate Hypersensitivity Reactions. In *StatPearls*, StatPearls Publishing: Treasure Island (FL), 2022.
78. Lieberman, P.; Nicklas, R. A.; Randolph, C.; Oppenheimer, J.; Bernstein, D.; Bernstein, J.; Ellis, A.; Golden, D. B. K.; Greenberger, P.; Kemp, S.; et al. Anaphylaxis--a Practice Parameter Update 2015. *Ann. Allergy Asthma Immunol. Off. Publ. Am. Coll. Allergy Asthma Immunol.* **2015,** *115* (5), 341–384. https://doi.org/10.1016/j.anai.2015.07.019.
79. 19.1 Hypersensitivities - Microbiology | OpenStax https://openstax.org/books/microbiology/pages/19-1-hypersensitivities (accessed Feb 13, 2023).
80. Bajwa, S. F.; Mohammed, R. H. Type II Hypersensitivity Reaction. In *StatPearls*, StatPearls Publishing: Treasure Island (FL), 2022.
81. Namikawa, A.; Shibuya, Y.; Ouchi, H.; Takahashi, H.; Furuto, Y. A Case of ABO-Incompatible Blood Transfusion Treated by Plasma Exchange Therapy and Continuous Hemodiafiltration. *CEN Case Rep.* **2018,** *7* (1), 114–120. https://doi.org/10.1007/s13730-018-0307-4.
82. Rhesus (Rh) Factor: Incompatibility, Complications & Pregnancy https://my.clevelandclinic.org/health/diseases/21053-rh-factor (accessed Feb 12, 2023).
83. Usman, N.; Annamaraju, P. Type III Hypersensitivity Reaction. In *StatPearls*, StatPearls Publishing: Treasure Island (FL), 2022.
84. Marwa, K.; Kondamudi, N. P. Type IV Hypersensitivity Reaction. In *StatPearls*, StatPearls Publishing: Treasure Island (FL), 2022.
85. Hypersensitivity Reactions - Types - T cell - TeachMePhysiology https://teachmephysiology.com/immune-system/immune-responses/hypersensitivity-reactions/ (accessed Feb 12, 2023).
86. Mountain, A. Gene Therapy: The First Decade. *Trends Biotechnol.* **2000,** *18* (3), 119–128. https://doi.org/10.1016/S0167-7799(99)01416-X.
87. Schäfer, R. Advanced Cell Therapeutics Are Changing the Clinical Landscape: Will Mesenchymal Stromal Cells Be a Part of It? *BMC Med.* **2019,** *17* (1), 53. https://doi.org/10.1186/s12916-019-1289-6.
88. Razi Soofiyani, S.; Baradaran, B.; Lotfipour, F.; Kazemi, T.; Mohammadnejad, L. Gene Therapy, Early Promises, Subsequent Problems, and Recent Breakthroughs. *Adv. Pharm. Bull.* **2013,** *3* (2), 249–255. https://doi.org/10.5681/apb.2013.041.

89. Somatic Gene Therapy - an overview | ScienceDirect Topics https://www.sciencedirect.com/topics/medicine-and-dentistry/somatic-gene-therapy (accessed Feb 10, 2023).
90. Germino-Watnick, P.; Hinds, M.; Le, A.; Chu, R.; Liu, X.; Uchida, N. Hematopoietic Stem Cell Gene-Addition/Editing Therapy in Sickle Cell Disease. *Cells* **2022,** *11* (11), 1843. https://doi.org/10.3390/cells11111843.
91. Kaufmann, K. B.; Büning, H.; Galy, A.; Schambach, A.; Grez, M. Gene Therapy on the Move. *EMBO Mol. Med.* **2013,** *5* (11), 1642–1661. https://doi.org/10.1002/emmm.201202287.
92. Chavda, V. P.; Pandya, A.; Kumar, L.; Raval, N.; Vora, L. K.; Pulakkat, S.; Patravale, V.; Salwa; Duo, Y.; Tang, B. Z. Exosome Nanovesicles: A Potential Carrier for Therapeutic Delivery. *Nano Today* **2023,** *49,* 101771. https://doi.org/10.1016/j.nantod.2023.101771.
93. Wallen M, Aqil F, Spencer W, Gupta RC. Exosomes as an Emerging Plasmid Delivery Vehicle for Gene Therapy. Pharmaceutics. **2023** Jun 27; *15*(7): 1832. doi: 10.3390/pharmaceutics15071832.
94. Krishnan, A.; Gangadaran, P.; Chavda, V. P.; Jogalekar, M. P.; Muthusamy, R.; Valu, D.; Vadivalagan, C.; Ramani, P.; Laishevtcev, A.; Katari, N. K.; et al. Convalescent Serum-Derived Exosomes: Attractive Niche as COVID-19 Diagnostic Tool and Vehicle for MRNA Delivery. *Exp. Biol. Med. Maywood.* **2022,** *247* (14), 1244–1252. https://doi.org/10.1177/15353702221092984.
95. Research, C. for B. E. and. Approved Cellular and Gene Therapy Products. *FDA* **2022**.
96. Sarsaiya, S.; Shi, J.; Chen, J. Bioengineering Tools for the Production of Pharmaceuticals: Current Perspective and Future Outlook. *Bioengineered* **2019,** *10* (1), 469–492. https://doi.org/10.1080/21655979.2019.1682108.
97. Vukovic, N.; van Elsas, A.; Verbeek, J. S.; Zaiss, D. M. W. Isotype Selection for Antibody-Based Cancer Therapy. *Clin. Exp. Immunol.* **2021,** *203* (3), 351–365. https://doi.org/10.1111/cei.13545.
98. Barragán-Ocaña, A.; Silva-Borjas, P.; Olmos-Peña, S.; Polanco-Olguín, M. Biotechnology and Bioprocesses: Their Contribution to Sustainability. *Processes* **2020,** *8* (4), 436. https://doi.org/10.3390/pr8040436.

CHAPTER 2

Advancement in Genetic Engineering

TOSHIKA MISHRA[1], SATHVIKA KAMARAJ[1], NIKITA SHARMA[1], SWATI PUNETHA[1], SUNEETHA VUPPU[1*], and YUHAO XIE[2]

[1]Department of Biotechnology, School of Bio Sciences and Technology (SBST), Vellore Institute of Technology, Vellore, India

[2]Department of Pharmaceutical Sciences, College of Pharmacy and Health Sciences, St. John's University, Queens, New York, USA

**Corresponding author.*

ABSTRACT

In recent times, alteration or modification of specific genes is responsible for unique desired properties and characteristics of organisms. Genetic engineering is a popular approach widely used in various domains of biology to manipulate the genes of an organism directly, using the tools of biotechnology. It has applications in the field of agriculture, medicine, research, industry, environment, pharmaceutical sector, etc. The last decade saw the development of innovative, novel genome modification approaches that are efficient in the effective manipulation and regulation of genes of different types of genomes. The scientific community all around the world is constantly working to bring out innovative techniques in genetic engineering. Some popular advancements in genetic engineering that are widely used globally include recombinant DNA (rDNA) technology, fluorescence in-situ hybridization (FISH), DNA microarray, Clustered Regularly Interspaced Short Palindromic Repeats-(CRISPR) associated protein 9 (CRISPR-Cas 9) system, Promoter Engineering, Metabolic Engineering, RNA interference (RNAi), etc. However, with the growing popularity of genetic engineering techniques, the ethical issues linked to it cannot be ignored, as it interferes with nature and living forms. This chapter focuses on the design and development of advanced, innovative, novel techniques in genetic engineering along with their advantages and limitations and also newly developed in-silico tools.

Advances in Pharmaceutical Biotechnology, Volume 1: Exploring Cutting-Edge Innovations and Applications.
Vivek P. Chavda and Zhe-Sheng (Jason) Chen (Eds.)

DOI: 10.1201/9781774919798-2

2.1 INTRODUCTION

Genetic engineering is among the recent developments in the domain of Biology. It includes several techniques that focus on the manipulation of genetic material (predominantly DNA) for the alternation, repair, or enhancement of its function and structure.[1] The techniques involve the incorporation of nucleic acid molecules, coding for the selected traits into the host genome using vectors like plasmids, bacteriophages, etc.[2]

Genetic engineering is a relatively new technology, which emerged from the knowledge of Genetics and Molecular Biology. It all started when the first civilizations began artificial selection to domesticate plants and animals. However, major scientific work in this field was initiated by Mendel's experiments and his proposal of "factors" of inheritance followed by a detailed study of inheritance principles and genetic mapping.[3] This was followed by confirming DNA as the genetic factor and a comprehensive understanding of the process of transfer of genetic material in bacteria. Then came the breakthrough discovery of the double-helical DNA conformation by Watson and Crick which initiated the study of genetics at the molecular level. Further, a base for the new field of genetics was set with the determination of genetic code in the year 1966.[4] This new field was Genetic Engineering which provided tools and techniques to manipulate the genome of organisms and insert the desired gene of interest from foreign sources in them. The discovery of DNA Ligases, restriction enzymes, the development of DNA sequencing, and DNA cloning techniques gave scientists the idea to alter or manipulate chromosomal DNA in a cell of any organism.[5] Further, the completion of the Human Genome Project in 2003 facilitated the identification of any human gene of interest.[6] The timeline for discovery and advancements in genetic engineering is shown in Figure 2.1.

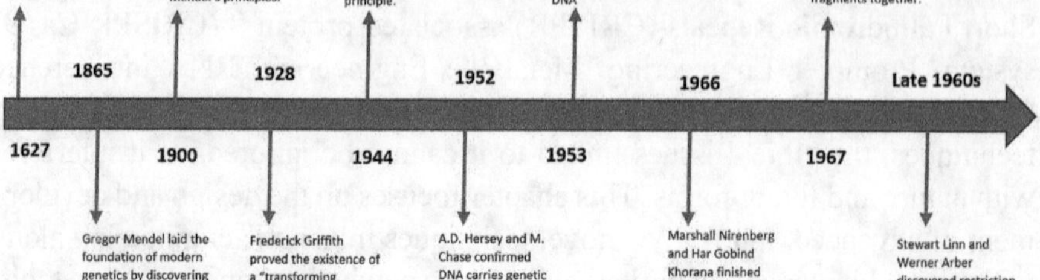

FIGURE 2.1 Timeline of genetic engineering research.

Source: Created with Biorender.com

In 1985, scientists discovered zinc-binding regions in proteins with finger-like conformations to hold the DNA, which are now known as Zinc-finger proteins (ZFP).[7] In 1994, work on ZFP proved that they could be modified to identify and suppress an oncogene, can be used as a promoter for other genes, and can also regulate gene expression by binding to effector proteins.[8] Based on this knowledge, the scientists fused ZFP with a restriction enzyme, Fok1, at a nonspecific cleavage domain, naming this heterodimeric complex as Zinc-finger nuclease (ZFN).[9] This can recognize a sequence and introduce double-stranded breaks which can be used to insert repaired or modified genes. In 2009, scientists discovered a similar DNA-binding motif called Transcription activator-like effector (TALE) which was obtained from the plant pathogen, *Xanthomonas*. These can also be used for the regulation of genes by combining with effectors or endonucleases. However, they have more target-specificity and efficiency than ZFN.[6]

The prokaryotic defensive mechanism against pathogens known as the Clustered Regularly Interspaced Short Palindromic Repeats-CRISPR associated protein 9 (CRISPR-Cas 9) system was found in 2007.[10] With continuous research on this system, by 2013, scientists acclaimed CRISPR-Cas 9 as an efficient, simple, and multiplexable technique for gene editing including genes of the human genome. It is a highly preferred technique for genetic and epigenetic alteration.[6] The details of this technique are discussed further in the chapter.

Further development of genetic engineering techniques resulted in various advancements to reach its present form. Various techniques have been developed to transfer the DNA to the host cell. These can be categorized into physical, chemical, and biological methods.

The widely established and used physical methods include techniques like Electroporation, Ultrasonication, Microinjection, Gene gun or Biolistics, Liposome fusion, silicon carbide fibers, etc.[11] Chemical methods of gene transfer include Poly-ethylene glycol (PEG)-mediated, diethyl amino ethyl (DEAE) dextran-mediated, calcium phosphate precipitation, etc.[12] The physical and chemical methods are called direct gene transfer methods. However, the biological methods largely comprise vector systems for the transfer. Plasmids and viruses are the most often utilized vectors.

Genetic manipulation is a tool that can be utilized in various fields and facilitate the design and development of solutions. It has numerous applications in the pharmaceutical industry and is crucial to the creation of innovative products like vaccines. It is also used in gene therapy for treatment of serious medical conditions. Microbiology-based industries use microbial culture as a primary raw material for strain improvement. These techniques also find applications in agriculture to incorporate desired traits like high yield, disease, pest or stress resistance, etc., in the plants.[13]

Over time, a lot of work has been done in this field giving rise to numerous novel techniques and procedures with greater efficiency. This chapter aims to discuss some major advancements that are breakthrough discoveries for society and have important utilization in numerous applications which include the study of phenotypic characteristics (observable features) on development of genotypic variations (alterations in the genetic material that are transferred via gametes) caused by insertion and deletion (knockout). This approach is useful in determining the molecular aspects of physiological properties and behavioral alterations in the conditions of health or disease.[14]

2.2 RECOMBINANT DNA TECHNOLOGY

The phenotype of an organism is altered when a genetically modified vector is inserted and integrated into the genomic makeup of the organism (host). In essence, this procedure involves splicing a foreign piece of DNA structure into the genome that houses our target gene. The gene inserted is termed recombinant gene while the technology is named recombinant DNA technology.[15] The identification of a new family of enzymes called "restriction endonucleases" served as the impetus for the advancement of recombinant DNA technology.[16] As DNA is ubiquitous in its presence, it is possible to clone genes from any organism, whether they are of an animal, plant, bacterial, or viral origin. Cloning genes in hosts from plants and animals became feasible with the investigation of additional cloning vectors, such as plasmids that could integrate into host genomes. Thereby, "designer" chimeric organisms comprising specific genes that confer desired characteristics can potentially be created in the laboratory.[16]

This can be accomplished using one of four methods:

- Transformation
- Transduction
- Nonbacterial transformation
- Bacterial conjugation

2.2.1 TRANSFORMATION

Step 1: Choose a desired DNA of interest to be incorporated into a vector.

Step 2: The restriction enzyme is then used to cut that fragment of DNA.

Step 3: Using DNA Ligase, the vector is ligated with the DNA. A selectable marker is included, which allows recombinant molecules to be identified. Following transformation, when exposed to a specific antibiotic, a host cell without a vector dies, whereas the host with the vector survives as it is resilient.

Step 4: Then the vector is incorporated into the host cell and this process is termed Transformation.[17] Under ideal conditions, the recombinant DNA multiplies and is introduced into the host cell, where it is synthesized as the produced protein. The steps of vector-based recombination are demonstrated in Figure 2.2.

To accept the foreign DNA, the host cells must be particularly prepared. Antibiotic resistance, color changes, or any other characteristic that distinguishes transformed from untransformed hosts can all be used as selectable markers.[18] Various vectors have unique features that make them ideal for a wide range of applications. Asymmetrical cloning sites, size, and a high copy number are examples of properties.[17]

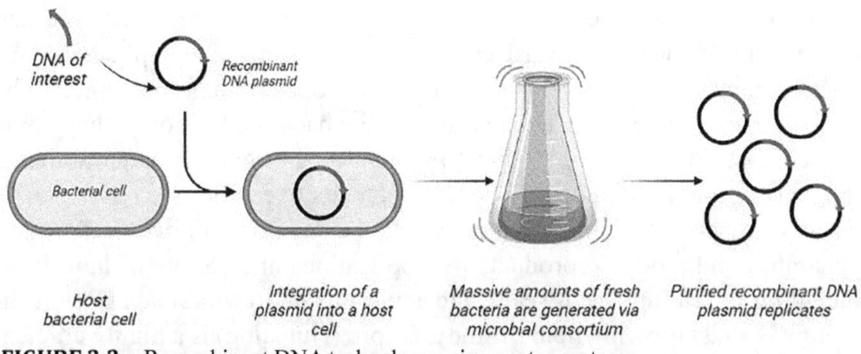

FIGURE 2.2 Recombinant DNA technology using vector systems.

Source: Created with Biorender.com

2.2.2 NONBACTERIAL TRANSFORMATION

Nonbacterial transformation is quite similar to that of the transformation procedure. This is a method where it does not require bacteria as a host. Using microinjection, biolistics, or other methods, the DNA gets directly inserted into the genome of the host cell.[17]

2.2.3 TRANSDUCTION

This method is similar to that of transfection, except that instead of bacteria, phages are being used.[17] The vector is packaged in vitro. This method employs lambda or M13 phages to generate phage plaques comprising recombinants. Recombinants can be distinguished from non-recombinants through differences brought about by various selection techniques.[17]

2.2.4 BACTERIAL CONJUGATION

Horizontal gene transfer is the exchange of genetic material between living systems (HGT). It is the main mechanism by which bacteria have acquired virulence and antibiotic resistance, and it is also important for bacterial evolution.[19] When Joshua Lederberg and Edward Tatum noticed genetic recombination between two nutritionally deficient *E. coli* in 1946, they knew they had found the first method of gene transfer that had been thoroughly studied. *E. coli* strains produced wild-type *E. coli*.[20]

A donor and a recipient bacterium are brought into direct contact to exchange genetic material during conjugation. The donor bacterium contains a segment of DNA recognized as the Fertility Factor (F-factor). The donor bacterium can create sex pilus (tiny "bridge") that adheres to the recipient cell and pulls the cell toward episome, which can either replicate on its own or combine with other pieces to form a bacterial chromosome.[20] The donor can exchange genetic material with the bacterial host once they come into contact. Plasmids are the most frequent type of genetic material that is transferred, and they can confer genetic benefits like antibiotic resistance.[20]

Researchers from all around the world are developing novel techniques, apparatus, and modified products for applications in a variety of industries, including agriculture, health care, and environment. Recombinant DNA technology is a fast growing field of study. Lispro (Humalog) is a highly efficient and quick-acting recombinant insulin that is superior to conventional human insulin.[21,22] On the other hand, a distinctive and popular recombinant protein called epoetin alfa can be utilized to effectively cure anemia.[21,23] Recombinant human growth hormone (rhGH) therapy has been demonstrated to have significant benefits for children who cannot produce enough hGH.[21] Recombinant DNA Technology has made significant contributions to the agricultural sciences in the areas of nutritional enrichment, tolerant plants, and enhanced crop yield output.[24]

2.2.5 *APPLICATIONS*

- The advancement of recombinant DNA (rDNA) technology emerged as a result of the efficient application of conventional techniques and instruments in inventive ways, which led to a wide range of alterations and analyses of the organization and structure of complex genomes' genes.[25]
- rDNA technology gives scientists the techniques they need to develop gene therapy to cure diseases, propagate livestock and crops with superior agronomic traits, as well as generate "synthetic" living creatures to overcome ecological problems like oil spillage and plastic waste management.[16]
- An efficient methodology for examining foodborne and phyto-pathogenic pathogens on fruits and vegetables has been made possible by the emergence of recombinant DNA sequencing techniques.[26]
- Through the application of recombinant DNA (rDNA) technology, a variety of peptides, proteins, and metabolites can be produced by cells that are not usually capable of generating them.[27]
- The biological research field has been radically transformed by recombinant DNA, which has also had a profound effect on the agricultural sector.[28] With the cloning and integration of bacterial genes into plant cells, recombinant DNA has boosted plant production in agriculture by enhancing the efficiency of fixing atmospheric nitrogen.[28]
- GMOs or transgenic organisms are organisms with desired characteristics produced using rDNA technology.[16]
- To produce bioactive molecules, vaccines, gene therapy, molecular diagnostics, and other medical supplies, rDNA technology is essential.[24]
- Bacterial cells are now used for the manufacturing of fine chemicals such as insulin, somatostatin, somatotropin, and p-endorphin due to the advancement of rDNA technology. Humulin is the first therapeutic component developed using it.[17] Do refer Figure 2.3 for a pictorial presentation of rDNA.

2.3 CRISPR CAS9

One of the greatest threats to prokaryotic life is posed by bacterial and archaeal viruses, which are among the most abundant organisms on Earth.

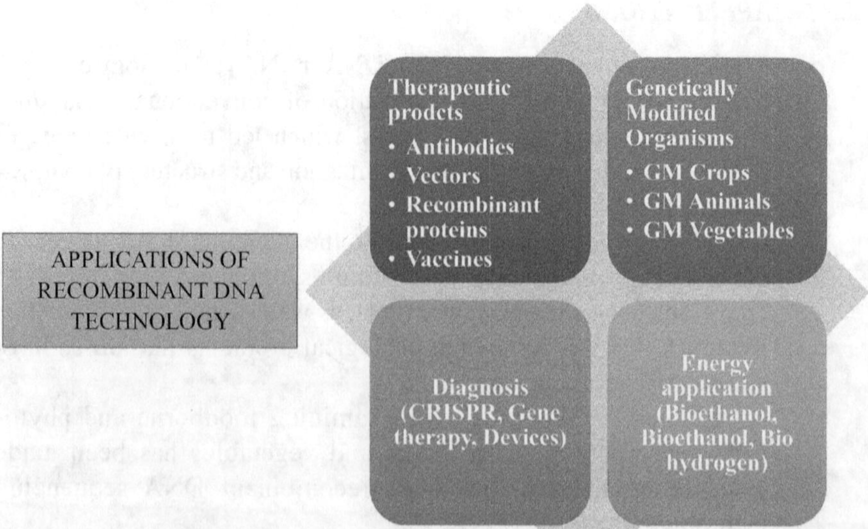

FIGURE 2.3 Applications of recombinant DNA technology.
Source: Created with Biorender.com

To combat phages, prokaryotes have developed several defence strategies. Due to its unique inherent flexibility and therapeutic potential, CRISPR-Cas (Clustered Regularly Interspaced Short Palindromic Repeats-CRISPR-associated protein) has gained recognition in the scientific community over the past ten years.[29] For sequence-specific interference with added nucleic acids, small guide RNAs (sgRNA) are employed in this system. The CRISPR genomic locus, which is made up of brief repetitive elements and unique sequences known as spacers, can be found in mobile genetic elements such as bacteriophages, transposable elements, and plasmids. This region is the basis for CRISPR-Cas.[29]

The CRISPR-Cas systems are generally classified into Class I (types I, III, and IV) and Class II on the basis of structure and functions of Cas-proteins (type II, V, and VI).[30] Class I systems are made up of multi-subunit Cas-protein complexes, whereas class II systems are made up of a single Cas-protein.[31] The two essential components of the CRISPR-Cas-9 system are guiding RNA (gRNA) and CRISPR-associated (Cas-9) proteins. Streptococcus pyogenes was used to retrieve the Cas-9 protein, the very first Cas protein used during genome editing (SpCas-9).[31] The bacteria-based CRISPR-Cas 9 system is represented in Figure 2.4.

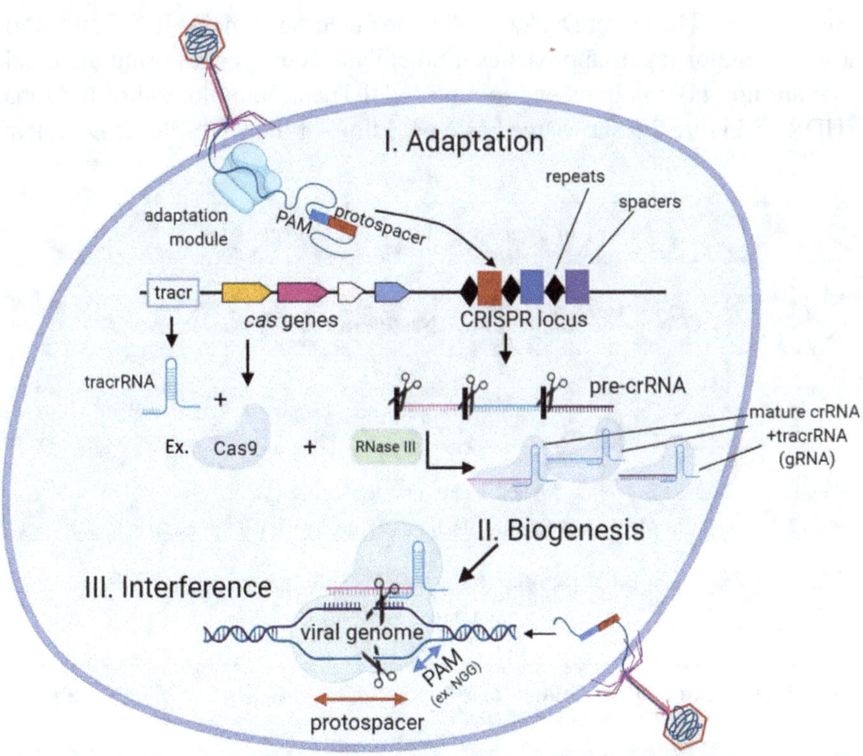

FIGURE 2.4 CRISPR system in bacteria. I-Adaption II-Biogenesis III-Interference.

Source: Reprinted from Ref. [32]. https://creativecommons.org/licenses/by/4.0/

The following domains make up the Cas9 DNA cleaver: RuvC, REC I, REC II, the bridge helix, the protospacer adjacent motif (PAM)-interacting domain, and the REC I and REC II domains.[33] The REC I domain aids the binding of the guide RNA to the target sequence. Refer to Figure 2.5 for a better understanding.

Once the target DNA is bound, the arginine-rich bridge helix can activate the cleavage function, and the PAM-interacting domain is then in charge of starting the interaction with the target sequence. For effective genome engineering, a crispr RNA (crRNA)–trans-activating crispr (tracrRNA) duplex can be concatenated to produce a chimeric single-guide RNA (sgRNA). A 20-nucleotide guide sequence that complements the target region makes up the sgRNA.[34] As soon as the sgRNA locates the target sequence, it directs Cas9 to dissolve the DNA strand, causing a double-stranded break (DSB) at the target location. This is accomplished via binding through Watson–Crick base-pairing. The target DNA is cut by the nuclease domains RuvC and HNH. The two major repair approaches ultimately account for restoring the breaks that are non-homologous end joining (NHEJ) and homology-directed repair (HDR).[35] Figure 2.5 shows the basic workflow of the CRISPR-Cas9 system.

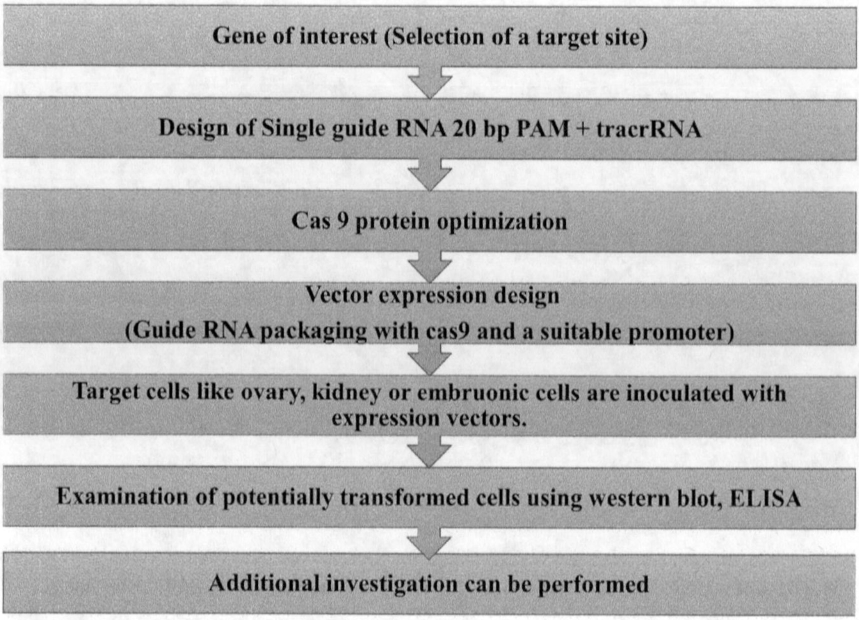

FIGURE 2.5 Flowchart for CRISPR technology.

CRISPR-Cas9 is a genome-editing technique with good potential in the treatment of a wide range of diseases. Nevertheless, developing a reliable and effective delivery system remains a key obstacle which must be addressed to translate this new technology into application areas. Figure 2.6 lists the important delivery methods of CRISPR-Cas9.

2.3.1 APPLICATIONS

- A CRISPR can specifically reach any other particular gene sequence to rectify mutations and incorporate genetic elements into target DNA areas in cells and tissue.
- Applications for the CRISPR-Cas9 system are quite diverse. It has been used in scientific research on pathogen identification, viral infections, cancer, and hereditary illnesses. For example, the

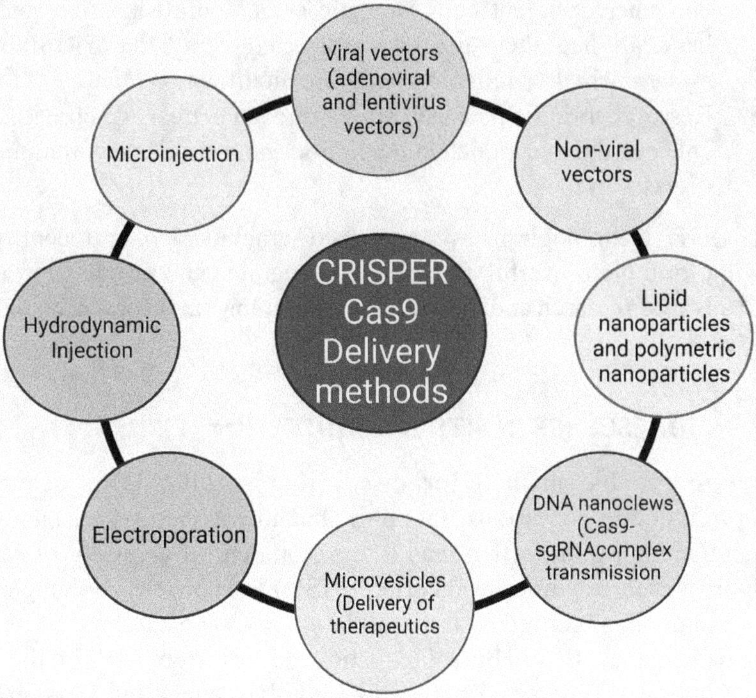

FIGURE 2.6 Various delivery methods of CRISPR-Cas9.

Source: Created with Biorender.com

mutation causing Duchenne muscular dystrophy (DMD) has indeed been effectively corrected using this technique in mice.[36]

- CRISPR-Cas9 genome engineering allows for the modification of any gene in any of the other plant varieties. It provides for speedier genetic alteration than other approaches due to its accessibility, productivity, relative inexpensiveness, and ability to target many genes.[37]
- With the help of CRISPR-Cas9, remarkable genetic alterations have been implemented to improve metabolic processes, resistance to biotic (pathogenic fungi, bacteria, or viruses) or abiotic stimuli (cold, drought, salt), nutritive values, productivity and crop quality, obtain haploid seeds, herbicide tolerance, and other traits.[38]
- The primary and significant application of CRISPR-Cas systems in the medical world is the generation of genetically engineered animal and cell models of numerous human diseases, such as gene knockout models, exogenous gene knock-in models, and site-directed mutagenesis models.[39]

- Human pluripotent cells, somatic cells, zebrafish, mice, and pigs have all had their genomes corrected using the CRISPR-Cas9 system, which is reliable in genome modification.[35]
- Researchers anticipate using this approach to treat and cure illnesses, enhance the nutritional value of crops, and eradicate communicable diseases in the future.[30]

CRISPR technologies have advanced dramatically in recent years, allowing for more powerful, flexible, and accurate genome engineering.[40, 41] It will advance research and therapy for immunological disorders and allergy diseases.[30]

2.4 FLUORESCENCE IN SITU HYBRIDIZATION (FISH)

The most reliable method for discovering specific DNA sequences, detecting genetic conditions, mapping and identifying novel oncogenes or genetic abnormalities that lead to development of a variety of cancers is Fluorescence in situ hybridization (FISH).[42] The field of cytogenetics has undergone a complete transformation as a result, and it is now acknowledged as a reliable diagnostic and research tool in the fight against inherited illnesses. This method involves annealing DNA or RNA probes to a specific target sequence in sample DNA using a fluorescent reporter molecule. The result may then be analyzed through fluorescence microscopy.[43]

In metaphase or interphase cells, FISH can identify particular sites of the target DNA sequence. The authors were the first to apply this technique to plant chromosomes, which were originally established for mammalian chromosomes.[44] The method relies on the potential of single-stranded DNA to anneal to complementary DNA, similar to any other DNA hybridization technique.[45]

In this process, a fluorescent compound absorbs one wavelength of light and then emits light of a different wavelength, which is then detected using a fluorescent microscope.[46] The protocol of FISH is represented as a flowchart in Figure 2.7.

This technology has a wide range of applications like gene mapping, chromosome aberration characterization, diagnosis of Prader–Willi syndrome, Angelman syndrome, 22q13 deletion syndrome, chronic myelogenous leukemia, acute lymphoblastic leukemia, Cri-du-Chat syndrome, velocardiofacial syndrome, and down syndrome.[47] It can also be used for diagnostic testing, prognosis, and remission evaluation of diseases including cancer. The technique has significantly aided cancer cytogenetics, and as a result, clinical laboratories have benefited from the approach, which is swift and may be executed on tissues.[42] Refer Figure 2.8.

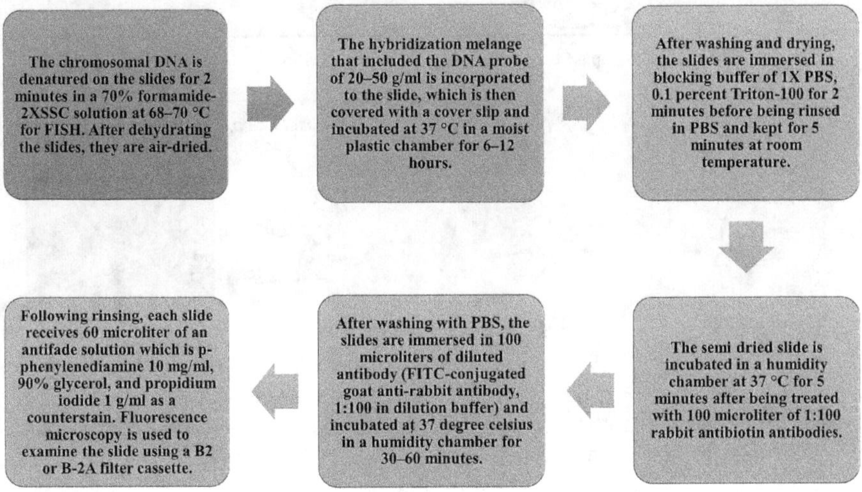

(a)

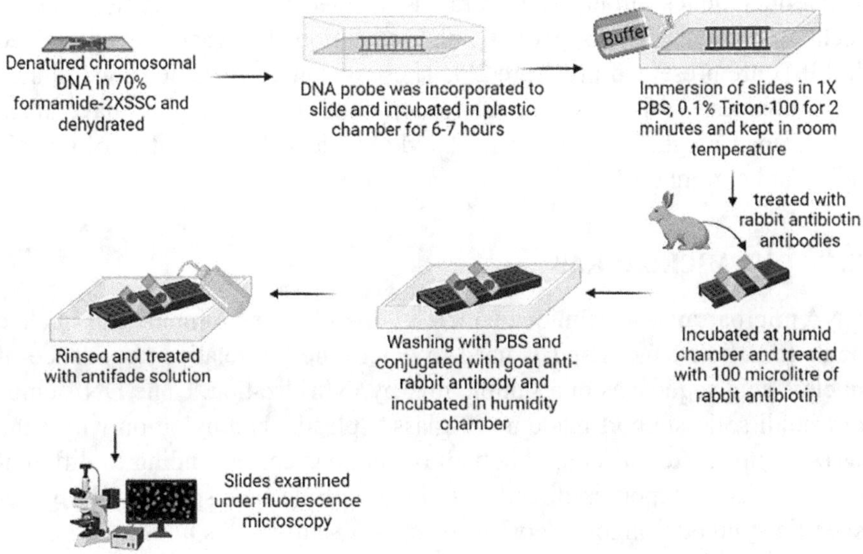

(b)

FIGURE 2.7 Protocol for the FISH technique (a) flowchart, (b) schematic representation.

Source: Created with Biorender.com

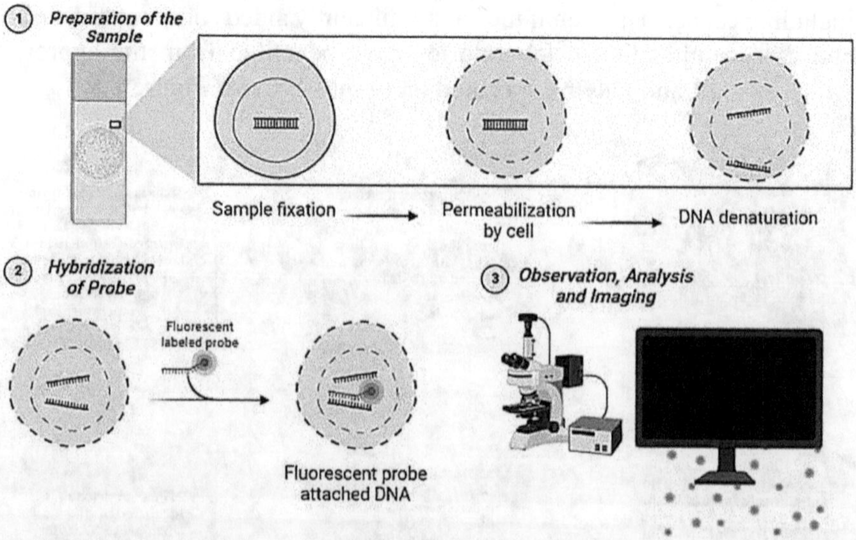

FIGURE 2.8 Steps involved in the FISH assay.
Source: Created with Biorender.com

The evolution of the original FISH methodology into a plethora of innovative techniques through the years has resulted from advancements in the technique's sensitivity, specificity, and resolution. The basic steps involved in FISH are presented in Figure 2.8. These recent advancements, combined with advances in fluorescence microscopy and digital imaging, have aided in a deeper insight into the physical and chemical characteristics of nucleic acids and chromatin.[42]

2.5 DNA MICROARRAY

DNA microarray is a technique which involves binding of numerous nucleic acids to solid surfaces and is used to determine the relative abundance of nucleic acid sequences in a sample mix by hybridization.[48] The DNA chips are small solid support made up of glass,[49] plastic, and nylon on which the DNA samples are loaded and bound to locations corresponding to different genes. [50] The support is directly produced, printed, or spotted with gene-specific synthetic oligonucleotides or cDNA samples. As a result, each dot on the array carries a DNA sequence specific to a particular gene, which will be hybridized into mRNA for that gene.[51]

The principle behind microarrays is the hybridization between the complementary sequences; the binding of complementary sequences occurs due to hydrogen bonding between the nucleotide base pairs.[52] mRNA is regarded as the intermediary molecule that transports genetic information from the cell nucleus to the cytoplasm for protein synthesis, according to central dogma.[53] Consequently, by evaluating different mRNAs, one may evaluate genetic information or gene expression, and hence mRNA serves as a marker.[54] Total RNA was isolated from samples of two tissues, and then reverse transcription was carried out on each sample to create cDNA copies in order to analyze the differences between the transcriptomes of healthy and diseased tissue, dNTPs as a substrate was added and labeling of cDNA is done by fluorochrome dyes Cy3 (green) and Cy5 (red), the further sample is mixed and allowed to hybridize to the microarray.[55] Figure 2.9 presents the steps involved in DNA microarray.

2.5.1 APPLICATION

2.5.1.1 DISEASE DIAGNOSTICS

Microarray offers a great platform for testing a broad range of genes.[56] It includes the recognition of SNPs, classification, and identification of cancer tumors, suppressors, and biomarkers. Furthermore, it can be used to compare the various gene expression levels between healthy individuals and cancer patients. Gene microarrays are based on comparative genomic hybridization where the genomic DNA is fluorescently labeled and utilized to find gene loss or amplification.[57,58] It has been also used to depict genetic abnormalities in tumors like bladder cancer,[59] breast cancer,[58] fallopian tube carcinoma,[60] melanoma,[61] and gastric carcinoma.[62]

2.5.1.2 COMPARISON OF DISEASED AND HEALTHY CELLS

It can be used for comparison of similar cell types like RNA samples from a brain tumor cell and the healthy cell can be compared by tagging this with a fluorescently labeled probe that binds to the tumor sample but not healthy cell indicating the genes that are uniquely associated with diseases.[63]

Several other applications of microarray technologies involve the classification of diseases, or molecular phenotyping, to study the gene function concerning gene regulatory networks, or functional genomics, evolution analysis,[64] molecular phylogenetics, pharmacogenomics,[65] monitoring of effects of drugs and toxins in the cellular system and developmental biology.

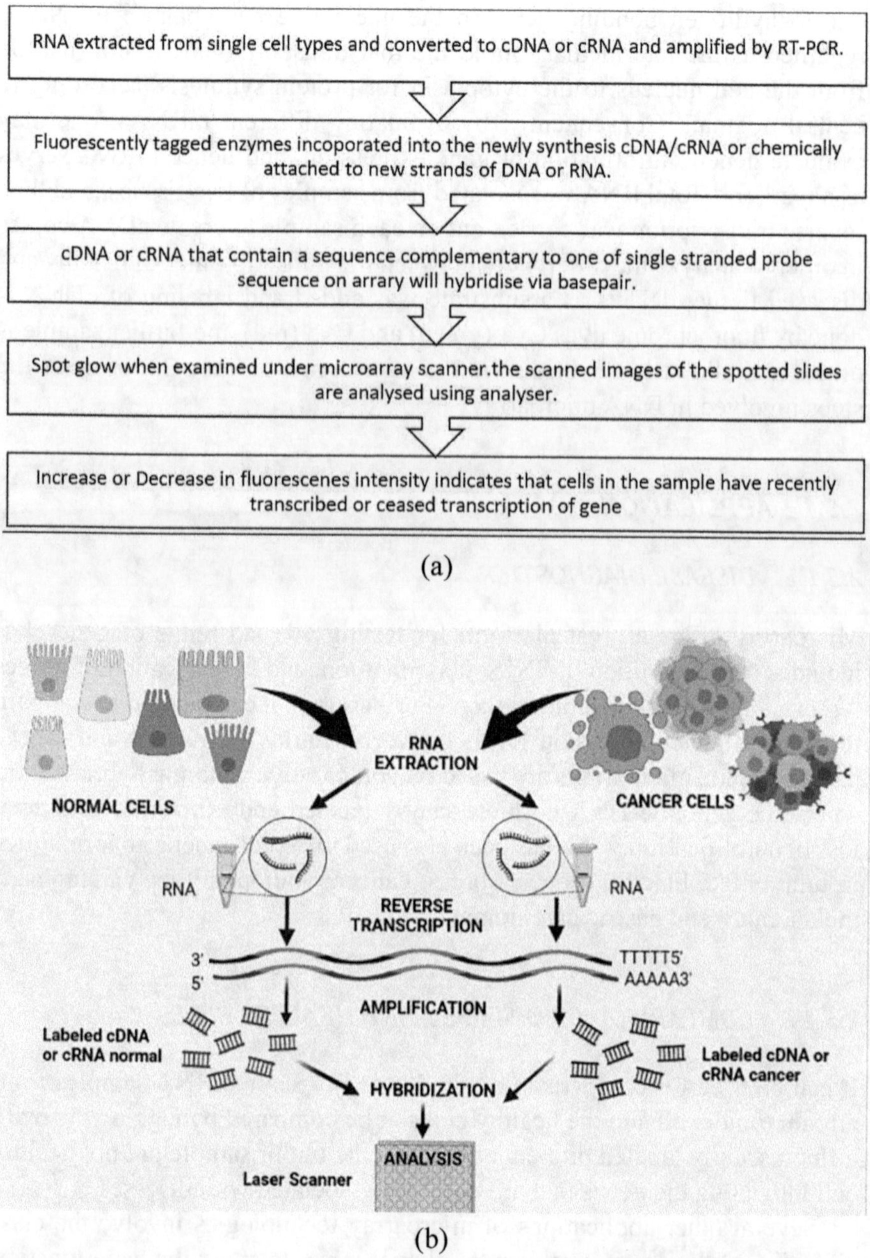

FIGURE 2.9 Steps involved in DNA microarray (a) flowchart and (b) schematic diagram.
Source: Created with Biorender.com

2.6 METABOLIC ENGINEERING

Metabolic engineering is the simple technique which is involved in the production of appropriate compounds after alteration or moderation of cellular metabolism. With recombinant DNA technology, we can obtain altered metabolic pathways of various organisms. As host organisms, bacteria, fungi, plant, and animal cells are all used, and the properties of interesting paths primarily drive selection.[66,67] Various secondary metabolites are generated using pathways including terpenes, phenolics, S-containing compounds, and N-containing compounds are illustrated in Figure 2.10.

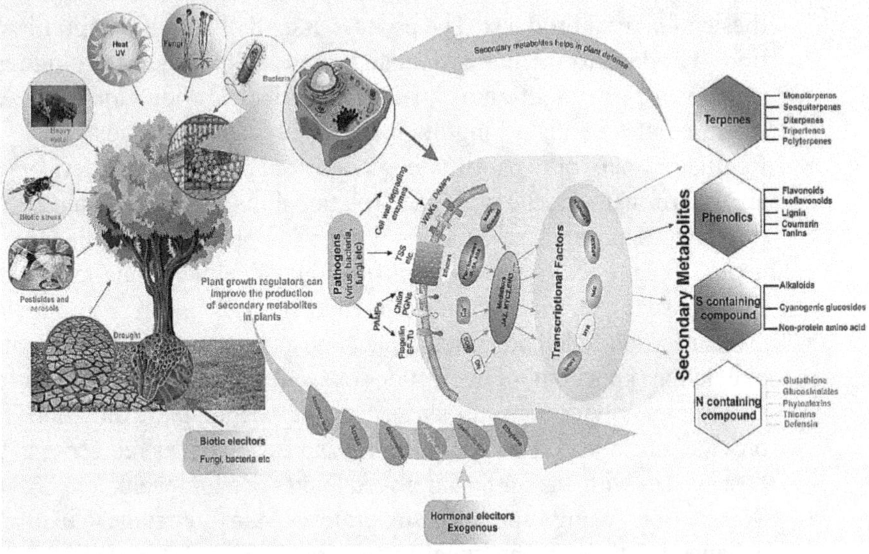

FIGURE 2.10 Production of secondary metabolites through different pathways.

Source: Reprinted from Ref. [68]. https://creativecommons.org/licenses/by/4.0/

Metabolic pathways generally involved are altered and then used for the production of chemicals, amino acids, biofuels, pharmaceuticals, and medicine.[69] In order to comprehend and exploit biological pathways more effectively, this strategy entails functional change of metabolic pathways. On the basis of industrial applications, one can create useful metabolites by enhancing the evolving strains. In order for cells to manufacture a novel product using this approach, certain proteins in a metabolic pathway must be overexpressed or downregulated.[70] For effective engineering, a thorough understanding of the host cell for genetic alterations is a prerequisite. Examining the impact of gene modification on growth is also necessary along with genetic modification which inhibits phosphoenolpyruvate overexpression and interrupts the heat-shock response and nitrogen regulation.[71] To change the host cell for enhancing the production of substances required can be achieved by removing hazardous byproducts and competing routes. Genetic alteration in the traditional metabolic engineering method relies on enzyme network route and its kinetics.[72]

Several approaches used for secondary metabolites improvement in bacteria are

- Heterologous gene clusters expression: Numerous microbes can be utilized for the expression of gene clusters responsible for biosynthesis of natural products. The process essentially requires optimization of codons to facilitate regulated expression in the heterologous hosts. A major limitation of this approach is the production of toxic intermediates by disturbing the metabolic regulation in hosts.[73]
- Regulatory networks pathway engineering: The regulatory networks are determinants of the organized relationships among the transcription factors, proteins associated with the sample as well as the target genes. The process involves modifications of cellular metabolism as a response to genetic and environmental stimuli.[74]
- Gene insertion or deletion: This application involves addition (insertion) or deletion (knockout) of nucleotides or sequences which consequently alter the genotypic makeup as well as the phenotypic expression. It provides an approach for obtaining the expression of desired genes in a host system with an aim for improved expression and yield.[75]
- Stimulation using specific substrates: Many enzymes exhibit enhanced activity using specific substrates. These stimulations are often linked to the development of substrate-enzyme complex, which favors reactions that result in increased yield or expression of the desired characteristics. [76]

Fundamental approaches for evolutionary and classical process engineering provide contributions to both strains and bioprocess development until we fully understand the cell complexity for metabolic and regulatory networks. [77]

Metabolic engineering can generate desirable products at high productivity. The basic approach for the biosynthesis of natural products in microbes using metabolic engineering is presented in Figure 2.11. One significant interrogation of metabolic engineering is identifying optimal organisms and determining the objective for changes in individual genes, whole pathways, or even transcriptional and translational control elements.[78]

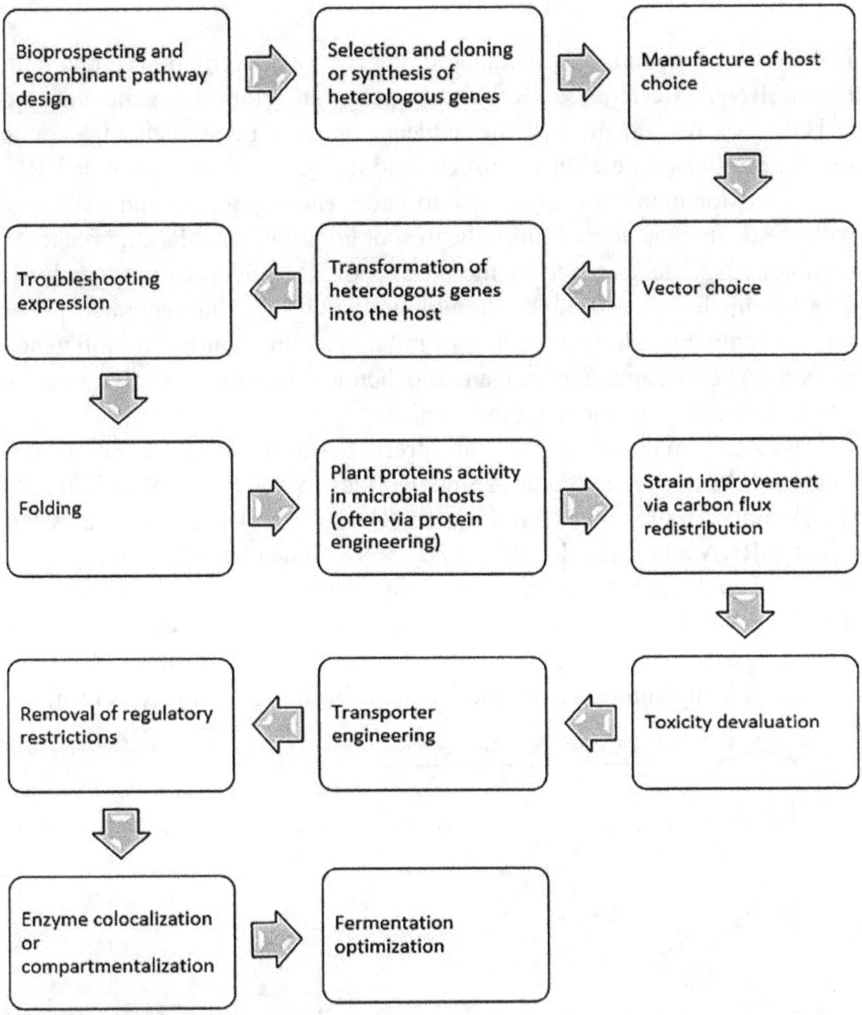

FIGURE 2.11 Steps involved in the biosynthesis of natural products in microbes using metabolic engineering.

Source: Created with Biorender.com

Although many regulatory control systems in nature are still poorly understood, the process for metabolic engineering is structured and defined. Therefore, the use of methodical, bioinformatics-based methodologies integrated with genomic, proteomic, and metabolomic investigations is becoming a common strategy.[79] Natural product biosynthesis may produce enormous amounts of desired substances by using metabolic engineering in bacteria.[78]

2.7 RNA INTERFERENCE

In functional genomics, therapeutic intervention, agriculture, and other fields, RNA interference (RNAi) or post-transcriptional gene silencing (PTGS) is a natural method for sequence-specific gene silencing. For the regulation and expression of protein-coding genes, double-stranded RNA is utilized for mediating immunity to endogenous parasite and exogenous pathogenic nucleic acids.[80] Initially, to spot homologous mRNA, breakdown of small RNAs takes place by the process of RNA degradation which triggers ribonucleases and targets homologous mRNA. The generated phenotypes resemble an allelic sequence of mutants or are identical to null genetic mutants. The in-depth genomic and biochemical research reveals a two-step RNAi-induced gene silencing mechanism.[81]

The steps involved in RNAi are presented in Figure 2.12. Small interfering RNAs (siRNAs) are 21–25 nucleotides long. The siRNAs join RISC (RNA-induced silencing complex), an RNase complex that interacts with target mRNA and degrades it.[82] Dicer, RNA-dependent RNA polymerase, helicases, and dsRNA endonucleases are important parts that are crucial to RNAi. Some of these elements also control the growth of various species by managing a large number of noncoding RNAs, or "micro-RNAs." In general, micro-RNA biosynthesis and function are related to RNAi mechanism.[83]

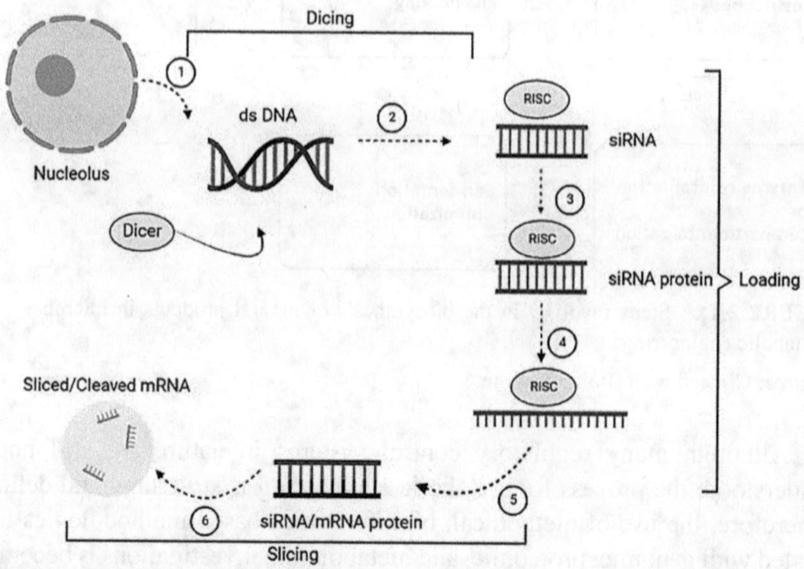

FIGURE 2.12 Schematic diagram of the mechanism involved in RNA interference.
Source: Created with Biorender.com

Recent studies indicated that in the context of RNAi, the genome also undergoes alterations in DNA methylation, heterochromatin formation, and programmed DNA elimination.[84]

Because of its exquisite specificity and efficiency, RNAi is considered an essential tool for functional genomics and gene-specific therapeutic activities targeting the mRNAs of disease-related genes.[85]

This experiment paved the way for the effortless production of null mutants and the process of silencing a functional gene by exogenously using dsRNA, termed RNA interference (RNAi).[86] This powerful method requires only catalytic amounts of dsRNA per cell to silence gene expression.

2.8 PROMOTER ENGINEERING

The process of creating novel promoters with characteristics not found in native sequences, such as increased strength or distinctive inducibility, is known as promoter engineering. By using methods like machine learning and other design paradigms, novel promoters may be made. From metabolic engineering to cellular therapies, promoter engineering has emerged as a crucial technique for a variety of synthetic biology applications.[87]

The primary mechanism by which cell metabolism is controlled is transcription, and promoters are thought to be crucial regulatory components that initiate transcription. In metabolic engineering and synthetic biology, the main uses of promoter engineering are to control gene expression and enhance metabolite production.[88] By logically and precisely controlling the expression of genes in multiple-gene pathways, it has a striking impact on the distribution of metabolic flux and promotes the synthesis of certain metabolites.[89] Simultaneously, several initiatives have focused on discovering natural promoters, creating inducible or hybrid promoters, and creating artificial promoters to fine-tune certain gene expressions at the transcriptional level and enhance the degree of synthesis of significant production metabolites.

An important stage in regulating the flow of cellular metabolism is the fine-tuned gene expression of metabolic pathways.[90,91] Since the promoters have a significant impact on gene expression levels, promoter engineering is one of the most effective methods for optimizing transcriptional regulation.[90,92] But there are frequent bottlenecks for precise transcriptional regulation because of the small number of endogenous promoters that are readily available. At the moment, the methods of promoter engineering are more heavily involved in the creation of a promoter library, the replacement of the native promoter, the alteration of promoter architecture, and the rational design of the hybrid promoter.[90,92–94] *E. coli*[95] and *Corynebacterium glutamicum*[96] and *Saccharomyces cerevisiae*.[97] are important industrial model microorganisms that may be engineered using promoters. Promoter engineering in *Saccharomyces cerevisiae* for the development of synthetic promoters with active ranges is depicted in Figure 2.13.[98]

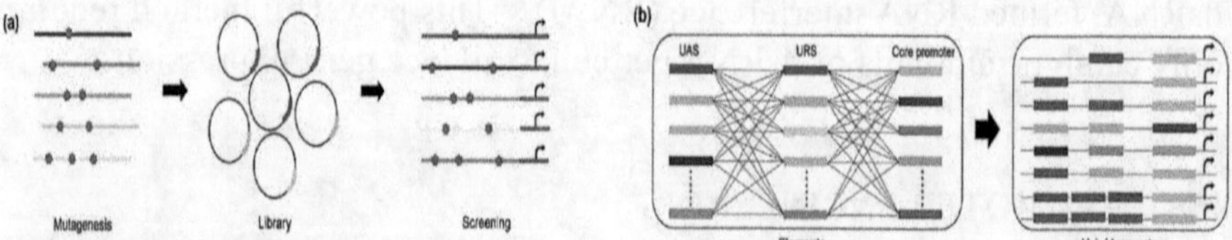

FIGURE 2.13 Development of synthetic promoters with active ranges in *Saccharomyces cerevisiae*.

Source: Reprinted from Ref. [98]. https://creativecommons.org/licenses/by/4.0/

2.9 EDIBLE VACCINE

A lot of diseases can be prevented via vaccination. The conventional approaches of vaccine development and immunisation, however, have drawbacks. Consequently, the idea of edible vaccines emerged in the 1990s by Charles Arntzen.[99] To create an edible vaccine, selected desired genes are introduced into plants, and the changed plants are then stimulated to produce the desired proteins.[100] Plants can incorporate the antigen-encoding genes of bacterial and viral pathogens without losing their original immunogenic characteristics.[101] Generally speaking, it refers to plants that generate vitamins, proteins, or other nutrients that serve as a preventative measure against certain diseases.[102] On ingestion, the plant-based vaccine elicits both mucosal and humoral immune responses.[103] Presently edible vaccines are being produced using various plant sources such as tobacco, potato, banana, tomato, rice, lettuce, soybean, carrots, peanuts, corn, wheat,[104] etc., which are cost-effective, easily stored, and can be readily acceptable for all age groups. The edible vaccine provides various possibilities for the significant reduction of pathogenic diseases including measles, hepatitis B, diphtheria, tetanus, and cholera.[105]

The majority of plant cell breakdown happens in the intestine as a result of the activity of the digestive or bacterial enzymes when an edible vaccine is consumed orally. Plasma cells that produce IgA are concentrated in Peyer's patches.[106] They have the capacity to occupy mucosal tissue because they also function as mucosal immune effector sites. The intestinal surface has 30 to 40 lymphoid nodules that include follicles, and these follicles operate as the place through which antigen enters the intestinal epithelium cell in the breakdown of edible vaccines close to Peyer's patches.[106] The antigen then comes into contact with M-cells, which are specialized epithelial cells of the mucosa-associated lymphoid tissues. The antigen is then passed on to macrophages, which then stimulate the T-cell to generate an immune response.[107,108]

The production of the edible vaccine involves either direct or indirect gene delivery of selected desired genes into the host plant. Two transformations are primarily used to express the transgene in the host plant.[109]

2.9.1 *DIRECT GENE DELIVERY METHOD*

The chosen DNA or RNA is directly inserted into the plant host cell using a biolistic gene gun or micro-projectile bombardment technique in this vector-independent approach.[110] The process is depicted in Figure 2.14.

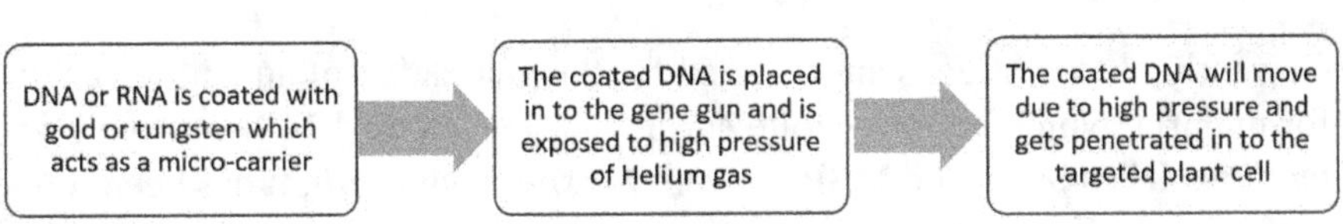

FIGURE 2.14 Gene delivery by biolistic method.

Gene gun methods can also help in nuclear and chloroplast transformation; vaccines produced by gene gun methods are Lyme disease, anthrax, tetanus, plague, cholera, and rotavirus[111].

2.9.2 *INDIRECT GENE DELIVERY METHOD*

This vector-mediated approach involves the targeting of plant cells with bacteria or viruses to express the desired protein. It includes the Agrobacterium Mediated Gene Transfer method' some examples of this vaccine are diarrhea, tuberculosis, dengue, avian flu, and ebolavirus[112].

2.9.3 APPLICATIONS

2.9.3.1 DIARRHEAL DISEASES

The vaccine against enterotoxigenic *Escherichia coli* (ETEC), cholera, and norovirus was developed by transferring gene-encoding LT-B to potato and tobacco leaves through *Agrobacterium tumefaciens* gene transfer methods and found that after ingestion it developed a serum IgG and mucosal IgA anti-LT-B.[103]

2.9.3.2 AUTOIMMUNE DISEASES

Type 1 diabetes is an example of an autoimmune disease; it has been seen that potatoes contain insulin or glutamic acid decarboxylase with innocuous B subunit of the *Vibrio cholera* toxin exhibits improvement in diabetic mice by suppression of immune responses and maintenance of the level of insulin.[113]

2.9.3.3 MALARIA

Currently, three antigens are selected for the generation of an edible vaccine for malaria, which are merozoite surface protein (MSP4/MSP5) from *Plasmodium falciparum*, and MSP4/5 from *Plasmodium yoelii*. It has been found that when MSP 4, MSP 4/5, and MSP1 are orally administrated with cholera toxin B (CTB) in mice they induce antibody responses.[103]

2.9.3.4 HEPATITIS

According to WHO, millions of infections and death are caused by the hepatitis B virus and related diseases like liver cirrhosis, and hepatocellular carcinoma. HBsAg is a hepatitis B surface antigen utilized for the generation of edible hepatitis B vaccines mainly from potatoes.[114]

The edible vaccine provides various possibilities for the significant reduction of pathogenic diseases. Moreover, it can be a solution to decrease some potential hazards associated with traditional vaccines such as allergy and the risk of attenuated vaccine reverting to toxicity.

2.10 IN-SILICO STUDIES

Genetic engineering requires precise editing at the desired site. With advancements in technology, several computerized and machine-learning-based programs have been developed.[115] These tools or techniques facilitate the easier performance of the experiments. Some *in-silico* software like Gene Designer enables rapid designing of synthetic genes. It facilitates easy addition, editing, and combination of genetic elements, in addition to optimization of codons to study protein expression in the host. Similarly, various software is designed to simplify and increase the accuracy of genetic engineering experiments.[116] These software target different aspects of genetic engineering like synthetic primer or gene designing (e.g., Gene Designer),[116] editing (e.g., Easy-Prime, Prime Induced Nucleotide Engineering Creator of New Edits [PINE-CONE], PrimeDesign),[117–119] virtual PCR (e.g., FastPCR),[120] gene silencing (e.g., OptKnock),[121] etc. Some of such recently designed in-silico methods are explained below.

2.10.1 EASY-PRIME

This is a machine-learning-based program that contains a variety of published data as sources. It includes features like folding of RNA conformation and combinations to optimize features to enhance the efficacy of editing.[122] This program works in collaboration with the method, Prime Editing which facilitates all types of transitions, transversion, customized insertions, and deletions. It involves Cas9 nickase attached with reverse transcriptase and prime editing guide RNA that facilitates the movement of the complex to the target site using a single-guide RNA sequence.[123] The Easy-Prime tool helps to evaluate the effect of position and sequence on the activity of prime editors. The tool also integrates prime editor-associated features and estimates the efficiency of editing using the collection of published data. Further, the tool also incorporates RNA folding features that are associated with the efficiency of prime editors.[124] The tool offers features in various categories like spCas9 activity, position, oligo, target mutation, RNA folding, etc. that suggest experimental designs.[117]

2.10.2 FASTPCR

Polymerase chain reaction (PCR) is an important technique in molecular biology and genetic engineering. The efficiency of this technique however depends on the accuracy of in-silico analysis of the nucleotide sequence before the initiation of the wet-lab experiment.[125] The software, FastPCR, is an online Java-based program for virtual PCR and fast search for primer and probe from all types of databases.[126] The prediction of sensitivity and specificity of the probe and primer is dependent on the database search for sequences having optimal values for mismatches, similarity, and stability. Thus, the software efficiently locates the primer position, orientation, and binding efficiency in addition to calculating the melting temperature of primer for standard and degenerate nucleotides. It is appropriate for processing batch files which facilitates the analysis of large quantities of data.[120,126]

2.10.3 OPTKNOCK

With recent advancements and a focus on computational methods, several new algorithms are being developed to identify gene knockout strategies for obtaining the desired phenotype. One such algorithm is OptKnock which suggests knockout strategies that result in the overproduction of the desired metabolite.[121, 127] It searches for the set of gene deletions that are capable of elevating the flux toward the desired product; however, all other objectives of biological significance are optimized by the operation of internal flux distribution. Thus, the microbe produces the desired metabolite to achieve maximal growth.[128] The algorithm benefits from the inherent characteristic of the metabolic pathway of the microbes to optimize the production of the desired metabolite. Since the algorithm is precisely related to the biological objectives of microorganisms, it provides an efficient framework for in-silico metabolic engineering.[121]

2.10.4 OPTFLUX

It is software for in-silico metabolic engineering.[129,130] This is open-source and modular software that is a reference computational application and is available for free to the users (precisely researchers who require the software for their study and benefit from its easy availability). It was the first tool to include strain optimization programs, for instance, to identify the targets for metabolic engineering by using Evolutionary or OptKnock algorithms. It facilitates the utilization of stoichiometric models for phenotypic simulation of wild and mutant organisms, for performing metabolic flux analysis, and for conducting pathway analysis.[131] It also examines methods for simplification of models and preprocessing operations that can lower the search space for optimization algorithms. Additionally, it also has a module for visualization which supports the study of the structure of the model compatible with the layout data that facilitates superimposition of the results of simulation and the graph of the model.[129]

2.11 ALTERNATIVES OF ANTIMICROBIALS

Genetic engineering provides an efficient approach to developing novel solutions to curb the global problem of antimicrobial resistance (AMR). The ability of bacteria to flourish in the presence of compounds (often antibiotics) that have the capability to destroy or suppress their growth is known as antimicrobial resistance. This renders the treatment against the pathogens ineffective.[132] It has become a serious health concern for humans and animals. Due to this problem, numerous widely used antibiotics have lost their ability and efficiency in the treatment of infections. Furthermore, the rate of development of new antibiotics is too slow and not up to mark with the requirement.[133] Thus, arises the need to develop new alternatives for which genetic engineering proves to be a beneficial approach. Some of the genetic engineering-based alternatives are:

2.11.1 BACTERIOPHAGE THERAPY

Bacteriophages are viruses that infect and kill bacterial cells.[134] It is a safer option with no harm to host cells as it lacks the machinery for interference with eukaryotic systems.[135] It is not a novel approach but has been rediscovered for its application as an alternative to antibiotics. The role genetic engineering plays in the rediscovery of this approach is important as the bacteriophages can be engineered to specifically target a particular microbe in the system, without harming beneficial flora residing in the system. They can also be bioengineered for the purification of lytic proteins. Both phages and lytic proteins are largely being used presently to specifically target multi-drug-resistant strains.[136]

2.11.2 LYSINS

Lysins are synthesized in host cells that lead to their lysis for releasing progenies after being introduced by bacteriophages. The structure of lysins is significantly important in its mechanism, which is different for different strains but contains a common N-terminal and cell-binding domain (CBD) connected with a linker.[137] It digests peptidoglycan to create pores, disrupting the bacterial cell wall structure and leading to hypotonic lysis of the bacterium. Using genetic engineering tools recombinant lysins can be produced using bacteriophages and result in restricting the pathogenicity of the microbe and biofilm formation, in cases of biofilm-producing pathogens.[138]

2.12 ETHICAL IMPLICATIONS OF GENETIC ENGINEERING

Although Genetic Engineering has helped us to find solutions for numerous problems, there are a large number of associated ethical issues that need to be addressed. The ethical implications of genetic engineering are highly debated questions. Some of the major ethical issues are:

- The major argument against the use of genetically modified organisms (GMOs) or transgenics is the concern related to the safety of the people as well as its probable impact on the environment. This is the reason that many societies are not accepting it.[139]
- In the case of genetically modified animals, the key concerns related to the welfare of animals were the invasiveness of the techniques, the requirement of numerous animals for the study, the implementation of ethical guidelines and limits in the experiment, and other unexpected concerns regarding their welfare.[140]
- Genetic engineering of animals was designed as a tool to make animals suitable for the environment by manipulating their genetic makeup. However, many people do not favor this notion and believe that instead of changing the animals' telos or functions (based on Aristotle's teleology), the focus should be on modifying their rearing conditions.[141]
- Patenting of gene sequences is also a controversial ethical issue associated with genetic engineering technologies. Many developing countries like the European Union favor the grant of patents to gene sequences.[142] The USA had approved patents for around 30,000 human genes.[143] However, the US Supreme Court in 2013 concluded the proposition that DNA segments and encoded information cannot be patented.[142] The patenting of genes especially for diagnostic purposes requires serious consideration of several aspects associated with it. The disease-specific gene sequences can be easily used as markers for efficient diagnosis, yet there have been cases where the patent owner has taken action against clinical diagnostic labs for using the technology for diagnosis.[143] This is a big obstruction in improving the healthcare status of people and making the services available to the masses. As a consequence, only a few people with a greater economic situation can have a monopoly on these services.

- Germline therapy is accepted by many in cases where it deals with the transfer of severe genetic disorders from parents to the child. However, it is completely unacceptable when it is not for medical aspects but for genetic enhancement to get desired characteristics. Furthermore, unnatural modifications of human nature can have long-term consequences, which cannot be predicted presently.[144]
- Cloning of human genes for research purposes is accepted in some countries and rejected and opposed in other countries. Also, reproductive cloning is universally unaccepted.[144]
- There is always the threat of genetically modified organisms (animals, microbes, and plants) being mixed or cross-bred with the natural wild types, which can lead to severe consequences. Thus, it becomes very crucial to restrict the GMOs in a facility and follow all the guidelines by ethical committees to prevent any unintentional exposure to wild types.

ACKNOWLEDGMENT

We want to express our sincere gratitude to our hon'ble chancellor Dr. G Viswanathan, for his constant motivation and encouragement and we also want to express special thanks to VPC for his support. Authors are also thankful to Prof. Shamlan Reshamwala and Dr. Jayesh Mewada from ICT Bombay for their valuable suggetions in the refinement of the Chapter.

KEYWORDS

- **genetic engineering**
- **rDNA technology**
- **CRISPR Cas-9**
- **DNA microarray**
- **edible vaccine**
- **metabolic engineering**
- **antimicrobial resistance**
- ***in-silico* tools**

REFERENCES

1. Robert, J. S.; Baylis, F. Genetic Engineering. *Int. Encycl. Public Health*, **2008,** 35–39. https://doi.org/10.1016/B978-012373960-5.00133-7.
2. B Primrose, S.; M Twyman, R.; W Old, R. Principles of Gene Manipulation—Old & Primrose.pdf—Google Drive [Online] https://docs.google.com/file/d/0B0VT-PXQdjb7MUNzNk1HZmxNYzQ/edit?resourcekey=0-aG4gmGLXm7tuZn0i8nwgNg (accessed Aug 29, 2022).
3. Nicholl, D. S. T. *An Introduction to Genetic Engineering*; Cambridge University Press, 2023.
4. S.T. Nicoll, D. An Introduction to Genetic Engineering—Desmond S. T. Nicholl—Google Books [Online]. https://books.google.co.in/books?id=g1v6WMHVkTgC&printsec=frontcover#v=onepage&q&f=false (accessed Aug 29, 2022).
5. Lanigan, T. M.; Kopera, H. C.; Saunders, T. L. Principles of Genetic Engineering. *Genes* **2020,** *11* (3). https://doi.org/10.3390/GENES11030291.
6. Rothschild, J. Ethical Considerations of Gene Editing and Genetic Selection. *J. Gen. Fam. Med.* **2020,** *21* (3), 37. https://doi.org/10.1002/JGF2.321.
7. Klug, A. The Discovery of Zinc Fingers and Their Applications in Gene Regulation and Genome Manipulation. *Annu. Rev. Biochem.* **2010,** *79,* 213–231. https://doi.org/10.1146/annurev-biochem-010909-095056.
8. Juárez-Moreno, K.; Erices, R.; Beltran, A. S.; Stolzenburg, S.; Cuello-Fredes, M.; Owen, G. I.; Qian, H.; Blancafort, P. Breaking through an Epigenetic Wall. *Epigenetics* **2013,** *8* (2), 164–176. https://doi.org/10.4161/epi.23503.
9. Chandrasegaran, S.; Carroll, D. Origins of Programmable Nucleases for Genome Engineering. *J. Mol. Biol.* **2016,** *428* (5 Pt B), 963–989. https://doi.org/10.1016/j.jmb.2015.10.014.
10. Richter, C.; Chang, J. T.; Fineran, P. C. Function and Regulation of Clustered Regularly Interspaced Short Palindromic Repeats (CRISPR) / CRISPR Associated (Cas) Systems. *Viruses* **2012,** *4* (10), 2291–2311. https://doi.org/10.3390/v4102291.
11. Physical methods of nucleic acid transfer: general concepts and applications - PMC https://www.ncbi.nlm.nih.gov/pmc/articles/PMC2697808/ (accessed Feb 16, 2023).
12. Fus-Kujawa, A.; Prus, P.; Bajdak-Rusinek, K.; Teper, P.; Gawron, K.; Kowalczuk, A.; Sieron, A. L. An Overview of Methods and Tools for Transfection of Eukaryotic Cells in Vitro. *Front. Bioeng. Biotechnol.* **2021,** *9*.
13. Kowalczyk, T.; Merecz-Sadowska, A.; Picot, L.; Brčić Karačonji, I.; Wieczfinska, J.; Śliwiński, T.; Sitarek, P. Genetic Manipulation and Bioreactor Culture of Plants as a Tool for Industry and Its Applications. *Molecules* **2022,** *27* (3), 795. https://doi.org/10.3390/molecules27030795.
14. Taylor, P.; Lewontin, R. The Genotype/Phenotype Distinction. In *The Stanford Encyclopedia of Philosophy*; Zalta, E. N., Ed.; Metaphysics Research Lab, Stanford University, 2021.
15. Alberts, B.; Johnson, A.; Lewis, J.; Raff, M.; Roberts, K.; Walter, P. Isolating, Cloning, and Sequencing DNA. *Mol. Biol. Cell* **2002**.
16. Nambisan, P. Recombinant DNA Technology and Genetically Modified Organisms. *Introd. Ethical Saf. Intellect. Prop. Rights Issues Biotechnol.* **2017,** 83–126. https://doi.org/10.1016/B978-0-12-809231-6.00004-1.

17. Shinde, S.; Society, I. E.; Sapkal, S. Recombinant DNA Technology and Its Applications : A Review Recombinant DNA Technology and Its Applications: A Review 2018, No. August.
18. Genetic Engineering of Novel Products of Health Significance: Recombinant DNA Technology [Online]. https://link.springer.com/chapter/10.1007/978-3-030-42319-3_26 (accessed Feb 16, 2023).
19. Burmeister, A. R. Horizontal Gene Transfer. *Evol. Med. Public Health* **2015,** *2015* (1), 193–194. https://doi.org/10.1093/emph/eov018.
20. Cecchetelli, A. Plasmids 101: Transformation, Transduction, Bacterial Conjugation, and Transfection.
21. Khan, S.; Ullah, M. W.; Siddique, R.; Nabi, G.; Manan, S.; Yousaf, M.; Hou, H. Role of Recombinant DNA Technology to Improve Life. *Int. J. Genom.* **2016,** *2016,* 2405954. https://doi.org/10.1155/2016/2405954.
22. Lomedico, P. T. Use of Recombinant DNA Technology to Program Eukaryotic Cells to Synthesize Rat Proinsulin: A Rapid Expression Assay for Cloned Genes. *Proc. Natl. Acad. Sci. U. S. A.* **1982,** *79* (19), 5798–5802. https://doi.org/10.1073/pnas.79.19.5798.
23. Masson, E. A.; Patmore, J. E.; Brash, P. D.; Baxter, M.; Caldwell, G.; Gallen, I. W.; Price, P. A.; Vice, P. A.; Walker, J. D.; Lindow, S. W. Pregnancy Outcome in Type 1 Diabetes Mellitus Treated with Insulin Lispro (Humalog). *Diabet. Med. J. Br. Diabet. Assoc.* **2003,** *20* (1), 46–50. https://doi.org/10.1046/j.1464-5491.2003.00840.x.
24. Vipra, M.; Patil, N.; Sivaram, A. Applications of Recombinant DNA Technology. In *A Complete Guide to Gene Cloning: From Basic to Advanced*; Patil, N., Sivaram, A., Eds.; Springer International Publishing: Cham, 2022; pp 143–157. https://doi.org/10.1007/978-3-030-96851-9_9.
25. A Revolution by Recombinant DNA Technology to Improve the Quality of Life. *Biomed. Lett.* **2021,** *7* (1), 12–25. https://doi.org/10.47262/BL/7.1.20201223.
26. DNA Sequencing, Genomes and Genetic Markers of Microbes on Fruits and Vegetables [Online]. https://ami-journals.onlinelibrary.wiley.com/doi/epdf/10.1111/1751-7915.13560?src=getftr (accessed Feb 16, 2023).
27. Porro, D.; Gasser, B.; Fossati, T.; Maurer, M.; Branduardi, P.; Sauer, M.; Mattanovich, D. Production of Recombinant Proteins and Metabolites in Yeasts. *Appl. Microbiol. Biotechnol.* **2011,** *89* (4), 939–948. https://doi.org/10.1007/s00253-010-3019-z.
28. Wall, D. Recombinant DNA, Basic Procedures. In *Encyclopedia of Microbiology*, 3rd ed.; Schaechter, M., Ed.; Academic Press: Oxford, 2009; pp 271–280. https://doi.org/10.1016/B978-012373944-5.00031-6.
29. Hille, F.; Charpentier, E. CRISPR-Cas : Biology, Mechanisms and Relevance. **2016**.
30. Asmamaw, M.; Zawdie, B. Mechanism and Applications of CRISPR/Cas-9-Mediated Genome Editing. *Biol. Targets Ther.* **2021,** *15,* 353–361. https://doi.org/10.2147/BTT.S326422.
31. Asmamaw, M. Biologics : Targets and Therapy Mechanism and Applications of CRISPR / Cas-9-Mediated Genome Editing. **2021,** 353–361.
32. Arroyo-Olarte, R. D.; Bravo Rodríguez, R.; Morales-Ríos, E. Genome Editing in Bacteria: CRISPR-Cas and Beyond. *Microorganisms* **2021,** *9* (4), 844. https://doi.org/10.3390/microorganisms9040844.
33. Globyte, V.; Lee, S. H.; Bae, T.; Kim, J.; Joo, C. CRISPR/Cas9 Searches for a Protospacer Adjacent Motif by Lateral Diffusion. *EMBO J.* **2019,** *38* (4), e99466. https://doi.org/10.15252/embj.201899466.

34. Babu, K.; Amrani, N.; Jiang, W.; Yogesha, S. D.; Nguyen, R.; Qin, P. Z.; Rajan, R. Bridge Helix of Cas9 Modulates Target DNA Cleavage and Mismatch Tolerance. *Biochemistry* **2019,** *58* (14), 1905–1917. https://doi.org/10.1021/acs.biochem.8b01241.
35. Barman, N. C.; Khan, N. M.; Islam, M.; Nain, Z.; Roy, R. K.; Haque, A.; Barman, S. K. CRISPR-Cas9: A Promising Genome Editing Therapeutic Tool for Alzheimer's Disease-A Narrative Review. *Neurol. Ther.* **2020,** *9* (2), 419–434. https://doi.org/10.1007/s40120-020-00218-z.
36. Cai, L.; Fisher, A. L.; Huang, H.; Xie, Z. CRISPR-Mediated Genome Editing and Human Diseases. *Genes Dis.* **2016,** *3* (4), 244–251. https://doi.org/10.1016/j.gendis.2016.07.003.
37. El-Mounadi, K.; Morales-Floriano, M. L.; Garcia-Ruiz, H. Principles, Applications, and Biosafety of Plant Genome Editing Using CRISPR-Cas9. *Front. Plant Sci.* **2020,** *11.*
38. Borrelli, V. M. G.; Brambilla, V.; Rogowsky, P.; Marocco, A.; Lanubile, A. The Enhancement of Plant Disease Resistance Using CRISPR/Cas9 Technology. *Front. Plant Sci.* **2018,** *9,* 1245. https://doi.org/10.3389/fpls.2018.01245.
39. Xu, Y.; Li, Z. CRISPR-Cas Systems: Overview, Innovations and Applications in Human Disease Research and Gene Therapy. *Comput. Struct. Biotechnol. J.* **2020,** *18,* 2401–2415. https://doi.org/10.1016/j.csbj.2020.08.031.
40. Wang, B.; Yang, W.; McKittrick, J.; Meyers, M. A. Keratin: Structure, Mechanical Properties, Occurrence in Biological Organisms, and Efforts at Bioinspiration. *Prog. Mater. Sci.* **2016,** *76,* 229–318. https://doi.org/10.1016/j.pmatsci.2015.06.001.
41. Yang, Y.; Xu, J.; Ge, S.; Lai, L. CRISPR/Cas: Advances, Limitations, and Applications for Precision Cancer Research. *Front. Med.* **2021,** *8* (March), 1–11. https://doi.org/10.3389/fmed.2021.649896.
42. Shakoori, A. R. Fluorescence In Situ Hybridization (FISH) and Its Applications. In *Chromosome Structure and Aberrations*; Bhat, T. A., Wani, A. A., Eds.; Springer India: New Delhi, 2017; pp 343–367. https://doi.org/10.1007/978-81-322-3673-3_16.
43. Shakoori, A. R. Fluorescence In Situ Hybridization (FISH) and Its Applications. *Chromosome Struct. Aberrations* **2017,** 343–367. https://doi.org/10.1007/978-81-322-3673-3_16.
44. Schwarzacher, T.; Leitch, A.R.; Bennett, M.D.; Heslop Harrison, J. S. In Situ Localization of Parental Genomes in a Wide Hybrid. *Ann. Bot.* **1989,** *64* ((3)), 315–324.
45. Joanna M. Bridger, E. V. V. *Fluorescence in Situ Hybridization (FISH) Protocols and Applications*, 1st ed.; Joanna M. Bridger, E. V. V., Ed.; Humana Totowa, NJ, 2010.
46. Singh, B. D. *Genetics*, 2nd ed.; Kalyani Publishers, 2009.
47. Liu, C.; Zhang, X.; Wang, J.; Zhang, Y.; Wang, A.; Lu, J.; Huang, Y.; Liu, S.; Wu, J.; Du, L.; et al. Genetic Testing for Prader-Willi Syndrome and Angelman Syndrome in the Clinical Practice of Guangdong Province, China. *Mol. Cytogenet.* **2019,** *12,* 7. https://doi.org/10.1186/s13039-019-0420-x.
48. Cuzin, M. DNA Chips: A New Tool for Genetic Analysis and Diagnostics. *Transfus. Clin. Biol. J. Soc. Francaise Transfus. Sang.* **2001,** *8* (3), 291–296. https://doi.org/10.1016/S1246-7820(01)00141-0.
49. Brazma, A.; Vilo, J. Gene Expression Data Analysis. *FEBS Lett.* **2000,** *480* (1), 17–24. https://doi.org/10.1016/S0014-5793(00)01772-5.
50. Fodor, S. P. A.; Read, J. L.; Pirrung, M. C.; Stryer, L.; Lu, A. T.; Solas, D. Light-Directed, Spatially Addressable Parallel Chemical Synthesis. *Science* **1991,** *251* (4995), 767–773. https://doi.org/10.1126/SCIENCE.1990438.

51. Lockhart, D. J.; Winzeler, E. A. Genomics, Gene Expression and DNA Arrays. *Nature* **2000,** *405* (6788), 827–836. https://doi.org/10.1038/35015701.
52. Ducray, F.; Honnorat, J.; Lachuer, J. [DNA Microarray Technology: Principles and Applications to the Study of Neurological Disorders]. *Rev. Neurol. (Paris)* **2007,** *163* (4), 409–420. https://doi.org/10.1016/S0035-3787(07)90417-8.
53. Alberts, B.; Johnson, A.; Lewis, J.; Raff, M.; Roberts, K.; Walter, P. From DNA to RNA. *Mol. Biol. Cell.* **2002.**
54. Govindarajan, R.; Duraiyan, J.; Kaliyappan, K.; Palanisamy, M. Microarray and Its Applications. *J. Pharm. Bioallied Sci.* **2012,** *4* (Suppl 2), S310-2. https://doi.org/10.4103/0975-7406.100283.
55. Schena, M.; Shalon, D.; Davis, R. W.; Brown, P. O. Quantitative Monitoring of Gene Expression Patterns with a Complementary DNA Microarray. *Science* **1995,** *270* (5235), 467–470. https://doi.org/10.1126/SCIENCE.270.5235.467.
56. Lockhart, D. J.; Dong, H.; Byrne, M. C.; Follettie, M. T.; Gallo, M. V.; Chee, M. S.; Mittmann, M.; Wang, C.; Kobayashi, M.; Horton, H.; et al. Expression Monitoring by Hybridization to High-Density Oligonucleotide Arrays. *Nat. Biotechnol.* **1996,** *14* (13), 1675–1680. https://doi.org/10.1038/NBT1296-1675.
57. Pollack, J. R.; Perou, C. M.; Alizadeh, A. A.; Eisen, M. B.; Pergamenschikov, A.; Williams, C. F.; Jeffrey, S. S.; Botstein, D.; Brown, P. O. Genome-Wide Analysis of DNA Copy-Number Changes Using CDNA Microarrays. *Nat. Genet.* **1999,** *23* (1), 41–46. https://doi.org/10.1038/12640.
58. Pinkel, D.; Segraves, R.; Sudar, D.; Clark, S.; Poole, I.; Kowbel, D.; Collins, C.; Kuo, W. L.; Chen, C.; Zhai, Y.; et al. High Resolution Analysis of DNA Copy Number Variation Using Comparative Genomic Hybridization to Microarrays. *Nat. Genet.* **1998,** *20* (2), 207–211. https://doi.org/10.1038/2524.
59. Veltman, J. A.; Fridlyand, J.; Pejavar, S.; Olshen, A. B.; Korkola, J. E.; DeVries, S.; Carroll, P.; Kuo, W. L.; Pinkel, D.; Albertson, D.; et al. Array-Based Comparative Genomic Hybridization for Genome-Wide Screening of DNA Copy Number in Bladder Tumors. *Cancer Res.* **2003,** *63* (11), 2872–2880. https://doi.org/10.2/JQUERY.MIN.JS.
60. Snijders, A. M.; Nowee, M. E.; Fridlyand, J.; Piek, J. M. J.; Dorsman, J. C.; Jain, A. N.; Pinkel, D.; Van Diest, P. J.; Verheijen, R. H. M.; Albertson, D. G. Genome-Wide-Array-Based Comparative Genomic Hybridization Reveals Genetic Homogeneity and Frequent Copy Number Increases Encompassing CCNE1 in Fallopian Tube Carcinoma. *Oncogene* **2003**, *22* (27), 4281–4286. https://doi.org/10.1038/sj.onc.1206621.
61. Buckley, P. G.; Mantripragada, K. K.; Benetkiewicz, M.; Tapia-Páez, I.; Diaz de Ståhl, T.; Rosenquist, M.; Ali, H.; Jarbo, C.; de Bustos, C.; Hirvelä, C.; et al. A Full-Coverage, High-Resolution Human Chromosome 22 Genomic Microarray for Clinical and Research Applications. *Hum. Mol. Genet.* **2002,** *11* (25), 3221–3229. https://doi.org/10.1093/HMG/11.25.3221.
62. Weiss, M. M.; Snijders, A. M.; Kuipers, E. J.; Ylstra, B.; Pinkel, D.; Meuwissen, S. G. M.; van Diest, P. J.; Albertson, D. G.; Meijer, G. A. Determination of Amplicon Boundaries at 20q13.2 in Tissue Samples of Human Gastric Adenocarcinomas by High-Resolution Microarray Comparative Genomic Hybridization. *J. Pathol.* **2003,** *200* (3), 320–326. https://doi.org/10.1002/PATH.1359.
63. Chrzanowska, N. M.; Kowalewski, J.; Lewandowska, M. A. Use of Fluorescence In Situ Hybridization (FISH) in Diagnosis and Tailored Therapies in Solid Tumors. *Molecules* **2020,** *25* (8), 1864. https://doi.org/10.3390/molecules25081864.

64. Gibson, G. Microarrays in Ecology and Evolution: A Preview. *Mol. Ecol.* **2002,** *11* (1), 17–24. https://doi.org/10.1046/J.0962-1083.2001.01425.X.
65. Chin, K. V.; Kong, A. N. T. Application of DNA Microarrays in Pharmacogenomics and Toxicogenomics. *Pharm. Res.* **2002,** *19* (12), 1773–1778. https://doi.org/10.1023/A:1021425004264.
66. Wuest, D. M.; Hou, S.; Lee, K. H. Metabolic Engineering. *Compr. Biotechnol.* **2011,** *3,* 617–628. https://doi.org/10.1016/B978-0-08-088504-9.00229-4.
67. Raghavendran, V.; Asare, E.; Roy, I. Bacterial Cellulose: Biosynthesis, Production, and Applications. *Adv. Microb. Physiol.* **2020,** *77,* 89–138. https://doi.org/10.1016/BS.AMPBS.2020.07.002.
68. Jan, R.; Asaf, S.; Numan, M.; Lubna; Kim, K. M. Plant Secondary Metabolite Biosynthesis and Transcriptional Regulation in Response to Biotic and Abiotic Stress Conditions. *Agron.* **2021,** *11* (5), 968. https://doi.org/10.3390/AGRONOMY11050968.
69. Jarboe, L. R.; Zhang, X.; Wang, X.; Moore, J. C.; Shanmugam, K. T.; Ingram, L. O. Metabolic Engineering for Production of Biorenewable Fuels and Chemicals: Contributions of Synthetic Biology. *J. Biomed. Biotechnol.* **2010,** *2010,* 761042. https://doi.org/10.1155/2010/761042.
70. Pickens, L. B.; Tang, Y.; Chooi, Y.-H. Metabolic Engineering for the Production of Natural Products. *Annu. Rev. Chem. Biomol. Eng.* **2011,** *2,* 211–236. https://doi.org/10.1146/annurev-chembioeng-061010-114209.
71. Yadav, D.; Tanveer, A.; Malviya, N.; Yadav, S. Overview and Principles of Bioengineering: The Drivers of Omics Technologies. In *Omics Technologies and Bio-Engineering: Towards Improving Quality of Life*; Azevedo, V., Debmalya, B., Eds.; Academic Press, 2018; Vol. 1, pp 3–23.
72. Oberhardt, M. A.; Palsson, B. Ø.; Papin, J. A. Applications of Genome-Scale Metabolic Reconstructions. *Mol. Syst. Biol.* **2009,** *5* (1), 320. https://doi.org/10.1038/msb.2009.77.
73. Luo, Y.; Enghiad, B.; Zhao, H. New Tools for Reconstruction and Heterologous Expression of Natural Product Biosynthetic Gene Clusters. *Nat. Prod. Rep.* **2016,** *33* (2), 174–182. https://doi.org/10.1039/c5np00085h.
74. Engineering regulatory networks for complex phenotypes in E. coli [Online]. Nature Communications https://www.nature.com/articles/s41467-020-17721-4 (accessed Feb 17, 2023).
75. Fan, Y.; Wang, W.; Ma, G.; Liang, L.; Shi, Q.; Tao, S. Patterns of Insertion and Deletion in Mammalian Genomes. *Curr. Genomics* **2007,** *8* (6), 370–378. https://doi.org/10.2174/138920207783406479.
76. Bazzi, M. D.; Nelsestuen, G. L. Substrate-Specific Stimulation of Protein Kinase C by Polyvalent Anions. *Biochem. Biophys. Res. Commun.* **1987,** *147* (1), 248–253. https://doi.org/10.1016/S0006-291X(87)80113-4.
77. Yang, S. T.; Liu, X.; Zhang, Y. Metabolic Engineering – Applications, Methods, and Challenges. *Bioprocess. Value-Added Prod. Renew. Resour.* **2007,** 73–118. https://doi.org/10.1016/B978-044452114-9/50005-0.
78. Joo, E. J.; Cress, B. F.; Koffas, M. A. G. Using Recombinant Microorganisms for the Synthesis and Modification of Flavonoids and Stilbenes. *Polyphenols Hum. Health Dis.* **2014,** *1,* 483–488. https://doi.org/10.1016/B978-0-12-398456-2.00036-0.
79. Haas, R.; Zelezniak, A.; Iacovacci, J.; Kamrad, S.; Townsend, S.; Ralser, M. Designing and Interpreting 'Multi-Omic' Experiments That May Change Our Understanding of Biology. *Curr. Opin. Syst. Biol.* **2017,** *6,* 37–45. https://doi.org/10.1016/j.coisb.2017.08.009.

80. Agrawal, N.; Dasaradhi, P. V. N.; Mohmmed, A.; Malhotra, P.; Bhatnagar, R. K.; Mukherjee, S. K. RNA Interference: Biology, Mechanism, and Applications. *Microbiol. Mol. Biol. Rev.* **2003,** *67* (4), 657–685. https://doi.org/10.1128/MMBR.67.4.657-685.2003.
81. Ying, S.-Y.; Chang, D. C.; Lin, S.-L. The MicroRNA (MiRNA): Overview of the RNA Genes That Modulate Gene Function. *Mol. Biotechnol.* **2008,** *38* (3), 257–268. https://doi.org/10.1007/s12033-007-9013-8.
82. Xu, W.; Jiang, X.; Huang, L. RNA Interference Technology. *Compr. Biotechnol.* **2019,** 560–575. https://doi.org/10.1016/B978-0-444-64046-8.00282-2.
83. Ambesajir, A.; Kaushik, A.; Kaushik, J. J.; Petros, S. T. RNA Interference: A Futuristic Tool and Its Therapeutic Applications. *Saudi J. Biol. Sci.* **2012,** *19* (4), 395–403. https://doi.org/10.1016/j.sjbs.2012.08.001.
84. Grewal, S. I. S. RNAi-Dependent Formation of Heterochromatin and Its Diverse Functions. *Curr. Opin. Genet. Dev.* **2010,** *20* (2), 134–141. https://doi.org/10.1016/j.gde.2010.02.003.
85. Fire, A.; Xu, S.; Montgomery, M. K.; Kostas, S. A.; Driver, S. E.; Mello, C. C. Potent and Specific Genetic Interference by Double-Stranded RNA in Caenorhabditis Elegans. *Nat.* **1998,** *391* (6669), 806–811. https://doi.org/10.1038/35888.
86. Timmons, L.; Fire, A. Specific Interference by Ingested DsRNA. *Nature* **1998,** *395* (6705), 854–854. https://doi.org/10.1038/27579.
87. Cazier, A. P.; Blazeck, J. Advances in Promoter Engineering: Novel Applications and Predefined Transcriptional Control. *Biotechnol. J.* **2021,** *16* (10), 2100239. https://doi.org/10.1002/BIOT.202100239.
88. Tang, H.; Wu, Y.; Deng, J.; Chen, N.; Zheng, Z.; Wei, Y.; Luo, X.; Keasling, J. D. Promoter Architecture and Promoter Engineering in Saccharomyces Cerevisiae. *Metabolites* **2020,** *10* (8), 320. https://doi.org/10.3390/metabo10080320.
89. Transcription control engineering and applications in synthetic biology [Online]. PMC https://www.ncbi.nlm.nih.gov/pmc/articles/PMC5655343/ (accessed Feb 16, 2023).
90. Chen, X.; Gao, C.; Guo, L.; Hu, G.; Luo, Q.; Liu, J.; Nielsen, J.; Chen, J.; Liu, L. DCEO Biotechnology: Tools To Design, Construct, Evaluate, and Optimize the Metabolic Pathway for Biosynthesis of Chemicals. *Chem. Rev.* **2018,** *118* (1), 4–72. https://doi.org/10.1021/ACS.CHEMREV.6B00804.
91. Raman, K.; Chandra, N. Flux Balance Analysis of Biological Systems: Applications and Challenges. *Brief. Bioinform.* **2009,** *10* (4), 435–449. https://doi.org/10.1093/BIB/BBP011.
92. Blazeck, J.; Alper, H. S. Promoter Engineering: Recent Advances in Controlling Transcription at the Most Fundamental Level. *Biotechnol. J.* **2013,** *8* (1), 46–58. https://doi.org/10.1002/BIOT.201200120.
93. Chae, T. U.; Choi, S. Y.; Kim, J. W.; Ko, Y. S.; Lee, S. Y. Recent Advances in Systems Metabolic Engineering Tools and Strategies. *Curr. Opin. Biotechnol.* **2017,** *47,* 67–82. https://doi.org/10.1016/J.COPBIO.2017.06.007.
94. Deaner, M.; Alper, H. S. Promoter and Terminator Discovery and Engineering. *Adv. Biochem. Eng. Biotechnol.* **2018,** *162,* 21–44. https://doi.org/10.1007/10_2016_8/COVER.
95. Hwang, H. J.; Lee, S. Y.; Lee, P. C. Engineering and Application of Synthetic Nar Promoter for Fine-Tuning the Expression of Metabolic Pathway Genes in Escherichia Coli. *Biotechnol. Biofuels* **2018,** *11* (1), 1–13. https://doi.org/10.1186/S13068-018-1104-1/TABLES/2.

96. Yim, S. S.; An, S. J.; Kang, M.; Lee, J.; Jeong, K. J. Isolation of Fully Synthetic Promoters for High-Level Gene Expression in Corynebacterium Glutamicum. *Biotechnol. Bioeng.* **2013,** *110* (11), 2959–2969. https://doi.org/10.1002/BIT.24954.
97. Portela, R. M. C.; Vogl, T.; Kniely, C.; Fischer, J. E.; Oliveira, R.; Glieder, A. Synthetic Core Promoters as Universal Parts for Fine-Tuning Expression in Different Yeast Species. *ACS Synth. Biol.* **2017,** *6* (3), 471–484. https://doi.org/10.1021/ACSSYNBIO.6B00178/SUPPL_FILE/SB6B00178_SI_002.ZIP.
98. Tang, H.; Wu, Y.; Deng, J.; Chen, N.; Zheng, Z.; Wei, Y.; Luo, X.; Keasling, J. D. Promoter Architecture and Promoter Engineering in Saccharomyces Cerevisiae. *Metabolites* **2020,** *10* (8), 320. https://doi.org/10.3390/METABO10080320.
99. Aswathi, P. B. Plant Based Edible Vaccines against Poultry Diseases: A Review. *Adv. Anim. Vet. Sci.* **2014,** *2* (5), 305–311. https://doi.org/10.14737/JOURNAL.AAVS/2014/2.5.305.311.
100. Gunasekaran, B.; Gothandam, K. M. A Review on Edible Vaccines and Their Prospects. *Braz. J. Med. Biol. Res.* **2020,** *53* (2). https://doi.org/10.1590/1414-431X20198749.
101. Edible Vaccines: Promises and Challenges [Online]. PMC https://www.ncbi.nlm.nih.gov/pmc/articles/PMC7090473/ (accessed Feb 16, 2023).
102. Concha, C.; Cañas, R.; Macuer, J.; Torres, M. J.; Herrada, A. A.; Jamett, F.; Ibáñez, C. Disease Prevention: An Opportunity to Expand Edible Plant-Based Vaccines? *Vaccines* **2017,** *5* (2). https://doi.org/10.3390/VACCINES5020014.
103. Mishra, N.; Gupta, N.; Khatri, K.; Goyal, A. K.; Vyas, S. P. Edible Vaccines: A New Approach to Oral Immunization. *Indian J. Biotechnol.* **2008,** *7,* 283–294.
104. Lal, P.; Ramachandran, V.; Goyal, R.; Sharma, R. Edible Vaccines: Current Status and Future. *Indian J. Med. Microbiol.* **2007,** *25* (2), 93–102. https://doi.org/10.4103/0255-0857.32713.
105. Jain, A.; Saini, V.; Veer Kohli, D. Edible Transgenic Plant Vaccines for Different Diseases. *Undefined,* **2013,** *14* (6), 594–614. https://doi.org/10.2174/138920101131400225.
106. Kunisawa, J.; Kurashima, Y.; Kiyono, H. Gut-Associated Lymphoid Tissues for the Development of Oral Vaccines. *Adv. Drug Deliv. Rev.* **2012,** *64* (6), 523–530. https://doi.org/10.1016/J.ADDR.2011.07.003.
107. Kurup, V. M.; Thomas, J. Edible Vaccines: Promises and Challenges. *Mol. Biotechnol.* **2020,** *62* (2), 79–90. https://doi.org/10.1007/S12033-019-00222-1.
108. Mildner, A.; Jung, S. Development and Function of Dendritic Cell Subsets. *Immunity* **2014,** *40* (5), 642–656. https://doi.org/10.1016/J.IMMUNI.2014.04.016.
109. Clarke, J. L.; Paruch, L.; Dobrica, M. O.; Caras, I.; Tucureanu, C.; Onu, A.; Ciulean, S.; Stavaru, C.; Eerde, A.; Wang, Y.; et al. Lettuce-produced Hepatitis C Virus E1E2 Heterodimer Triggers Immune Responses in Mice and Antibody Production after Oral Vaccination. *Plant Biotechnol. J.* **2017,** *15* (12), 1611. https://doi.org/10.1111/PBI.12743.
110. Altpeter, F.; Baisakh, N.; Beachy, R.; Bock, R.; Capell, T.; Christou, P.; Daniell, H.; Datta, K.; Datta, S.; Dix, P. J.; et al. Particle Bombardment and the Genetic Enhancement of Crops: Myths and Realities. *Mol. Breed.* **2005,** *15* (3), 305–327. https://doi.org/10.1007/S11032-004-8001-Y.
111. Wu, L.; Jiang, L.; Zhou, Z.; Fan, J.; Zhang, Q.; Zhu, H.; Han, Q.; Xu, Z. Expression of Foot-and-Mouth Disease Virus Epitopes in Tobacco by a Tobacco Mosaic Virus-Based Vector. *Vaccine* **2003,** *21* (27–30), 4390–4398. https://doi.org/10.1016/S0264-410X(03)00428-6.

112. Obembe, O. O.; Popoola, J. O.; Leelavathi, S.; Reddy, S. V. Advances in Plant Molecular Farming. *Biotechnol. Adv.* **2011,** *29* (2), 210–222. https://doi.org/10.1016/j.biotechadv.2010.11.004.
113. Richter, L. J.; Thanavala, Y.; Arntzen, C. J.; Mason, H. S. Production of Hepatitis B Surface Antigen in Transgenic Plants for Oral Immunization. *Nat. Biotechnol.* **2000,** *18* (11), 1167–1171. https://doi.org/10.1038/81153.
114. Giddings, G.; Allison, G.; Brooks, D.; Carter, A. Transgenic Plants as Factories for Biopharmaceuticals. *Nat. Biotechnol.* **2000,** *18* (11), 1151–1155. https://doi.org/10.1038/81132.
115. Mushtaq, M.; Ahmad Dar, A.; Skalicky, M.; Tyagi, A.; Bhagat, N.; Basu, U.; Bhat, B. A.; Zaid, A.; Ali, S.; Dar, T.-U.-H.; et al. CRISPR-Based Genome Editing Tools: Insights into Technological Breakthroughs and Future Challenges. *Genes* **2021,** *12* (6), 797. https://doi.org/10.3390/genes12060797.
116. Villalobos, A.; Ness, J. E.; Gustafsson, C.; Minshull, J.; Govindarajan, S. Gene Designer: A Synthetic Biology Tool for Constructing Artificial DNA Segments. *BMC Bioinf.* **2006,** *7,* 285. https://doi.org/10.1186/1471-2105-7-285.
117. Li, Y.; Chen, J.; Tsai, S. Q.; Cheng, Y. Easy-Prime: A Machine Learning–Based Prime Editor Design Tool. *Genome Biol.* **2021,** *22* (1), 1–11. https://doi.org/10.1186/S13059-021-02458-0/FIGURES/4.
118. Standage-Beier, K.; Tekel, S. J.; Brafman, D. A.; Wang, X. Prime Editing Guide RNA Design Automation Using PINE-CONE. *ACS Synth. Biol.* **2021,** *10* (2), 422–427. https://doi.org/10.1021/ACSSYNBIO.0C00445/SUPPL_FILE/SB0C00445_SI_001.PDF.
119. Hsu, J. Y.; Grünewald, J.; Szalay, R.; Shih, J.; Anzalone, A. V.; Lam, K. C.; Shen, M. W.; Petri, K.; Liu, D. R.; Keith Joung, J.; et al. PrimeDesign Software for Rapid and Simplified Design of Prime Editing Guide RNAs. *Nat. Commun.* **2021,** *12* (1), 1–6. https://doi.org/10.1038/s41467-021-21337-7.
120. Kalendar, R.; Khassenov, B.; Ramankulov, Y.; Samuilova, O.; Ivanov, K. I. FastPCR: An in Silico Tool for Fast Primer and Probe Design and Advanced Sequence Analysis. *Genomics* **2017,** *109* (3–4), 312–319. https://doi.org/10.1016/J.YGENO.2017.05.005.
121. Patil, K. R.; Rocha, I.; Förster, J.; Nielsen, J. Evolutionary Programming as a Platform for in Silico Metabolic Engineering. *BMC Bioinf.* **2005,** *6,* 308. https://doi.org/10.1186/1471-2105-6-308.
122. Li, Y.; Chen, J.; Tsai, S. Q.; Cheng, Y. Easy-Prime: A Machine Learning–Based Prime Editor Design Tool. *Genome Biol.* **2021,** *22,* 235. https://doi.org/10.1186/s13059-021-02458-0.
123. Petri, K.; Zhang, W.; Ma, J.; Schmidts, A.; Lee, H.; Horng, J. E.; Kim, D. Y.; Kurt, I. C.; Clement, K.; Hsu, J. Y.; et al. CRISPR Prime Editing with Ribonucleoprotein Complexes in Zebrafish and Primary Human Cells. *Nat. Biotechnol.* **2022,** *40* (2), 189–193. https://doi.org/10.1038/s41587-021-00901-y.
124. Improved prime editors enable pathogenic allele correction and cancer modelling in adult mice [Online]. Nature Communications https://www.nature.com/articles/s41467-021-22295-w (accessed Feb 17, 2023).
125. Liu, H. Y.; Hopping, G. C.; Vaidyanathan, U.; Ronquillo, Y. C.; Hoopes, P. C.; Moshirfar, M. Polymerase Chain Reaction and Its Application in the Diagnosis of Infectious Keratitis. *Med. Hypothesis Discov. Innov. Ophthalmol.* **2019,** *8* (3), 152–155.
126. Kalendar, R.; Tselykh, T. V.; Khassenov, B.; Ramanculov, E. M. Introduction on Using the FastPCR Software and the Related Java Web Tools for PCR and Oligonucleotide

Assembly and Analysis. *Methods Mol. Biol.* **2017,** *1620,* 33–64. https://doi.org/10.1007/978-1-4939-7060-5_2.

127. Burgard, A. P.; Pharkya, P.; Maranas, C. D. Optknock: A Bilevel Programming Framework for Identifying Gene Knockout Strategies for Microbial Strain Optimization. *Biotechnol. Bioeng.* **2003,** *84* (6), 647–657. https://doi.org/10.1002/BIT.10803.
128. Machado, D.; Herrgård, M. J.; Rocha, I. Stoichiometric Representation of Gene–Protein–Reaction Associations Leverages Constraint-Based Analysis from Reaction to Gene-Level Phenotype Prediction. *PLoS Comput. Biol.* **2016,** *12* (10), e1005140. https://doi.org/10.1371/journal.pcbi.1005140.
129. Rocha, I.; Maia, P.; Evangelista, P.; Vilaça, P.; Soares, S.; Pinto, J. P.; Nielsen, J.; Patil, K. R.; Ferreira, E. C.; Rocha, M. OptFlux: An Open-Source Software Platform for in Silico Metabolic Engineering. *BMC Syst. Biol.* **2010,** *4,* 45–45. https://doi.org/10.1186/1752-0509-4-45.
130. Vilaca, P.; Maia, P.; Giesteira, H.; Rocha, I.; Rocha, M. Analyzing and Designing Cell Factories with OptFlux. *Metab. Netw. Reconstr. Model.* **2017,** 37–76.
131. Copeland, W. B.; Bartley, B. A.; Chandran, D.; Galdzicki, M.; Kim, K. H.; Sleight, S. C.; Maranas, C. D.; Sauro, H. M. Computational Tools for Metabolic Engineering. *Metab. Eng.* **2012,** *14* (3), 270–280.
132. Aslam, B.; Wang, W.; Arshad, M. I.; Khurshid, M.; Muzammil, S.; Rasool, M. H.; Nisar, M. A.; Alvi, R. F.; Aslam, M. A.; Qamar, M. U.; et al. Antibiotic Resistance: A Rundown of a Global Crisis. *Infect. Drug Resist.* **2018,** *11,* 1645–1658. https://doi.org/10.2147/IDR.S173867.
133. Sengupta, S.; Chattopadhyay, M. K.; Grossart, H.-P. The Multifaceted Roles of Antibiotics and Antibiotic Resistance in Nature. *Front. Microbiol.* **2013,** *4,* 47. https://doi.org/10.3389/fmicb.2013.00047.
134. Gordillo Altamirano, F. L.; Barr, J. J. Phage Therapy in the Postantibiotic Era. *Clin. Microbiol. Rev.* **2019,** *32* (2). https://doi.org/10.1128/CMR.00066-18.
135. Harada, L. K.; Silva, E. C.; Campos, W. F.; Del Fiol, F. S.; Vila, M.; Dąbrowska, K.; Krylov, V. N.; Balcão, V. M. Biotechnological Applications of Bacteriophages: State of the Art. *Microbiol. Res.* **2018,** *212–213,* 38–58. https://doi.org/10.1016/J.MICRES.2018.04.007.
136. Lin, D. M.; Koskella, B.; Lin, H. C. Phage Therapy: An Alternative to Antibiotics in the Age of Multi-Drug Resistance. *World J. Gastrointest. Pharmacol. Ther.* **2017,** *8* (3), 162. https://doi.org/10.4292/WJGPT.V8.I3.162.
137. Schmelcher, M.; Donovan, D. M.; Loessner, M. J. Bacteriophage Endolysins as Novel Antimicrobials. *Future Microbiol.* **2012,** *7* (10), 1147–1171. https://doi.org/10.2217/fmb.12.97.
138. Heselpoth, R. D.; Euler, C. W.; Fischetti, V. A. PaP1, a Broad-Spectrum Lysin-Derived Cationic Peptide to Treat Polymicrobial Skin Infections. *Front. Microbiol.* **2022,** *13.* https://doi.org/10.3389/FMICB.2022.817228.
139. Carter, L. Re-Interpreting Some Common Objections to Three Transgenic Applications: GM Foods, Xenotransplantation and Germ Line Gene Modification (GLGM). *Transgenic Res.* **2004,** *13* (6), 583–591. https://doi.org/10.1007/S11248-004-2835-2.
140. Ormandy, E. H.; Dale, J.; Griffin, G. Genetic Engineering of Animals: Ethical Issues, Including Welfare Concerns. *Can. Vet. J.* **2011,** *52* (5), 544.

141. Rollin, B. E. Telos, Conservation of Welfare, and Ethical Issues in Genetic Engineering of Animals. *Curr. Top. Behav. Neurosci.* **2015,** *19,* 99–116. https://doi.org/10.1007/7854_2014_279.
142. Cole, P. Patentability of Genes: A European Union Perspective. *Cold Spring Harb. Perspect. Med.* **2015,** *5* (5). https://doi.org/10.1101/CSHPERSPECT.A020891.
143. Merz, J. F.; Cho, M. K. What Are Gene Patents and Why Are People Worried about Them? *Community Genet.* **2005,** *8* (4), 203. https://doi.org/10.1159/000087956.
144. Ardekani, A. M. Genetic Technologies and Ethics. *J. Med. Ethics Hist. Med.* **2009,** *2,* 11.

CHAPTER 3

Advancement in Protein Engineering

DIGNA PATEL[1], HIMALI PATEL[1], PRIYA SHARMA[1], BINAY DALAI[1], XIANG CHEN[2], and SARAT K. DALAI[1*]

[1]*Institute of Science, Nirma University, Ahmedabad, Gujarat, India*

[2]*Institute for Biotechnology, St. John's University, New York, USA*

**Corresponding author.*

ABSTRACT

Protein engineering is a technology that synthesizes new proteins by manipulating the native form of naturally existing proteins. The earliest articles on protein engineering date back to 1980 that were reviewed by Ulmer (1983). However, the aspects of engineering a protein are broad, and it is still in its infancy. Biopharmaceuticals are extensively using this approach to improvise drug response and develop novel drug therapies for different indications. Proteins are being modified to attain desirable qualities as smart therapeutic. Different types of proteins can be used for various applications but are not feasible to use due to several constraints (Common types of proteins include: Enzymes are proteins that assemble or disassemble compounds. Some antibodies, for example, are immune system proteins that protect against diseases. Signaling proteins and receptors are proteins that relay signals between cells and receive signals from other cells via proteins).[1] To overcome the limitations, protein engineering has been used to design, produce, purify, and improve modified proteins. Advanced protein engineering broadly depends on techniques related to molecular cell biology and principles of organic chemistry to efficiently design proteins with desired properties.[1] Desire properties include stability, receptor and target recognition, protein folding, improved half-life, enzyme activity, structure, and synthetic production to fulfill the necessities. This chapter will concentrate on protein engineering methods and their applications.

Advances in Pharmaceutical Biotechnology, Volume 1: Exploring Cutting-Edge Innovations and Applications. Vivek P. Chavda and Zhe-Sheng (Jason) Chen (Eds.)

DOI: 10.1201/9781774919798-3

3.1 PROTEIN ENGINEERING

Protein engineering is the method by which a scientist modifies a protein to fulfil a specific requirement and make it more suitable for a particular application or a particular purpose. Advanced protein engineering relies on concepts of organic chemistry and molecular cell biology approaches in order to effectively design proteins with desired features.[2] Stability, receptor and target recognition, better half-life, protein folding, enzyme activity, structure, and synthetic manufacturing are all desired features.[2]

Protein engineering is a rapidly developing field that has the potential to improve our understanding of how natural proteins work as well as to create artificial proteins that have been altered to improve human health. It now has a wider range of applications and high-throughput technology is emerging. There are many advancements in protein engineering.

3.2 METHODS OF PROTEIN ENGINEERING

Proteins are abundantly present in all living organisms in the form of cell membranes, enzymes, receptors, antibodies, cytoskeleton, etc. They are the most multifaceted macromolecules in living systems. A biological polymer is created when 20 amino acids are placed in a particular order and interact with one another through hydrogen bonds, disulphide bonds, Van Der Waals forces, and salt bridges.[3] About 100,000 human proteins were discovered based on coding genes discussed in next-generation sequencing projects.[3] Moreover, 25,000 enzymes are known now. Considering the full-length sequence of proteins, very few similarities are noted. For example, one of the most common enzymes alpha-amylase contains 500–600 amino acids; they have very well-conserved catalytic motifs among family members.[3] However, only 10% similarity is found in their whole sequence. Variation in the amino acid arrangement is responsible for the packing of polypeptides and structural folding. Thus, a change in even one amino acid is capable of affecting the properties of that protein.[4] There are several natural causes of mutations like radiation, transposition, mutagenic chemicals, and also errors that occur during meiosis or DNA replication. The possibility of favorable changes in protein gave rise to the field of protein engineering. This novel technique is applicable for basic research, to elucidate structure–function relations, and for the exploitation in the industry. Protein engineering so far has a boom for applications in pharmaceuticals, health, the food industry, and medical sciences. It encompasses the rapidly developing approved segment of the Food and Drug Administration which predicts revamping clinical outcomes in long run by improving the control over pharmacokinetics and pharmacodynamics.[5] This ever-expanding need for protein engineering in the therapeutics and healthcare sector demands the development of engineering technology.

Methods of protein engineering are adopted based on the application of the protein for its functional requirements. Major applications of protein engineering are in biomedical and related industries. For example, several decades ago the first recombinant technique was used for insulin production in large amounts. Likewise, other techniques were used to prepare advanced therapeutic proteins for bio-medical translational research or industrial products.[3] Refer Figure 3.1. Here are the two crucial steps involved in protein engineering at a genetic level and another at a protein level. There are different types of modifications, but applying both the level of modification techniques makes the desired product that serves pharmaceutical biotechnology for translational therapeutic and industrial purposes.[6]

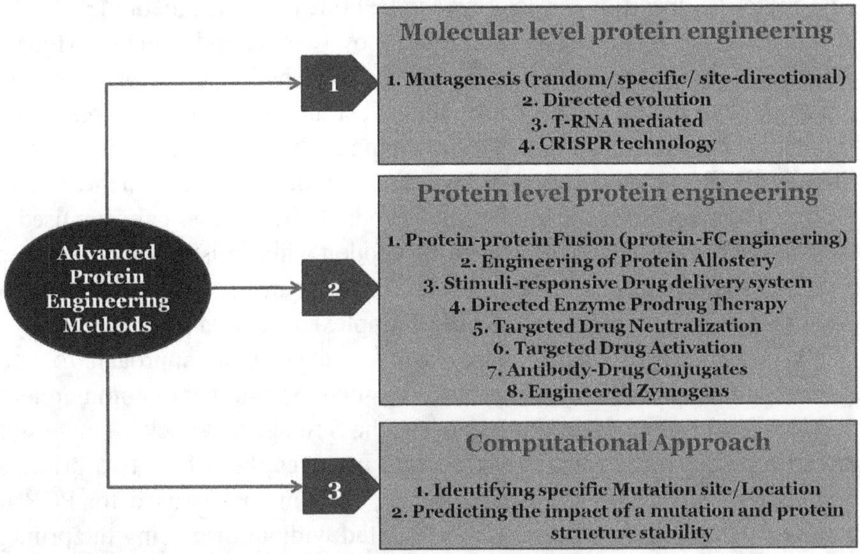

FIGURE 3.1 Different approaches for the protein engineering.

Source: Created with Biorender.com

3.2.1 GENETIC/MOLECULAR LEVEL PROTEIN ENGINEERING TECHNIQUES

In this technique, protein modification was done at the gene level for the betterment of protein function or for fulfilling the purpose of modification in the function of a particular protein.[1] It includes various techniques that we will discuss further with their application and examples.

3.2.1.1 MUTAGENESIS (RANDOM/SPECIFIC/SITE-DIRECTIONAL)

A plethora of techniques is available to engineer protein, owing to the advancements in recombinant DNA technology. The most classical method is called rational design. It comprises site-directed mutagenesis. It is a technique that originated in the 1970s. Michael Smith, the creator of site-directed mutagenesis, received the 1993 Nobel Prize in chemistry. This method comprises short single-stranded DNA fragments or oligonucleotides with desired mutations that act as a template after the addition of the DNA polymerase enzyme.[2] Site-directed mutagenesis of proteins is the conventional technique, commonly referred to as the "rational design" approach.[2] By using this technique, additional amino acids can be added to our target gene. The fundamental principle of the method requires a short-strand DNA primer complementary to the template DNA consisting of the mutation site. Point mutation, multiple base mutation, insertion, or deletion can be done using this method.[2]

Site-directed mutagenesis can be done by two methods, one of which is called "overlap extension." It is the most established method for generating mutations like deletions, insertions, and substitutions without any restriction sites and generates PCR products overlapping each other as the name suggests. Initially, PCR generates overlapping segments of the gene, which are later used as a template to make a full-length product. Two primer pairs are used in the approach, each of which has a mutant codon with a mismatched sequence. Two PCRs take place, in the first PCR all 4 primers are used to generate heteroduplexes, in the second PCR these heteroduplexes are used as a template.[2]

"Whole plasmid single round PCR" is a different approach to site-directed mutagenesis. This technique is applied in a number of commercially accessible mutagenesis products, such as the Stratagene, Quick-Change site-directed mutagenesis kit.[1] The technique involves the use of two primers, one containing the desired mutation. DNA polymerase is used for PCR to generate a copy of the plasmid to be mutated without displacing the primer. Hence, the plasmid product with breaks is obtained after PCR contains the desired mutation.[1] DpnImethylase is used to obtain a circular nicked vector with a mutated gene. After being transformed into competent cells, these nicks are repaired, and a circular mutant plasmid DNA is generated.[2] There is an alternate strategy for site-directed mutagenesis in which the region for mutation is replaced with a fragment obtained by the ligation of synthetic oligonucleotides.[2] These alternate methods also include Cassette mutagenesis, as the name suggests which comprises a synthetic double-stranded DNA i.e. cassette incorporated with requisite mutant sequence.[7] Only when the target site is located between two restriction endonuclease cleavage sites is it feasible. A synthetic oligonucleotide containing mutation can be used in place of this intervening sequence. There are several examples of engineered proteins by site-directed mutagenesis that we use in our day-to-day life.[7] The majority of regularly used detergents are subtilized, in which the protein's wild-type form has a methionine that bleach can oxidize, rendering the protein inactive. When alanine is used in its place, the protein still functions when bleach is present. The research of protein functions also frequently makes use of site-directed mutagenesis.[2]

3.2.1.2 DIRECTED EVOLUTION

The above-mentioned methods are only useful when the mechanism and structure of the target protein are well understood.[8] However, there is minimal information on the structure and function of a number of proteins in protein engineering.[8] In such cases "evolutionary methods" like "random mutagenesis and selection" are useful. Directed evolution mimics natural evolution in a laboratory setting. Do refer to Figure 3.2. It circumvents our profound incomprehension of how encoding of function is carried out by protein's sequence.[7] It expedites the evolution through recursive rounds of gene variations and library screening. Unlike natural evolution which has an ultimate goal of survival and mutation, directed evolution is carried out with higher mutations to obtain a protein with desired biological function Gene diversification by random mutagenesis and screening/selection of the appropriate phenotype are the two main elements of the overall procedure.[7] "Saturation mutagenesis" is a popular and straightforward method for random mutagenesis. It includes changing a single amino acid in a naturally occurring protein, and it is possible to know all the variations.[7] A technique called "localized or region-specific random mutagenesis" uses both rational and random techniques for protein engineering.[8] It requires a simultaneous change in a few amino acids altogether to obtain a new protein with different specificities.[8] Similar to site-directed mutagenesis, this method also uses overlap extension and whole plasmid single-round PCR mutagenesis; the main distinction is that in evolutionary methods, everything is random.[8] The other in-vitro methods for gene diversification involve error-prone PCR (EpPCR) and recombination-based DNA shuffling. The flawed replication process is mimicked in the Error-prone PCR method.[7] Naturally, the rate of flaw is 10–10 per base pair, while in EpPCR random mutations are introduced at a much higher rate, i.e., 10–4 per base pair. Similar to this, the natural process of homologous recombination is imitated in the DNA shuffling method to obtain a combination of favorable mutations.[7]

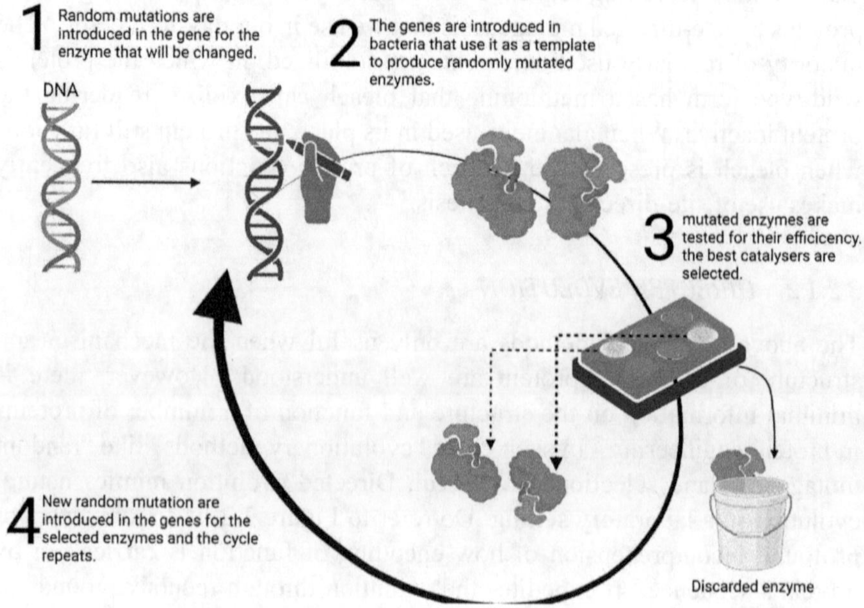

FIGURE 3.2 Method of directed evolution.

Source: Created with Biorender.com

3.2.1.3 T-RNA MEDIATED

A novel method for the incorporation of stable isotope labels and non-native amino acids using modified tRNA has opened new doors for basic biotechnology that defies and goes beyond conventional site-directed mutagenesis.[8] Genetic engineering has largely been constricted to 20 amino acids; on the other hand, TRAMPE allows variety in genetic code so that custom-made non-native amino acids can be introduced into proteins.[8] Previously, chemical modification of tRNAs was carried out which were enzymatically acylated with existing amino acids. Non-native amino acid replacement (NAAR) consists of introducing naturally occurring amino acids with derivatizedanalogs.[8] Misacylation of amino acids can be carried out by chemical aminoacylation or by the use of aminoacyl-tRNA-synthetase, firstly, a suppressor tRNA is misacylated using one of the above-mentioned methods, followed by targeting non-native amino acids to the appropriate position in existing protein through misaminoacylated-tRNA anticodon.[8] This ability of TRAMPE to produce novel amino acids apart from the already known 20 amino acids allows a plethora of futuristic applications in biotechnology.[8] One of these applications is Phototonic Proteins. Otherwise nascent proteins with no photo activity are endowed with customized photoactive amino acids which can be used in the production of novel biomaterials.[8] Refer Figure 3.3 for the diagrammatic preparation.

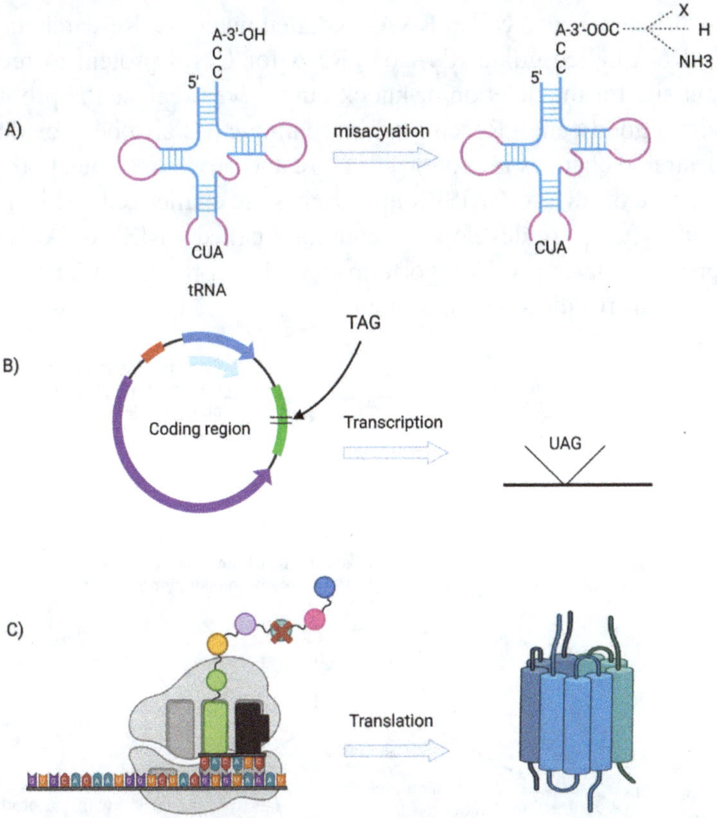

FIGURE 3.3 Stages of t-RNA mediated engineering.

Source: Created with Biorender.com

3.2.1.4 CRISPR TECHNOLOGY

The usage of the CRISPR gene-editing technique is well recognized. CRISPR stands for clustered regularly interspaced short palindromic repeats and CRISPR-associated protein 9.[9] It represented a molecular biology milestone. It was modified from a genome-editing system that is present in bacteria naturally as an immune function. When bacteria are attacked by viruses, the small segments of virus DNA are captured by the bacteria and assimilated into their DNA generating a particular pattern called CRISPR arrays.[9] This provides the bacteria to create a memory against these viruses or closely related ones. Upon subsequent exposure, RNA segments are produced by bacteria from CRISPR arrays to identify and adhere to specific regions in viruses' DNA.[9] This approach to immune defense was adapted by scientists to edit DNA and is also called RNA-mediated nuclease. Researchers design specific RNA called guide RNA or gRNA for Cas 9 protein to recognize the target site for the deletion or knock out or even replace the problematic gene with a good gene.[9] Recently, this technique has also been exploited to make desirable changes in protein structure and properties. Due to the gene-centric nature of gRNA, CRISPR approaches are challenged.[9] This has been addressed by a newly developed technology called CRISPR-TAPE, which has a protein-centric gRNA algorithm that allows the targeting of specific amino acids or residues within proteins.[9]

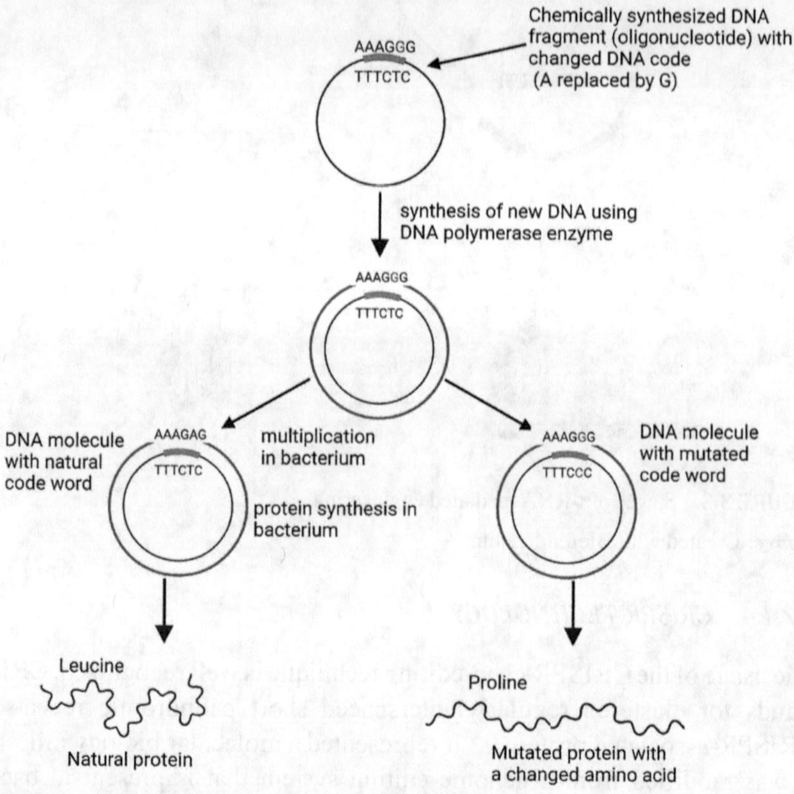

FIGURE 3.4 Process of CRISPR gene-editing technique.

Source: Created with Biorender.com

3.2.2 *PROTEIN LEVEL PROTEIN ENGINEERING TECHNIQUES*

This protein-based technique includes different methods which help mainly improve the protein stability, specificity, and half-life for its better therapeutic use in the field of biomedical or in industries. Engineering the protein or preparing a good quality drug is not enough in the field of pharmaceutical biotechnology. However, after designing engineered proteins or drugs it is equally important to deliver them to the target site with specificity and stability.[10] Therefore, there are several aspects to deliver the engineered product to become translational with more stability and specificity. Those aspects are also a part of protein engineering methods. Some of those techniques are mentioned in Figure 3.5.

Further, we will discuss those protein-based modification techniques with their application and examples.

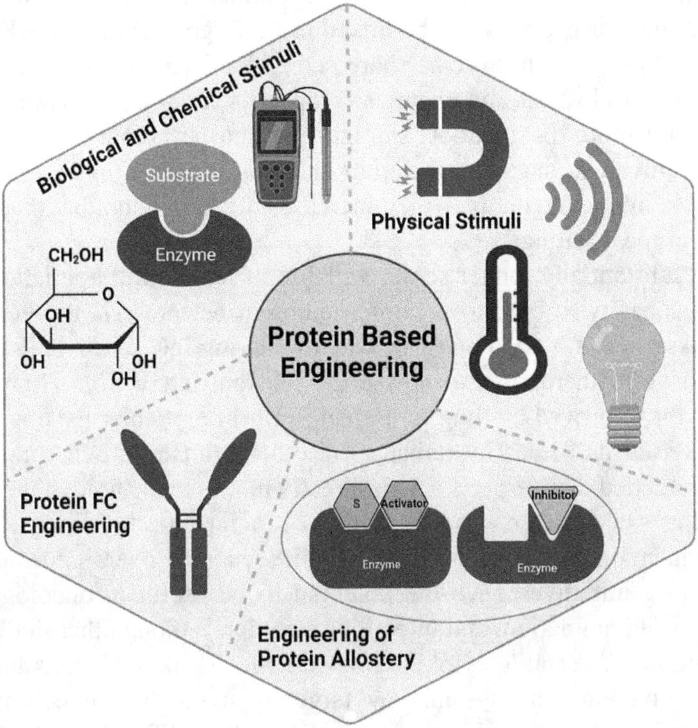

FIGURE 3.5 Protein-based protein engineering technique ("Figure Made in BioRender.com") The figure shows some of the majorly used protein-based protein engineering techniques such as physical stimuli includes magnetic, ultrasound, temperature, and light; biological and chemical stimuli includes pH, enzyme, and glucose; protein FC engineering, protein allostery, etc.

Source: Created with Biorender.com

3.2.2.1 PROTEIN–PROTEIN FUSION (PROTEIN-FC ENGINEERING)

The capacity of Fc to attach to immune-related molecules like Fc receptors can be used to regulate in vivo immune response.[10] The combined biological and pharmacological properties of an Fc-based therapeutic protein include (i) improved Fc-mediated effecter functions, (ii) prolonged half-life, (iii) enhanced stability and aggregation resistance, (iv) acquired multivalent binding to the target, and (v) modulated immunogenicity.[10]

Many endogenous proteins in our bodies as well as other unknown proteins may have the ability to take care of a number of illnesses because of their agonistic or antagonistic activities.[10] These proteins may have several drawbacks, which prevent further clinical use of them. They can be successfully fused to the antibody Fc fragment to address this problem. For example, extracellular receptor domains, soluble cytokines, ligands, enzymes, modified domains, or peptides can be linked to Fc fragments to create Fc-fusion proteins.[11,12] Fc-fusion proteins share some characteristics with antibodies, such as good physical and chemical properties for straightforward expression, purification, specification, storage, and transportation; they can also be bi- or multivalent, have a long serum half-life, improve function, and have adjustable immunogenicity, which increases the likelihood that they will be used in clinical settings.[11]

The substantially longer serum half-life of feature differentiation is due to their capacity to bind to a gamma globulin receptor via the Fc component.[12] As a result, a tried-and-true way of altering IgG serum half-life is Fc engineering to increase or decrease the relevant interaction. Human IgGs created for improved binding to human FcRn in particular have noticeably longer serum half-lives in humans and other primates when compared to their unmutated counterparts.[12] This directly relates to the field of therapeutic antibodies since longer serum half-lives of therapeutic IgGs may allow for less frequent dosing or administration while retaining overall efficacy. Such engineering initiatives have the potential to be useful in oncology, autoimmunity, inflammation, and infectious disorders, among other fields.[12]

Example: A versatile cytokine known as interleukin-15 links the innate and adaptive immune systems by targeting a number of cell types. It promotes T cell, natural killer cell, and B cell proliferation and differentiation. Among the cytokines, IL-15 has been identified as a top candidate for cancer immunotherapy. However, the major limitations of IL-15 to use as a therapeutic agent are its short-half life and poor bioavailability. To overcome these limitations, we designed and successfully made chimeric

IL-15 by fusing IL-15 with IgG2/IgG2a heavy-chain constant domain with an increased half-life of 40 h. An Indian patent has also been filed for this invention and published (Indian Patent Application No. 201721010096A). IgG2 presents IL-15 to CD8+ T cells via APC, and this action results in a potent Ag-specific memory T-cell response that is necessary for vaccine development and antitumor therapy.[13]

3.2.2.2 ENGINEERING OF PROTEIN ALLOSTERY

Engineering Allostery is involved in the modification or control of protein function.[14] It is the most significant application in the field of biotechnology and medicine. Changes at the dynamic or structural level lead to an increase or decrease in the protein catalytic activity.[14] This protein Allostery allows changes in the signaling pathways of protein. Because this kind of engineering makes the functional change of protein at one side by changing another site of the protein.[14]

Allosteric drugs are very specific, can activate as well as inhibit a protein function, and sometimes have a limit to their impact. Predicting Allosteric sites, Allosteric modulators, and residues implicated in the Allosteric signal propagation is essential for understanding and making use of Allostery. There are different methods involved in the engineering of protein Allostery; computational method, normal mode analysis method, molecular dynamic method, evolutionary method, Allosteric pathway prediction method, etc.[14] Preclinical compounds and probes known as Allosteric legends have been created for a variety of therapeutic targets, such as ion channels,[15,16] capsizes,[17] kinesis,[18] and phospholipases.[19]

Example of allosteric protein; An illustration of kinase inhibition is the regulation of cellular signaling by protein phosphorylation, which occurs in almost all known cellular signaling pathways.[20] Defects in kinase activity result in a variety of illnesses and disorders in humans. As a result, kinases are desirable therapeutic targets and a focus of pharmaceutical industry activity.[21]

3.2.2.3 STIMULI-RESPONSIVE DRUG DELIVERY SYSTEM

The drug delivery system (DDS) is a very important aspect of its action as well as the target specificity of the drug. Several different drug delivery systems have been used to deliver the drug with specific drug action and target purposes.[22] The major domain of the delivery system is to entrap the drug in nanoparticles and control its release by a simulative response called a stimuli-responsive drug delivery system (DDS), which is an internal or external stimulus. Further, it is divided into different parts such as physical and chemical.[22] Physical DDS includes the light, magnetic field, electric field, ultrasound, temperature, osmotic pressure, etc. Chemical DDS includes pH, ionic strength, glucose, etc. One more part called biological includes enzyme-based drug delivery.[22] According to the modified drug physiology used a specific type of stimulus which improves its efficacy at the specific site.[22] A schematic diagram of the same is explained in Figure 3.6.

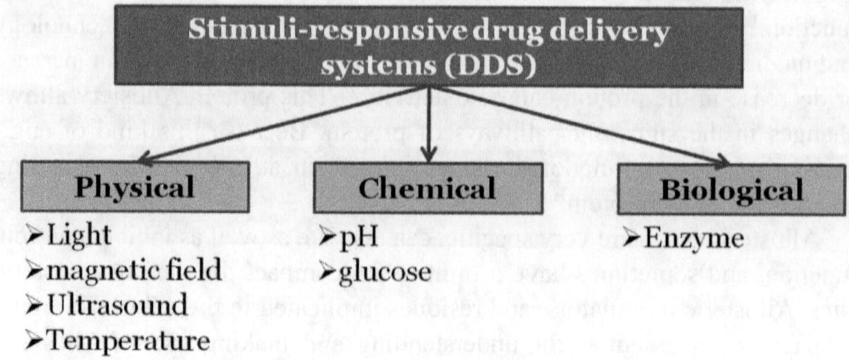

FIGURE 3.6 Different types of stimuli-responsive drug delivery systems.

Source: Created with Biorender.com

3.2.3.1 PHYSICAL STIMULI-RESPONSIVE DRUG DELIVERY SYSTEM

3.2.3.1.1 Light

By adding light-sensitive molecules to drug delivery platforms, light is exploited to improve medication distribution and release. Light is employed as an external stimulus for improved drug delivery, cargo release, imaging, and therapeutics.[22]

Drug release can be controlled by adjusting the wavelength, light intensity, exposure time, and duration.[22] Different wavelengths of light, such as infrared, UV, or laser are applied to release the drug. Examples include the use of near-infrared light (NIR) to control the delivery of protein payload. Silica-coated gold nanorods were created by Tang et al. to be used as a topical medicinal solution for the skin. Transdermal protein administration was facilitated by using the heat produced when NIR light was applied.[23] The amphiphilic titanium oxide nanotubular system for protein distribution invented by Song et al. is another example. This system involved capping hydrophilic titanium oxide nanotubes with hydrophobic monolayers. The precisely controlled removal of the hydrophobic cap, which prevents cargo leakage, was then accomplished upon exposure to the UV signal, resulting in a highly controlled release of the hydrophilic protein payload.[24] Another is continuous-wave laser light, which is also used to deliver drugs and was applied to the skin to raise its temperature. In a study, ovalbumin served as a model protein.[22] As a result, heat shock protein expression was elevated and inflammatory cell migration was facilitated. However, due to the limited penetration through the stratum corneum, pulsed laser treatment did not increase skin temperature. The same method could be used to administer proteins and vaccines transdermally.[22,24]

3.2.3.1.2 Ultrasound

Due to its ease of use, confidentiality, accessibility, and real-time applications, ultrasound has lately been exploited as an external stimulus in drug administration.[25] Insulin is a good example of this use. We produce insulin using recombinant technology, but its efficient release is a crucial component as well. Jin and colleagues developed an insulin-loaded ultrasound-responsive delivery device via an injectable polymeric nanonetwork.[25] A core/shell system was made with the double emulsion approach, consisting of an insulin-loaded matrix for the shell and a network of positively and negatively charged chitosan and alginate for the core.[25] Due to cavitation caused through the focused ultrasound system (FUS), the ultrasound's pulsatile profile caused the release of insulin. Adult diabetic mice given streptozotocin were the animal model utilized to investigate the pulsatile release characteristics of insulin.[25] Subsequent to the nanosystem having been administered for two days, focused ultrasound was used on the injection site for 30 s.[25] The blood glucose level dropped dramatically. Additionally, ultrasound therapy used at regular intervals demonstrated this FUS-triggered insulin administration strategy's capacity to control insulin delivery.[25]

3.2.3.1.3 Magnetic Field

Magnetic-responsive nanosystems can target tumors when an external magnetic field is applied.[26] For instance, by precipitating calcium oleate in ferrofluid using magnetite as the foundation, Chorny et al. were able to produce magnetically responsive nanoparticles (MNP). By using a magnetic field to combat severe oxidative stress, this newly developed nanosystem demonstrated notable therapeutic success.[26] It also demonstrated excellent antioxidant enzyme encapsulation efficiency, including superoxide dismutase and catalase.[26] In order to study the intracellular trafficking of their amino-functionalized mesoporous magnetic hollow nanoparticles (MMHs), Huang et al. used BSA as a model protein cargo. According to a previously described method, negatively charged polystyrene (PS) nanospheres were employed as a template to create the mesoporous hollow structure, which was subsequently removed to create modified MMHs.[27] The altered MMHs with amino groups demonstrated high intracellular protein transport and release targeting certain subcellular compartments, and they also demonstrated more protein-loading efficiency.[27]

3.2.3.1.4 Temperature

The characteristic of thermo-responsive materials is their nonlinear response to temperature and solubility. At a sharp temperature change, the solubility of the thermo-responsive compounds significantly either increases or decreases.[28] Both the hydrophilic and hydrophobic thermo-responsive polymers show a phase shift at a certain temperature. The capacity of thermo-responsive polymers to target many areas with certain physiological circumstances, such as malignant tissues, is its main advantage.[28] For instance, the tumor microenvironment, which has a temperature range of 38–39°C, can collapse long polymer chains. Hydrogels with covalent crosslinks are a significant class of thermo-responsive drug delivery methods. Combining already-formed protein and polymer molecules, the polymer matrix is then chemically crosslinked to keep the protein molecules inside its structure. Protein partitioning in an expanded hydrogel solution is a different technique.[28]

3.2.3.2 CHEMICAL STIMULI-RESPONSIVE DRUG DELIVERY SYSTEM

3.2.3.2.1 pH

In response to physiological pH gradients, stimuli-responsive nanosystems have been created that contain certain pH-sensitive moieties that can target particular areas of the human body. Two main approaches could be used to design such efficient pH-responsive nanosystems. Utilizing polymers with ionizable function groups[29] is the first technique, while including acid-cleavable bonds in polymers[30] is the second. Combining these two strategies has led to the development of numerous pH-responsive nanosystems for the delivery of protein therapies at the cellular or organ level.[31] Those systems have been carefully developed to target specific organs such as intracellular organelles, or even diseased areas like inflammatory tissues or cancer (Tumor) sites. Another application is, proteins can be preserved and released from acidic endo-lysosomal compartments with the aid of cellular-targeted pH-responsive nanosystems, which results in cytoplasmic protein targeting.[32] The benefit of endosomal acidic pH can be used as a targeted release stimulus using the endosomal cycle.

3.2.3.2.2 Glucose

A metabolic condition called diabetes is defined by chronically elevated blood glucose levels.[33] Insulin injections, which are painful for the patients, have been the standard diabetes treatment. Three typical forms have been investigated as drug delivery systems that release insulin upon exposure to glucose. Boronic acids, glucose oxidase, and glucose-binding proteins are some of these.[34] Many researchers suggested an alternative way to administer insulin, using glucose-responsive hydrogel delivery systems that can intelligently react to high blood glucose levels by releasing specific amounts of insulin under controlled circumstances as opposed to continuous invasive insulin administration.[35]

3.2.3.3 BIOLOGICAL STIMULI-RESPONSIVE DRUG DELIVERY SYSTEM

3.2.3.3.1 Enzyme

Certain concentrations of certain enzymes are accessible under specific physiological conditions to play a significant role in biological chemistry.[36] Polymers can be produced and programmed to target specific enzymes in particular physiological circumstances.[37] Furin is one of the enzymes that enzyme-responsive protein delivery systems use as a trigger. A transmembrane protein called furin cycles between the trans-Golgi network and the surface.[38] Furin's high level could be used as a stimulant for the delivery of chemotherapeutics to particular target cells, such as malignant cells, since overexpression of furin is linked to many disorders, including cancer.[37] Bisacrylated peptides were utilized by Biswas et al. to crosslink the polymeric matrix that furin had triggered. The set of enzymes known as matrix metalloproteinases (MMPs), which are involved in the oxidation and turnover of extracellular matrix during organogenesis, growth, and routine tissue turnover, serves as another example. MMPs are overexpressed in several colorectal diseases and cancers.[37] A class of enzymes known as hyaluronidases promotes the breakdown of hyaluronic acid. It is three to ten times higher in prostate cancer, 5-8 times higher in bladder cancer, and raised in breast cancer metastases.[38]

3.2.2.4 *DIRECTED ENZYME PRODRUG THERAPY*

An enzyme is administered specifically to the desired location and acts in directed enzyme prodrug therapy (DEPT).[39] The following step involves systemically delivering a non-bioactive prodrug, which the earlier delivered enzyme then activates locally. To guarantee that the prodrug's enzymatic activation takes place specifically at the desired site, both the prodrug and the activating enzyme often need to resemble natural human enzymes.[39] There are now several classes of DEPT being developed, each with a particular requirement for the engineering of the prodrug-activating enzymes.[39] These classes include antibody directed (ADEPT), gene directed (GDEPT), bacterial directed (BDEPT), and substrate mediated (SMEPT). An antibody–enzyme combination that acts as the effector is a biomolecular recognition (sensing) component of ADEPT.[39] A bacterial carboxypeptidase conjugated to a single-chain variable domain was used in a clinical trial that showed promise in stabilizing individuals with CEA-expressing tumors.[39] However, immunogenicity limited the trial's treatment options. By utilizing humanized antibodies and bacterial prodrug activation enzymes devoid of B- and T-cell epitopes, next-generation ADEPT attempts to circumvent this problem.[39,40] The second class of DEPT, known as GDEPT or "suicide gene therapy," is rapidly gaining popularity as a cancer treatment and normally distributes the therapeutic gene(s) by viral vectors (gene therapy).[41,42] BDEPT, an intriguing variant of GDEPT, uses bacteria that preferentially colonize cancer cells and are usually equipped with therapeutic genes to activate prodrugs locally.[43] The "bystander effect," or leakage of the activated drug to adjacent cells, is generally considered a beneficial component of GDEPT systems because conventional gene delivery methods cannot reach all cancer cells.[41] An enzyme through a substrate for better control of surface-mediated drug distribution, prodrug therapy integrates a prodrug activating enzyme in a hydrogel or other biomaterial. It is also used for anti-inflammatory and cancer therapy.[44]

3.2.2.5 TARGETED DRUG NEUTRALIZATION

As a result, antibodies may effectively serve as an antidote to pharmaceutical agents as a class that also includes other antibodies. Antibody-mediated immunity also provides a broadly applicable mechanistic basis for the control and prevention of both infectious and noninfectious diseases through passive or active immunization.[3]

Specific antidotes that specifically neutralize drugs or toxins are an essential group of therapy modalities to become more widely accessible for a number of targets as a result of recent advancements in the development of antibody treatments.[45] Careful monitoring is necessary for medications having a small therapeutic window, such as the anticoagulant warfarin and the cardiac glycoside digoxin, to keep drug levels within a range that is safe and effective.[45,46] Accidental overdoses of these medications still happen even under thorough medical supervision since every patient's reaction to a drug is different. The accessibility of an antidote can prevent death and improve the efficacy of treatment.[45] It has proven crucial for the treatment of both drugs to be able to counteract the effects of digoxin with immune globulins and warfarin with vitamin K injections.[45,47] For a large spectrum of drugs and toxins, such as anthrax, cocaine, methamphetamine, desipramine, and botulinum toxins, antibody-based antidotes have been created.[48,49] The development of antibodies prevents the anthrax toxin from attaching to cell-surface receptors or prevents the toxin from assembling.[45] The receptor-binding moiety of the botulinum and anthrax toxins undergoes conformational changes throughout their translocation mechanisms, which result in the creation of an ion-conductive pore and the translocation of the corresponding catalytic moieties into the cytosol under low pH conditions.[49] For these toxins, preventing hole formation and subsequent toxin transfer is an effective treatment approach. In addition to the more well-known class of antidotes that depend on molecular recognition for their activity, protein engineering is now used to produce enzymatic antidotes for therapeutic usage.[49] Particularly enzymatic antidotes are meant to treat drug overdoses and prevent addiction. Particularly important is the creation of hydrolases for the neutralization of cocaine.[5] The ability to profoundly alter a protein's function, such as the substrate selectivity of human butyrylcholinesterase, highlights the growing potential of protein engineering to have a wider impact on drug metabolism and medicine in general.[50]

3.2.2.6 TARGETED DRUG ACTIVATION

Drug activation means, when receptors are activated, either a specific response is immediately triggered in the body or the body releases hormones and/or other endogenous medicines to stimulate a certain response.[51] Protein treatments have a natural stimulus-responsive activity, which gives them an edge over conventional small-molecule medications. For example, tissue-type plasminogen activator (t-Pa) efficiently transforms plasminogen to plasmin only after binding to fibrin, a molecular recognition trigger (a proteolytic response).[51] Thus, in response to t-PA administration, clot disintegration takes place locally, close to thrombi (where fibrin is deposited), rather than systemically, leading to fewer adverse effects.[51]

3.2.2.7 ANTIBODY–DRUG CONJUGATES

Monoclonal antibodies (mAbs) have helped develop approved biopharmaceutical products in a number of therapeutic areas, including the treatment of cancer, infections, immunological disorders, and cardiovascular diseases. Antibody–drug conjugates (ADCs), which use a monoclonal antibody to drive the attached medication to specific areas, such as cancer cells, are the simplest stimulus-responsive protein therapeutics.[52,53] Although the idea behind antibody–drug conjugates (ADCs) may seem straightforward, creating effective products is frequently a significant challenge. Numerous variables, some specific to certain tumor types, must be optimized for ADC products. Some of these include ADC stability, pharmacokinetics, immunoreactivity retention, and efficient release of the drug's active form at the location of the disease. Release of the medication should not occur before the ADC has reached its in vivo goal. Selective drug release is made possible via a special linker that connects the antibody and the medication. High selectivity in binding to tumor cells and limited cross-reactivity with healthy tissues are crucial aspects to take into account when selecting a monoclonal antibody. High potency medicines are necessary since only a limited amount of ADC in each dose reaches the target area. Additionally, the linker must be stable enough to prevent early drug release.[52–54] However, vascular targeting and non-internalizing ADCs are also in development. Prior to drug release, the majority of ADCs that are currently on the market are designed for cell internalization via receptor-mediated endocytosis. Hydrazones, which cleave in the acidic cytosolic environment of the lysosomes and endosomes, disulfides, which promote drug release in the alkaline cytosolic environment, and cancer-specific protease susceptible peptides are examples of contemporary responsive linker approaches. To release the medication from "non-cleavable" linkers like thioethers, which are highly durable, the internalized antibody must be enzyme-degraded in lysosomes or endosomes. BR96-doxorubicin is a well-known ADC from this first generation. A chimeric mAb that targets the LewisY tetra-saccharide, which is frequently expressed in human carcinomas, has eight molecules of doxorubicin linked to the hinge cysteines. Doxorubicin, a DNA intercalator, has been recognized for the treatment of numerous varieties of tumors. A hydrazone linker that is acid-labile was used to engineer drug release. The conjugate would be taken up by the cells when the ADC attaches to the cell surface antigens, and the acidic endosomal and lysosomal vesicles would release doxorubicin. When BR96-doxorubicin was administered to tumor-bearing mice and rats, the tumors that exhibited the corresponding antigen were successfully treated.[53] About 35 ADCs are now being studied in clinical trials, and the US Food and Drug Administration (FDA) has approved the sale of two ADCs, brentuximabvedotin and adotrastuzumabemtansine.[4] Adotrastuzumabemtansine, a combination of the cytotoxic medication mertansine with the anti-Her2-mAb trotuzumab, has demonstrated encouraging outcomes in the treatment of advanced HER2-positive breast cancer.[53] Another example, a monoclonal antibody that specifically targets CD30, a protein expressed only in Hodgkin lymphoma, is combined with the anti-mitotic medication monomethylauristatin E to form the compound known as BrentuximabVedotin.[53]

3.2.2.8 ENGINEERED ZYMOGENS

Proteins that have been designed to be dormant until triggered by a particular signal, such as proteolytic cleavage, are known as engineered zymogens.[55] A class of therapeutic proteins known as engineered zymogens reacts to stimuli and acts as both a biosensor and an effector. Engineered zymogens are latent until an activating signal, like proteolytic cleavage, is provided, just like natural zymogens are. There are now several designed zymogens being developed, including zymoxins, dendronized proteases, and engineered coagulation cascade proteins.[55–57]

In order to build a proof of concept for potential revolutionary antiviral therapies that can precisely destroy virally infected cells, researchers created two chimeric toxins whose activation is regulated by HCV-NS3 protease cleavage (thus restricting virus production and distribution).[56] These chimeric toxins are formed by fusing the NS3-activable modified catalytic domains of the bacterial or plant toxins ricin toxin (RTA) or diphtheria toxin (DTA), respectively, with the binding and translocation domains of Pseudomonas exotoxin A.[56] The C terminus of these molecules, which we call "zymoxins," is fused to an inhibitory peptide or domain that has been properly designed (zymogenized toxins). The NS3-sensitive NS5A/B junction sequence, which the viral protease cleaves, is followed by the inhibitory domain. As a result, the toxin is separated from the inhibitory moiety.[57]

When administered to model cells that express NS3 (either cytosolic or membrane-bound NS3), as well as in vitro enzymatic studies, these toxins display a significant level of NS3 protease-dependent activation.[57] A less pronounced but substantial increase in cytotoxicity was seen when HCV-infected cells were used as test subjects. This finding suggests that viral protease-expressing cells can be eliminated at particular toxin concentration ranges.[57]

Synthetic pH-sensitive zymogens have also been produced by self-assembling dendrimers onto the surfaces of trypsin, papain, and DNase I. The dendrimers dissolve when the pH is lowered, allowing the zymogens to activate.[56] A drug's action must be further controlled using two-component systems, such as directed enzyme prodrug therapy or exogenous light-responsive control, to reach the next degree of therapeutic complexity.[56]

3.2.3 NEWER IN-SILICO APPROACH

Recent advancement in protein engineering has been done at two stages, one is computational and the next stage is a laboratory. All those basic protein engineering methods we discussed earlier are first applied to the protein through a computational approach and then developed in the laboratory. Those are the step-by-step process for engineering a protein for protein interaction with its receptor/ligand/binding site according to the functional requirements. Computational methods involve different software and algorithms which help in finding modification sites and predicting the structure and function of those modified proteins.[58]

3.2.3.1 *COMPUTATIONAL APPROACH FOR PROTEIN ENGINEERING*

The base for protein engineering has been established through the prominent development of techniques based on modifications at the molecular and protein levels.[58] The main issue in protein engineering, which residues to target and what modifications to introduce, was initially approached using either a rational or random approach. Expert knowledge is required for rational engineering, and protein structures are becoming more and more available.[59] The random technique introduces a large number of mutations to be examined in an effort to mimic spontaneous evolution.[1,7] A recent and newer approach "focused mutant protein engineering" is the current trend and combination of these two methods. Using computer predictions and rational components, this approach prioritizes the most likely protein locations for mutagenesis, resulting in smart-and-small mutant libraries with a high percentage of functional variations.[59] Refer to Figure 3.7.

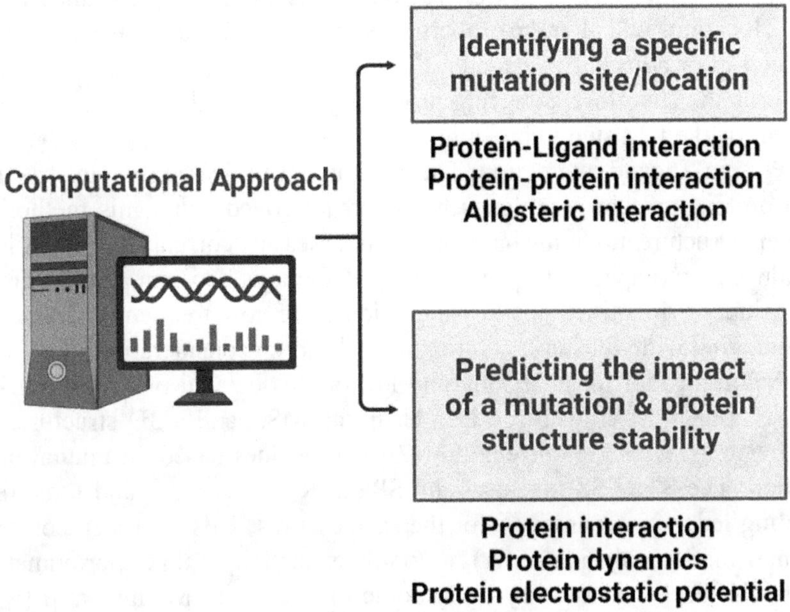

FIGURE 3.7 Two stages of *in silico* protein engineering technique.

Source: Created with Biorender.com

Computational tools have become essential for these types of engineering processes as a result of the enormous advancements in computer technology, which have led to the availability of protein structural models and algorithms.[6,60] Such strategies consistently aid in the effective distribution of proteins used in a variety of biotechnologies. Numerous new tools with various uses and objectives are created and released every year as a result of ongoing efforts to incorporate increasingly complex computational methodologies into targeted mutant protein editing.[60] We stated some tools for the sequential engineering stages that choose the most promising mutations with the highest interest function: (a) Identifying a specific mutation site and (b) predicting the impact of a mutation (Fig. 3.7)

3.2.3.1.1 Identifying a Specific Mutation Site/Location

Depending on the specific target attribute, a variety of methodologies utilizing protein structure for mutation site selection can be used, such as concentrating on ligand-binding pathways,[61] allosteric networks,[62] the proximity of binding sites, or flexible regions.[63] It is more likely that mutations at these locations will alter the properties of the protein because they are usually located in or adjacent to structurally or functionally significant portions of the protein. Therefore, selecting promising and crucial protein locations to modify is the first step in bioengineering.[63]

Protein–Ligand interaction: The accuracy of projected protein structures and protein–protein complexes has been improved using this method.[64,65] Protein structure modeling has a history of assessing correlated or coevolving residues to anticipate direct or indirect interactions between pairs or groups of residues.[66] Identified coevolving residues are now frequently targeted by engineering with allosteric control and stability in consideration.[67]

A current tool for analyzing and locating associated or coevolved sites is the visual CMAT web server.[68] Using an MSA and a 3D structure, this approach assesses the correlation between residues based on mutual information. The STRESS tool uses the SPACER web server[69] and a modified binding leverage technique[70] for the surface sites. This approach combines normal mode analysis (NMA), a computer technique that approximates a system's local dynamics by a harmonic motion, with Monte Carlo (MC)-based ligand docking.

Protein–protein interaction: Protein interactions are essential for the bulk of biological processes, such as substrate identification, metabolism, signaling, pathogen recognition, protein-enhanced activity, and protein deactivation. The implicated residues have drawn a lot of attention due to their ability to disrupt or enhance activity, know more about interactions, or serve as a guide for predicting protein structures.[71]

The PPI3D web server was made available as a user-friendly alternative that focuses on protein–protein interactions.[72] This web server uses a carefully selected local database to retrieve data and do analytics. PPI3D assesses and identifies three types of interactions: protein–protein, protein–peptide, and domain–domain. The objective is typically to improve the binding affinity or avoid modifying those residues, as opposed to aiming to destabilize the protein complex while searching for interface sites. Disruption is sometimes desired, though, which is why the DisruPPI program was created.[73] DisruPPI analyzes the stability of each monomer and their interactions to find site residues and disruptive mutations that prevent binding while preserving or enhancing the stability of the monomers.

Allosteric interaction: The AlloSigMA online service employs NMA to evaluate the effects of ligand binding or a mutation on allostery in addition to identifying sites of mutation.[74] The software CaverDock was built as a better option for expensive assessments of ligand transport conducted using MD simulations.[75] It requires tunnels for a specific macromolecule to be precalculated using software like CAVER 3.02.[76] The ligand is then docked with a modified AutoDockVina tool to systematically explore these tunnels.[77]

To automate these calculations, CaverDock has been connected to a Caver Web 1.0 server.[78] In addition to automatically calculating tunnels for a protein of interest, this server offers follow-up analyses of ligand transport through the user-selected tunnel. This server provides follow-up evaluations of ligand transport through the user-selected tunnel in addition to the automatic computation of tunnels for a protein of interest.[78]

3.2.3.1.2 Predicting the Impact of a Mutation and Protein Structure Stability

After the site of the mutation has been chosen, the next step is to determine how specific mutations in the region will affect the protein's desired qualities.

Protein interaction: Due to the significance of protein–protein interactions, many methods have been created to predict the impact of mutations in the residues creating these interactions.[79] The MutaBind online service was created to broaden the range of publicly accessible techniques for quantitatively forecasting the effects of single-point mutations on binding energy.[80]

Protein dynamics: Mutations can modify the flexibility or firmness of proteins, affecting their dynamics.[81] With the use of MD simulations, a potent technique that is computationally taxing and demands a high level of expertise, it is necessary to examine the impact of mutations on protein dynamics. As an alternative, rapid and ready-to-use techniques like NMA have been introduced. The DynaMut webserver[82] uses the NMA tools Bio3D and ENCoM[83] to make it easier to use NMA to assess the effects of mutation on protein dynamics and stability. To characterize the WT structure, the graph-based signature for the mCsm-PPI2, mCsm-NA, and mcsm-lig tools incorporate structural elements of the changed residue's surroundings, such as solvent-accessible surface area, residue depth, and secondary structure.

Protein electrostatic potential: Studies[84,85] have shown that proteins' electrostatic potential can affect protein adsorption, ligand binding, or the thermodynamic stability of the protein complex. There are numerous computer programs that can estimate the electrostatic potential of proteins.[86] The online server uses PDB2PQR to assign atom parameters, a Modeller to optimize mutant structures, and APBS[87] to determine electrostatic potential. As a result, each residue's electrostatic difference is shown as a histogram. The web server also enables the downloading of 3D structures and electrostatic maps for PyMOL or VMD visualization.[88]

3.3 CONCLUSION

The use of the protein for its functional requirements determines the protein engineering techniques that are used. Protein engineering is most commonly used in biomedical and associated fields. Three main techniques for protein engineering were covered here. (1) Genetic level protein engineering: In this method, protein modifications are made at the gene level to enhance protein function or to achieve specific protein function modification goals. It uses a variety of methods, including CRISPR, T-RNA-mediated, directed evolution, and mutagenesis. (2) Protein-level protein engineering: This protein-based methodology uses a variety of techniques to primarily enhance the stability, specificity, and half-life of proteins for their improved therapeutic usage in the biomedical or industrial fields. In the field of pharmaceutical biotechnology, creating a high-quality medication or engineering the protein is insufficient. However, it is similarly ineffective to deliver created proteins or medications to the target spot with specificity and stability after their construction. As a result, there are a number of factors that can be delivered to make the manufactured product more particular and stable throughout the translation. These features are also included in protein engineering techniques. (3)

In-silico computational approach: computational methods use a variety of tools and algorithms to locate modification sites and anticipate the structure and functionality of proteins that have undergone modification.

ACKNOWLEDGMENT

We acknowledge DBT (Department of Biotechnology), Govt. of India (BT/PR7857/BRB/10/1215/2013) and GSBTM (Gujarat State Biotechnology Mission), Govt. of Gujarat (GSBTM/JDR&D/608/2020/455) for research grants to SD; GSBTM for fellowship to DP; Nirma University for fellowship to HP and PS. Thanks to BioRender ("Figures Made in BioRender.com") and Dr. LalitVora, for help regarding the figures preparation of the book chapter.

KEYWORDS

- **protein engineering**
- **PCR**
- **CRISPR-Cas**
- **toxin**

REFERENCES

1. Arnold, F. H. Engineering Proteins for Nonnatural Environments. *FASEB J.* **1993,** *7* (9), 744–749.
2. Arnold, F. H. Engineering Proteins for Nonnatural Environments. *FASEB J.* **1993,** *7* (9), 744–749.
3. Tobin, P. H.; Richards, D. H.; Callender, R. A.; Wilson, C. J. Protein Engineering: A New Frontier for Biological Therapeutics. *Curr. Drug Metab.* **2014,** *15* (7), 743–756.
4. Beck, A.; Reichert, J. M. Antibody-Drug Conjugates: Present and Future. *mAbs* **2014,** *6* (1), 15–17.
5. Gorelick, D. A. Pharmacokinetic Strategies for Treatment of Drug Overdose and Addiction. *Future Med. Chem.* **2012,** *4* (2), 227–243.
6. Sun, M. G.; Seo, M.-H.; Nim, S.; Corbi-Verge, C.; Kim, P. M. Protein Engineering by Highly Parallel Screening of Computationally Designed Variants. *Sci Adv.* **2016,** *2* (7), e1600692.
7. Arnold, F. H. Innovation by Evolution: Bringing New Chemistry to Life (Nobel Lecture). *Angew. Chem. Int. Ed. Engl.* **2019,** *58* (41), 14420–14426.

8. Antikainen, N. M.; Martin, S. F. Altering Protein Specificity: Techniques and Applications. *Bioorg. Med. Chem.* **2005,** *13* (8), 2701–2716.
9. Anderson, D. P.; Benns, H. J.; Tate, E. W.; Child, M. A. CRISPR-TAPE: Protein-centric CRISPR Guide Design for Targeted Proteome Engineering. *Mol. Syst. Biol.* **2020,** *16* (6), e9475.
10. Levin, D.; Golding, B.; Strome, S. E.; Sauna, Z. E. Fc Fusion as a Platform Technology: Potential for Modulating Immunogenicity. *Trends Biotechnol.* **2015,** *33* (1), 27–34.
11. Czajkowsky, D. M.; Hu, J.; Shao, Z.; Pleass, R. J. Fc-fusion Proteins: New Developments and Future Perspectives. *EMBO Mol. Med.* **2012,** *4* (10), 1015–1028.
12. Benjamin, W.; Sun, Y.-N. Pharmacokinetics of Peptide–Fc Fusion Proteins. *J. Pharm. Sci.* **2014,** *103* (1), 53–64.
13. Patidar, M.; Yadav, N.; Dalai, S. K. Development of Stable Chimeric IL-15 for Trans-Presentation by the Antigen Presenting Cells. *Front. Immunol.* **2021,** *12*, 646159.
14. Wenthur, C. J.; Gentry, P. R.; Mathews, T. P.; Lindsley, C. W. Toxicology, Drugs for Allosteric Sites on Receptors. *Annu. Rev. Pharmacol. Toxicol.* **2014,** *54*, 165.
15. Monaghan, D. T.; Irvine, M. W.; Costa, B. M.; Fang, G.; Jane, D. E. Pharmacological Modulation of NMDA Receptor Activity and the Advent of Negative and Positive Allosteric Modulators. *Neurochem. Int.* **2012,** *61* (4), 581–592.
16. Taly, A.; Corringer, P.-J.; Guedin, D.; Lestage, P.; Changeux, J.-P. Nicotinic Receptors: Allosteric Transitions and Therapeutic Targets in the Nervous System. *Nat. Rev. Drug Discov.* **2009,** *8* (9), 733–750.
17. Häcker, H.-G.; Sisay, M. T.; Gütschow, M. Therapeutics, Allosteric Modulation of Caspases. Pharmacol. Ther. **2011,** *132* (2), 180–195.
18. Fang, Z.; Grütter, C.; Rauh, D. Strategies for the Selective Regulation of kinases with Allosteric Modulators: Exploiting Exclusive Structural Features. *ACS Chem. Biol.* **2013,** *8* (1), 58–70.
19. Selvy, P. E.; Lavieri, R. R.; Lindsley, C. W.; Brown, H. A. Phospholipase D: Enzymology, Functionality, and Chemical Modulation. *Chem. Rev.* **2011,** *111* (10), 6064–6119.
20. Johnson, L. N.; Lewis, R. J. Structural Basis for Control by Phosphorylation. *Chem. Rev.* **2001,** *101* (8), 2209–2242.
21. Pawson, T.; Scott, J. D. J. Protein Phosphorylation in Signaling–50 Years and Counting. *Trends Biochem. Sci.* **2005,** *30* (6), 286–290.
22. Fomina, N.; Sankaranarayanan, J.; Almutairi, A. J. A. D. D. R. Photochemical Mechanisms of Light-Triggered Release from Nanocarriers. **2012,** *64* (11), 1005–1020.
23. Tang, H.; Kobayashi, H.; Niidome, Y.; Mori, T.; Katayama, Y.; Niidome, T. J. J. O. C. R. CW/pulsed NIR Irradiation of Gold Nanorods: Effect on Transdermal Protein Delivery Mediated by Photothermal Ablation. **2013,** *171* (2), 178–183.
24. Song, Y.-Y.; Schmidt-Stein, F.; Bauer, S.; Schmuki, P. J. J. O. T. A. C. S. Amphiphilic TiO2 Nanotube Arrays: An Actively Controllable Drug Delivery System. **2009,** *131* (12), 4230–4232.
25. Ferrara, K. W. J. A. D. D. R. Driving Delivery Vehicles with Ultrasound. **2008,** *60* (10), 1097–1102.
26. Chorny, M.; Fishbein, I.; Forbes, S.; Alferiev, I. J. I. L. Magnetic Nanoparticles for Targeted Vascular Delivery. **2011,** *63* (8), 613–620.
27. Huang, Z.; Tang, F. J. J. O. C.; Science, I. Preparation, Structure, and Magnetic Properties of Mesoporous Magnetite Hollow Spheres. **2005,** *281* (2), 432–436.

28. Schmaljohann, D. J. A. D. D. R. Thermo-and pH-Responsive Polymers in Drug Delivery. **2006,** *58* (15), 1655–1670.
29. Gu, X.; Wang, J.; Wang, Y.; Wang, Y.; Gao, H.; Wu, G. J. C.; Biointerfaces, S. B. Layer-By-Layer Assembled Polyaspartamide Nanocapsules for pH-Responsive Protein Delivery. **2013,** *108*, 205–211.
30. Bae, Y.; Fukushima, S.; Harada, A.; Kataoka, K. J. A. C. Design of Environment-Sensitive Supramolecular Assemblies for Intracellular Drug Delivery: Polymeric Micelles that are Responsive to Intracellular pH Change. **2003,** *115* (38), 4788–4791.
31. Gao, W.; Chan, J. M.; Farokhzad, O. C. J. M. P. pH-Responsive Nanoparticles for Drug Delivery. **2010,** *7* (6), 1913–1920.
32. Dominska, M.; Dykxhoorn, D. M. J. J. O. C. S. Breaking Down the Barriers: siRNA Delivery and Endosome Escape. **2010,** *123* (8), 1183–1189.
33. Wu, Q.; Wang, L.; Yu, H.; Wang, J.; Chen, Z. J. C. R. Organization of Glucose-Responsive Systems and Their Properties. **2011,** *111* (12), 7855–7875.
34. Wu, W.; Zhou, S. J. M. B. Responsive Materials for Self-R Egulated Insulin Delivery. **2013,** *13* (11), 1464–1477.
35. Mo, R.; Jiang, T.; Di, J.; Tai, W.; Gu, Z. J. Emerging Micro-and Nanotechnology Based Synthetic Approaches for Insulin Delivery. *Chem. Soc. Rev.* **2014,** *43* (10), 3595–3629.
36. De La Rica, R.; Aili, D.; Stevens, M. M. Enzyme-Responsive Nanoparticles for Drug Release and Diagnostics. *Adv. Drug Deliv. Rev.* **2012,** *64* (11), 967–978.
37. Guo, D.-S.; Wang, K.; Wang, Y.-X.; Liu, Y. J. J. O. T. A. C. S. Cholinesterase-Responsive Supramolecular Vesicle. **2012,** *134* (24), 10244–10250.
38. Hajdin, K.; d'Alessandro, V.; Niggli, F. K.; Schäfer, B. W.; Bernasconi, M. J. P. O. Furin Targeted Drug Delivery for Treatment of Rhabdomyosarcoma in a Mouse Model. *PLoS One.* **2010,** *5* (5), e10445.
39. Bagshawe, K. D. Antibody-Directed Enzyme Prodrug Therapy (ADEPT) for Cancer. *Expert Rev. Anticancer Ther.* **2006,** *6* (10), 1421–1431.
40. Weidle, U. H.; Tiefenthaler, G.; Georges, G. Proteases as Activators for Cytotoxic Prodrugs in Antitumor Therapy. *Cancer Genom. Proteom.* **2014,** *11* (2), 67–79.
41. Dachs, G. U.; Tupper, J.; Tozer, G. M. From Bench to Bedside for Gene-Directed Enzyme Prodrug Therapy of Cancer. *Anticancer Drugs* **2005,** *16* (4), 349–359.
42. Duarte, S.; Carle, G.; Faneca, H.; de Lima, M. C.; Pierrefite-Carle, V. Suicide Gene Therapy in Cancer: Where Do We Stand Now? *Cancer Lett.* **2012,** *324* (2), 160–170.
43. Lehouritis, P.; Springer, C.; Tangney, M. Bacterial-Directed Enzyme Prodrug Therapy. *J. Control. Release* **2013,** *170* (1), 120–131.
44. Fejerskov, B.; Zelikin, A. N. Substrate Mediated Enzyme Prodrug Therapy. *PLoS ONE* **2012,** *7* (11), e49619.
45. Ehle, M.; Patel, C.; Giugliano, R. P. Digoxin: Clinical Highlights: A Review of Digoxin and Its Use in Contemporary Medicine. *Crit. Pathways Cardiol.* **2011,** *10* (2), 93–98.
46. Lenzini, P.; Wadelius, M.; Kimmel, S.; Anderson, J. L.; Jorgensen, A. L.; Pirmohamed, M.; Caldwell, M. D.; Limdi, N.; Burmester, J. K.; Dowd, M. B.; Angchaisuksiri, P.; Bass, A. R.; Chen, J.; Eriksson, N.; Rane, A.; Lindh, J. D.; Carlquist, J. F.; Horne, B. D.; Grice, G.; Milligan, P. E.; Eby, C.; Shin, J.; Kim, H.; Kurnik, D.; Stein, C. M.; McMillin, G.; Pendleton, R. C.; Berg, R. L.; Deloukas, P.; Gage, B. F. Integration of Genetic, Clinical, and INR Data to Refine Warfarin Dosing. *Clin. Pharmacol. Ther.* **2010,** *87* (5), 572–578.

47. Levi, M.; Eerenberg, E.; Kamphuisen, P. W. Bleeding Risk and Reversal Strategies for Old and New Anticoagulants and Antiplatelet Agents. *J. Thromb. Haemost.: JTH* **2011,** *9* (9), 1705–1712.
48. Rainey, G. J.; Young, J. A. Antitoxins: Novel Strategies to Target Agents of Bioterrorism. *Nat. Rev. Microbiol.* **2004,** *2* (9), 721–726.
49. Schiele, F.; van Ryn, J.; Canada, K.; Newsome, C.; Sepulveda, E.; Park, J.; Nar, H.; Litzenburger, T. A Specific Antidote for Dabigatran: Functional and Structural Characterization. *Blood* **2013,** *121* (18), 3554–3562.
50. Xue, L.; Ko, M.-C.; Tong, M.; Yang, W.; Hou, S.; Fang, L.; Liu, J.; Zheng, F.; Woods, J. H.; Tai, H.-H. Design, Preparation, and Characterization of High-Activity Mutants of Human Butyrylcholinesterase Specific for Detoxification of Cocaine. *Mol. Pharmacol.* **2011,** *79* (2), 290–297.
51. Craik, C. S.; Page, M. J.; Madison, E. L. Proteases as Therapeutics. *Biochem. J.* **2011,** *435* (1), 1–16.
52. Alley, S. C.; Okeley, N. M.; Senter, P. D. J. C. O. I. C. B. Antibody–Drug Conjugates: Targeted Drug Delivery for Cancer. *Curr. Opin. Chem. Biol.* **2010,** *14* (4), 529–537.
53. Casi, G.; Neri, D. Antibody-Drug Conjugates: Basic Concepts, Examples and Future Perspectives. *J. Control. Release* **2012,** *161* (2), 422–428.
54. Teicher, B. A.; Doroshow, J. H. The Promise of Antibody-Drug Conjugates. *N. Engl. J. Med.* **2012,** *367* (19), 1847–1848.
55. Ivanciu, L.; Toso, R.; Margaritis, P.; Pavani, G.; Kim, H.; Schlachterman, A.; Liu, J. H.; Clerin, V.; Pittman, D. D.; Rose-Miranda, R.; Shields, K. M.; Erbe, D. V.; Tobin, J. F.; Arruda, V. R.; Camire, R. M. A Zymogen-Like Factor Xa Variant Corrects the Coagulation Defect in Hemophilia. *Nat. Biotechnol.* **2011,** *29* (11), 1028–1033.
56. Ng, D. Y.; Arzt, M.; Wu, Y.; Kuan, S. L.; Lamla, M.; Weil, T. Constructing Hybrid Protein Zymogens Through Protective Dendritic Assembly. *Angew. Chem. (Int. ed. Engl.)* **2014,** *53* (1), 324–328.
57. Shapira, A.; Gal-Tanamy, M.; Nahary, L.; Litvak-Greenfeld, D.; Zemel, R.; Tur-Kaspa, R.; Benhar, I. Engineered Toxins "Zymoxins" are Activated by the HCV NS3 Protease by Removal of an Inhibitory Protein Domain. *PLoS One* **2011,** *6* (1), e15916.
58. Adli, M. J. N. C. The CRISPR Tool kit for Genome Editing and Beyond. *Nat. Commun.* **2018,** *9* (1), 1–13.
59. Thomas, P. G.; Russell, A. J.; Fersht, A. Tailoring the pH Dependence of Enzyme Catalysis Using Protein Engineering. *Nature* **1985,** *318* (6044), 375–376.
60. Pincus, D.; Pandey, J. P.; Feder, Z. A.; Creixell, P.; Resnekov, O.; Reynolds, K. A. Engineering Allosteric Regulation in Protein kinases. *Sci. Signal.* **2018,** *11* (555), eaar3250.
61. Morley, K. L.; Kazlauskas, R. J. Improving Enzyme Properties: When are Closer Mutations Better? *Trends Biotechnol.* **2005,** *23* (5), 231–237.
62. Rivalta, I.; Sultan, M. M.; Lee, N.-S.; Manley, G. A.; Loria, J. P.; Batista, V. S. Allosteric Pathways in Imidazole Glycerol Phosphate Synthase. **2012,** *Proc. Natl Acad. Sci. U. S. A. 109* (22), E1428–E1436.
63. Yu, H.; Huang, H. Engineering Proteins for Thermostability Through Rigidifying Flexible Sites. *Biotechnol. Adv.* **2014,** *32* (2), 308–315.
64. Ovchinnikov, S.; Kamisetty, H.; Baker, D. Robust and Accurate Prediction of Residue–Residue Interactions Across Protein Interfaces Using Evolutionary Information. *Elife* **2014,** *3*, e02030.

65. Kamisetty, H.; Ovchinnikov, S.; Baker, D. Assessing the Utility of Coevolution-Based Residue–Residue Contact Predictions in a Sequence-and Structure-Rich Era. Proc. Natl Acad. Sci. U. S. A. **2013,** *110* (39), 15674–15679.
66. Morcos, F.; Schafer, N. P.; Cheng, R. R.; Onuchic, J. N.; Wolynes, P. G. Coevolutionary Information, Protein Folding Landscapes, and the Thermodynamics of Natural Selection. *Proc. Natl Acad. Sci. U. S. A.* **2014,** *111* (34), 12408–12413.
67. Franceus, J.; Verhaeghe, T.; Desmet, T. Biotechnology, Correlated Positions in Protein Evolution and Engineering. *J. Ind. Microbiol. Biotechnol.* **2017,** *44* (4–5), 687–695.
68. Suplatov, D.; Sharapova, Y.; Timonina, D.; Kopylov, K.; Švedas, V. The visualcmat: A Web-Server to Select and Interpret Correlated Mutations/Co-Evolving Residues in Protein Families. *J. Bioinform. Comput. Biol.* **2018,** *16* (02), 1840005.
69. Goncearenco, A.; Mitternacht, S.; Yong, T.; Eisenhaber, B.; Eisenhaber, F.; Berezovsky, I. N. SPACER: server for Predicting Allosteric Communication and Effects of Regulation. *Nucleic Acids Res.* **2013,** *41* (W1), W266–W272.
70. Mitternacht, S.; Berezovsky, I. N. Binding Leverage as a Molecular Basis for Allosteric Regulation. **2011,** *7* (9), e1002148.
71. Kawabata, T. HOMCOS: An Updated Server to Search and Model Complex 3D Structures. *J. Struct. Funct. Genomics* **2016,** *17* (4), 83–99.
72. Dapkūnas, J.; Timinskas, A.; Olechnovič, K.; Margelevičius, M.; Dičiūnas, R.; Venclovas, Č. The PPI3D Web Server for Searching, Analyzing and Modeling Protein–Protein Interactions in the Context of 3D Structures. *Bioinformatics* **2017,** *33* (6), 935–937.
73. Choi, Y.; Furlon, J. M.; Amos, R. B.; Griswold, K. E.; Bailey-Kellogg, C. DisruPPI: Structure-Based Computational Redesign Algorithm for Protein Binding Disruption. *Bioinformatics* **2018,** *34* (13), i245–i253.
74. Guarnera, E.; Tan, Z. W.; Zheng, Z.; Berezovsky, I. N. AlloSigMA: Allosteric Signaling and Mutation Analysis Server. *Bioinformatics* **2017,** *33* (24), 3996–3998.
75. Vavra, O.; Filipovic, J.; Plhak, J.; Bednar, D.; Marques, S. M.; Brezovsky, J.; Stourac, J.; Matyska, L.; Damborsky, J. CaverDock: A Molecular Docking-Based Tool to Analyse Ligand Transport Through Protein Tunnels and Channels. *Bioinformatics* **2019,** *35* (23), 4986–4993.
76. Chovancova, E.; Pavelka, A.; Benes, P.; Strnad, O.; Brezovsky, J.; Kozlikova, B.; Gora, A.; Sustr, V.; Klvana, M.; Medek, P. CAVER 3.0: A Tool for the Analysis of Transport Pathways in Dynamic Protein Structures. *PLoS Comput. Biol.* **2012.**
77. Trott, O.; Olson, A. J. AutoDock Vina: Improving the Speed and Accuracy of Docking with a New Scoring Function, Efficient Optimization, and Multithreading. *J. Comput. Chem.* **2010,** *31* (2), 455–461.
78. Stourac, J.; Vavra, O.; Kokkonen, P.; Filipovic, J.; Pinto, G.; Brezovsky, J.; Damborsky, J.; Bednar, D. Caver Web 1.0: Identification of Tunnels and Channels in Proteins and Analysis of Ligand Transport. *Nucleic Acids Res.* **2019,** *47* (W1), W414–W422.
79. Brender, J. R.; Zhang, Y. Predicting the Effect of Mutations on Protein-Protein Binding Interactions Through Structure-Based Interface Profiles. *PLoS Comput. Biol.* **2015,** *11* (10), e1004494.
80. Boulenouar, S.; Doisne, J.-M.; Sferruzzi-Perri, A.; Gaynor, L. M.; Kieckbusch, J.; Balmas, E.; Yung, H. W.; Javadzadeh, S.; Volmer, L.; Hawkes, D. A. The Residual Innate Lymphoid Cells in NFIL3-Deficient Mice Support Suboptimal Maternal Adaptations to Pregnancy. *Front. Immunol.* **2016,** *7,* 43.

81. Otten, R.; Liu, L.; Kenner, L. R.; Clarkson, M. W.; Mavor, D.; Tawfik, D. S.; Kern, D.; Fraser, J. S. Rescue of Conformational Dynamics in Enzyme Catalysis by Directed Evolution. *Nat. Commun.* **2018,** *9* (1), 1–11.
82. Rodrigues, C. H.; Pires, D. E.; Ascher, D. B. DynaMut: Predicting the Impact of Mutations on Protein Conformation, Flexibility and Stability. *Nucleic Acids Res.* **2018,** *46* (W1), W350–W355.
83. Frappier, V.; Najmanovich, R. J. Coarse-Grained Elastic Network Atom Contact Model and Its Use in the Simulation of Protein Dynamics and the Prediction of the Effect of Mutations. *PLoS Comput. Biol.* **2014,** *10* (4), e1003569.
84. Borgia, A.; Borgia, M. B.; Bugge, K.; Kissling, V. M.; Heidarsson, P. O.; Fernandes, C. B.; Sottini, A.; Soranno, A.; Buholzer, K. J.; Nettels, D. Extreme Disorder in an Ultrahigh-Affinity Protein Complex. *Nature.* **2018,** *555* (7694), 61–66.
85. Pakulska, M. M.; Elliott Donaghue, I.; Obermeyer, J. M.; Tuladhar, A.; McLaughlin, C. K.; Shendruk, T. N.; Shoichet, M. S. Encapsulation-Free Controlled Release: Electrostatic Adsorption Eliminates the Need for Protein Encapsulation in PLGA Nanoparticles. *Sci. Adv.* **2016,** *2* (5), e1600519.
86. Walsh, I.; Minervini, G.; Corazza, A.; Esposito, G.; Tosatto, S. C.; Fogolari, F. Bluues Server: Electrostatic Properties of Wild-Type and Mutated Protein Structures. *Bioinformatics.* **2012,** *28* (16), 2189–2190.
87. Baker, N. A.; Sept, D.; Joseph, S.; Holst, M. J.; McCammon, J. A. Electrostatics of Nanosystems: Application to Microtubules and the Ribosome. *Proc. Natl. Acad. Sci. U. S. A.* **2001,** *98* (18), 10037–10041.
88. Humphrey, W.; Dalke, A.; Schulten, K. J. J. O. M. G. VMD: Visual Molecular Dynamics. *J. Mol. Graph.* **1996,** *14* (1), 33–38.

CHAPTER 4

Recombinant DNA Technology

PANKTI C. BALAR[1], JAGRUT SHAH[2], DIGNA PATEL[2], YESHA SHAH[3], and VIVEK P. CHAVDA[4*]

[1]*Pharmacy Section, L.M. College of Pharmacy, Ahmedabad, Gujarat, India*

[2]*Institute of Science, Nirma University, Ahmedabad, Gujarat, India*

[3]*PharmD Section, L.M. College of Pharmacy, Ahmedabad, Gujarat, India*

[4]*Department of Pharmaceutics and Pharmaceutical Technology, L.M. College of Pharmacy, Ahmedabad, Gujarat, India*

**Corresponding author.*

ABSTRACT

Recombinant DNA (rDNA) technology is an important tool of biochemistry that allows scientists to manipulate small fragments of deoxyribonucleic acid (DNA) and study them in detail. It is an important part of biochemistry where enzymes are utilized along with other laboratory techniques in order to isolate the DNA of interest and then combine it or splice it from a different species. This can lead to the formation of a gene with different functions. Amplification of this rDNA has a lot of applications such as the production of insulin, cancer therapeutics, vaccine production, and many more. This chapter summarizes the detailed process of rDNA technology along with its clinical applications.

4.1 INTRODUCTION

Biotechnology is the regulated use of biological agents for human benefit. It is a synthesis of biology and technology having the combined application of biochemistry, microbiology, and engineering disciplines to maximize the potential of microorganisms, and cultivated tissues/cells.

Advances in Pharmaceutical Biotechnology, Volume 1: Exploring Cutting-Edge Innovations and Applications.
Vivek P. Chavda and Zhe-Sheng (Jason) Chen (Eds.)

DOI: 10.1201/9781774919798-4

Biotechnology is currently used mostly in agriculture and medicine, with contemporary techniques permitting the manufacture of new and enhanced foods. Biotechnology is also being used to regulate pollution, mine, and produce energy. Toxic pollutants from industrial effluents and oil spills are cleaned up using genetically altered microorganisms and plants.[1]

Peter Lobban suggested the concept of rDNA in 1972 and 1973. Stanford University received a US patent in 1980 after applying it in 1974.[2] Human insulin, invented by Genentech and licensed by Eli Lilly and Company, was the first licensed medication produced utilizing rDNA technology. Since DNA molecules from all animals have the same chemical structure, the concept of rDNA is feasible.[3]

rDNA molecules are DNA molecules created in laboratories using genetic recombination techniques to combine genetic material from diverse sources, resulting in sequences not seen in biological organisms.[4] It is employed in the development of both live and dead vaccines with increased responsiveness and excellent specificity. It is also used to identify the protein component of a virus or microbial pathogen that can stimulate the generation of antibodies capable of neutralizing pathogen infectivity while possibly shielding the host from the pathogen.[4]

4.2 GENETIC MAKEUP

The term genetic comes from the Greek word genetikos meaning, "to generate." It is the study corresponding to gene and how effectively it passes information on to the next generation. It basically deals with DNA.[5] From the earlier research based on DNA, it carries genetic information, which helps in functioning of organism. Moreover, it is present in a twisted form that contains deoxyribose and a phosphate group. It contains four bases, i.e., Adenine (A), Guanine (G), Cytosine (C), and Thymine (T). As all the bases are present in a stranded manner, they are connected through a chemical bond. Hence, providing sufficient information on the transcriptional, translational, and post-translational modification of DNA, RNA, and protein (Figure 4.1).[6]

The genetic constitution, commonly known as the genotype, contains the complete set of hereditary information possessed by an organism. In the human species, the storage of genetic information occurs within DNA molecules, which are densely packed into 23 pairs of chromosomes located within the nucleus of each cell. The human genome is composed of around 3 billion base pairs, which contain a coding sequence for approximately

20,000–25,000 genes. These genetic factors govern a wide range of characteristics, encompassing physical traits such as stature and hair pigmentation, as well as inclinations toward specific ailments and the manner in which our bodies process pharmaceutical substances. Nonetheless, it is imperative to acknowledge that our genetic makeup does not only determine our future; rather, other external influences such as the environment, lifestyle choices, and other variables significantly contribute to shaping our overall well-being and other consequential results.[7] The complex interaction between our genetic predispositions and the external factors in our environment is responsible for shaping our distinct personal experiences. The completion of the Human Genome Project in 2003 provided a comprehensive depiction of the entirety of human genes, so initiating a new phase characterized by individualized medicine and enhanced understanding of human biology.[8]

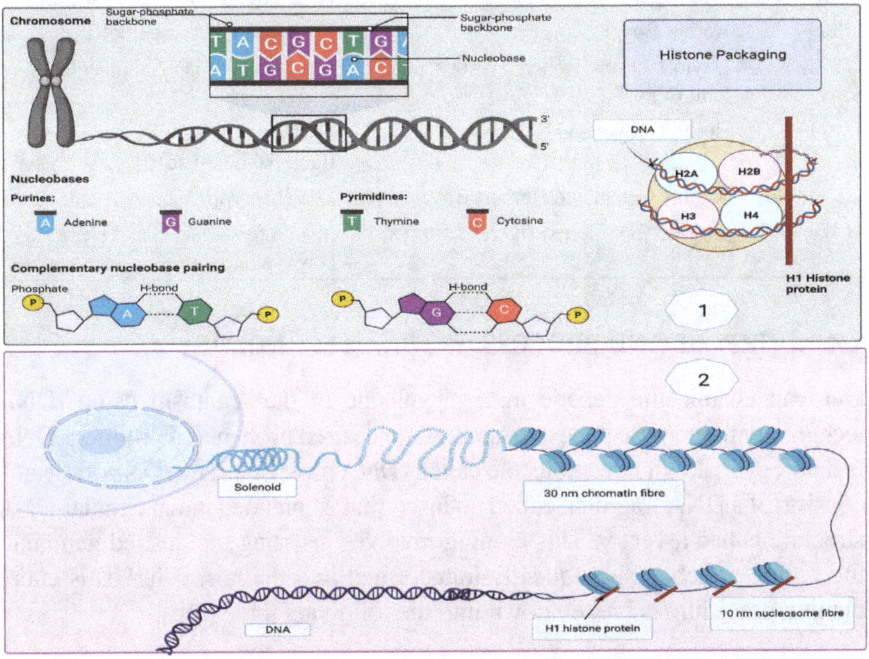

FIGURE 4.1 The components of a chromosome. (1) It is made up of DNA wrapped helically around histone proteins that are stacked to form a chromosome. DNA is a double helix structure that is made up of nucleobases, sugar, and a phosphate group. Nucleosomes adenosine and thiamine are complementary to each other whereas guanine and cytosine are complementary. The complementary ends of the DNA are paired through H-bonds via nucleobases. (2) This shows the histone proteins that make up the chromatin fiber and are stacked at a uniform 10 nm distance from each other.

Source: Created with Biorender.com

4.3 RDNA TECHNOLOGY—HISTORY

Stanley Cohen and Herbert Boyer initially introduced rDNA technology in 1972. They claim that rDNA technique involves combining DNA pieces from various species (hybrid DNA) and putting them into the host cells, which are often bacteria.[9] Prior to that, Werner Arber's discovery of the restriction enzyme was reported in 1969. Type I and Type II restriction enzymes (RE) are two different subtypes. In contrast to Type II RE, Type I RE only cleaves DNA at locations. As a result, Type II RE is used in the field of genetic engineering's rDNA technology for cloning.[10]

TABLE 4.1 History of rDNA Technology.

Year	Scientist	Discovery
Late 1960s	Werner Arbour and Hamilton Smith	Restriction Endonuclease Enzyme
1969	Herber Boyer	Restriction Enzyme EcoRI
1972	David Jackson, Robert Symons, and Paul Berg	Generation of DNA molecules
1973	Herbert Boyer and Stanley Cohen	First Recombinant Plasmid with Bacterial host-Psc101
1974	Stanley Cohen and Herbert Boyer	The eukaryotic gene was cloned
1979	Goeddel DV , Kleid DG, and Bolivar F	Insulin Synthesized and Hepatitis B was cloned

4.4 PURPOSE AND PROCESS OF rDNA TECHNOLOGY

One can change the genetic material outside of the organism using rDNA technology to increase the expression of the desired protein as a product. rDNA technology makes a new molecule called rDNA molecule. The rDNA molecule consists of a DNA fragment called an insert that is intended into a circular DNA structure called a vector. This cloning involves inserting the desired gene into the vector, which is subsequently transformed into the host cell.[11] This entire cloning procedure is broken down into the following three steps:

4.4.1 DESIRED GENE AMPLIFICATION AND VECTOR

Using rDNA technology, human hormones and enzymes with therapeutic uses can be produced in large quantities. The desired gene fragment from the genome can be identified and amplified using the PCR technique.[12] Denaturation, annealing, and extension are the three stages that make up the PCR procedure (Fig. 4.2). During the denaturation process, double-stranded DNA is reduced to a single strand, at which time primers anneal to certain genes based on their complementary sites and expand in the following step with the help of the enzyme *DNA polymerase*.[13] To identify the appropriate gene segment from the genome, primers are crucial. Purified the amplified gene fragment and, using primers, added restriction enzyme sites to both ends. The restriction enzyme sites should be identical in the vector and the targeted gene insert. Cloning vectors have a location that enables the insertion of DNA fragments, such as a multiple cloning site with a number of regularly used restriction sites that DNA fragments can ligate to.[13] Qualitative PCR is the term commonly employed when utilizing PCR for the purpose of identifying the presence or absence of a particular DNA product. The utilization of qualitative PCR proves to be a valuable approach in instances where PCR is conducted for the objectives of cloning or pathogen identification. Quantitative real-time polymerase chain reaction (qRT-PCR) offers insights beyond the simple identification of DNA. This measurement denotes the quantity of a particular DNA sequence or gene that is present within the given sample. Quantitative reverse transcription polymerase chain reaction (qRT-PCR) enables simultaneous real-time detection and quantification of the produced PCR product. There are two commonly employed techniques for the detection and quantification of the product. The first way involves the use of fluorescent dyes that have the ability to intercalate with double-stranded DNA in a non-specific manner. The second method utilizes sequence-specific DNA probes that are labeled with fluorescent dye.[14]

There exist numerous advantages associated with the utilization of PCR. Initially, this technique is characterized by its straightforward nature, facilitating comprehension and implementation, while also yielding prompt outcomes. The technology in question is characterized by its great sensitivity and has the capacity to generate a vast number of duplicates, ranging from millions to billions, of a certain product. These duplicates are subsequently utilized for purposes such as sequencing, cloning, and analysis. This statement holds true for quantitative reverse transcription polymerase chain reaction (qRT-PCR) as well. However, qRT-PCR offers the added benefit of enabling measurement of the generated product. Therefore, this technique can be employed to examine changes in the levels of gene expression in tumors, bacteria, or other pathological conditions.[12]

Despite its value, the PCR technology does possess certain limits. Due to the high sensitivity of the PCR technology, the presence of even minute quantities of DNA contamination in the sample might lead to the generation of false outcomes. Moreover, the acquisition of preceding sequence data is vital for the purpose of designing primers for PCR. Hence, PCR is limited to the detection of a specific disease or gene, confirming its existence or absence. An additional constraint arises from the potential for non-specific annealing of PCR primers to sequences that bear resemblance, albeit not total identity, to the target DNA. Furthermore, it is worth noting that the DNA polymerase has the potential to introduce erroneous nucleotides into the PCR sequence, albeit with a minimal occurrence.[15]

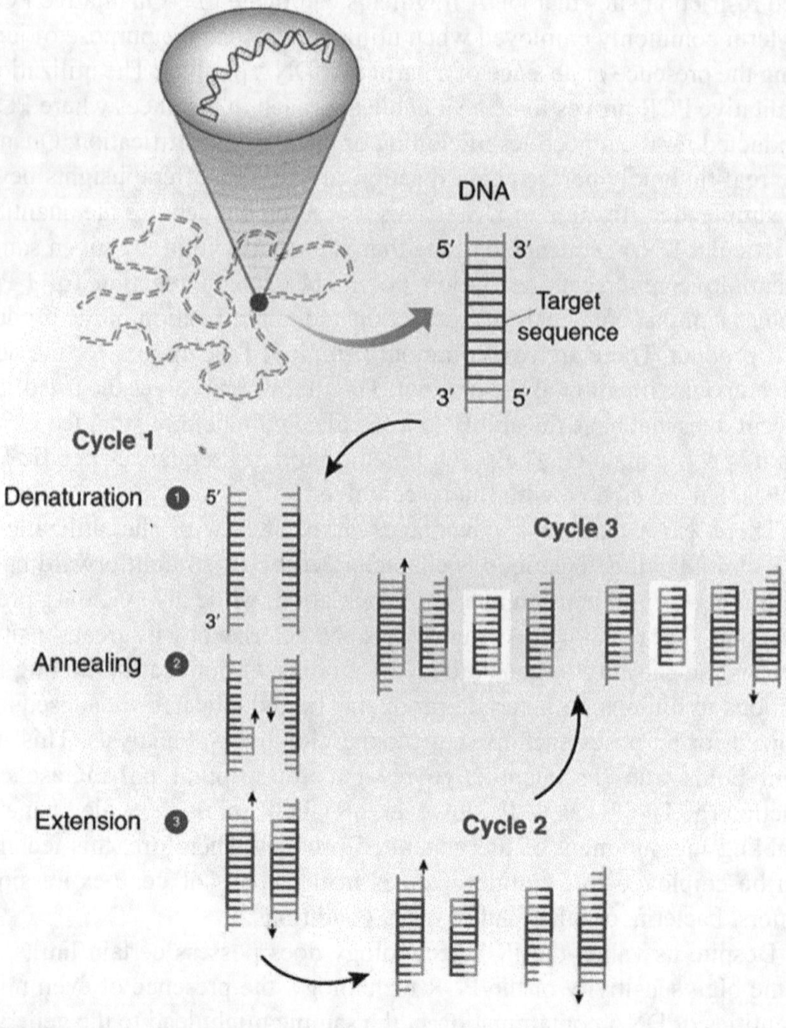

FIGURE 4.2 Summary of PCR process.

Source: Reprinted from Ref. [12]. https://creativecommons.org/licenses/by/4.0/

4.4.1.1 PLASMID VECTOR

Plasmids, which are typically observed in bacteria, are circular DNA molecules that exist independently from chromosomal DNA. Researchers utilize these inherent genetic structures as carriers to facilitate the introduction of exogenous DNA into recipient cells.[16] When employing a plasmid as a vector, the intended gene is incorporated into the plasmid, which is subsequently introduced into a bacterium. As the bacterium undergoes replication, the plasmid also undergoes replication, resulting in the generation of several copies of the inserted gene. Plasmids are of great significance in the field of rDNA technology, particularly in the areas of gene cloning, protein manufacturing, and gene therapy. The fundamental role of their manipulability and ability to propagate serves as a foundation for numerous operations in the field of molecular biology. The pBR322 vector is modified to produce pUC. pUC vectors are just 2.7 kb in size, which is less than pBR322. In contrast, it has a large number of copies. When the replication origin is altered, the plasmid replicates 500–600 times per cell without being amplified (Figure 4.3).[17]

4.4.1.1.1 pUC Vector

1. Replication Origin: It is generated from the replication origin of pBR322. A random mutation was performed on the ColE1 origin of replication of pBR322, resulting in 500–600 copies of the plasmid in each transformed *E. coli* cell.
2. Selectable Marker: An ampicillin resistance gene is present. Transformed host cells can thrive in an ampicillin-containing medium, but non-transformed cells perish.
3. Multi-cloning site (MCS) in the lac Z' gene.[17]

The pUC vector is a widely recognized plasmid vector commonly employed in the field of molecular biology for the purpose of cloning. The pUC class of plasmids, which originated from the pMB1 plasmid, are characterized by their ability to exist in high copy numbers within host cells of *Escherichia coli* (*E. coli*). This feature greatly enhances the production of DNA.[18] The pUC vector is distinguished by the inclusion of the lacZ gene, facilitating the implementation of blue–white screening. This approach entails the identification of recombinant bacteria that have successfully incorporated a plasmid harboring a specific insert of interest. When bacterial colonies are cultivated on agar plates supplemented with the substrate X-gal, the colonies with the unaltered lacZ gene exhibit a blue coloration, whereas those with a disrupted lacZ gene (caused by the insertion of exogenous DNA) manifest a white appearance. Moreover, pUC plasmids are furnished with a MCS, a concise DNA sequence encompassing numerous recognition sites for restriction enzymes. This feature greatly promotes the integration of exogenous DNA segments. In numerous genetic research applications, pUC vectors have demonstrated immense value.[19] It is used as cloning and expression vector.

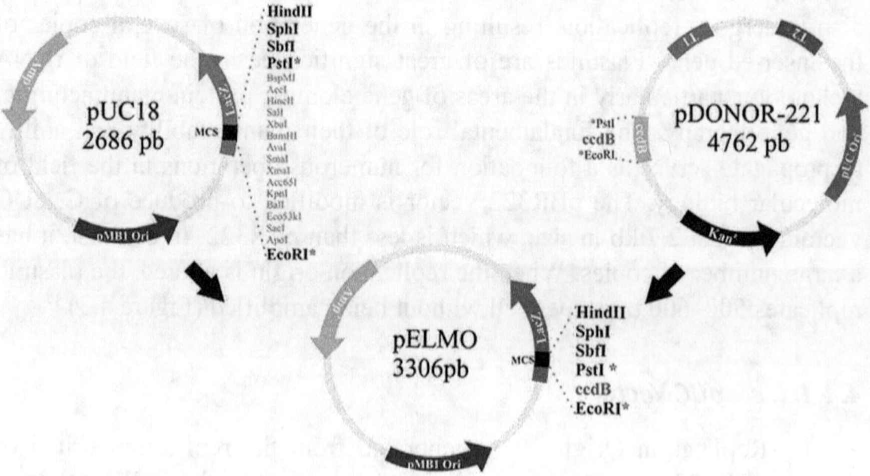

FIGURE 4.3 pUC18, pDONOR and pELMO Vector. The pUC vector is a frequently employed plasmid with a high copy number that is widely utilized in the process of cloning. The notable characteristics encompass the presence of an ampicillin resistance gene (Amp^R) to facilitate the selection of antibiotic-resistant cells, as well as the inclusion of the lacZ gene to enable blue–white screening. The insertion of a DNA fragment causes a disruption in the lacZ gene, resulting in the production of white colonies on X-gal plates, which facilitates the identification of recombinant clones. The MCS is a crucial component of its functionality, as it encompasses numerous recognition sites for restriction enzymes, hence enabling the insertion of various DNA fragments. The pUC plasmid possesses a very efficient and streamlined structure, rendering it highly suitable for a wide range of genetic research endeavors.

Source: Reprinted from Ref. [20]. https://creativecommons.org/licenses/by/4.0/

4.4.1.2 COSMID VECTOR

Cosmid vectors combine features of plasmids and bacteriophages, allowing them to package large DNA fragments (approximately 35–45 kb) into phage particles for efficient transformation into bacteria.[21] It is a lambda-based vector of the most complicated type. Cosmids are fusions of bacterial plasmids with phage DNA molecules.[22] They are designed this way because the enzymes that pack the lambda DNA molecule into the phage protein coat only need the cos sites to operate (Figure 4.4).[22] Hence, a plasmid with a cos site is known as a cosmid. Furthermore, needed are a plasmid replication origin and a selectable marker, such as an ampicillin resistance gene. It is important to note that because cosmid lacks all the lambda genes, it does not produce plaques. Instead, colonies grow on selective medium, much like plasmid vectors.[23] The target gene is put into the cosmid's specific restriction site once the cosmid is opened. These fragments are frequently created *via* partial digestion with a restriction endonuclease since complete digestion nearly always results in fragments that are too tiny to be cloned using a cosmid. Ligation leads to the formation of catenanes. The next step is to infect an *E. coli* culture with the lambda phages. All colonies are recombinant because non-recombinant lambda phages cannot fit within the lambda bacteriophage's head.[23] It is used for the construction of eukaryotic genomic library. Cosmids are widely utilized in the creation of genomic libraries, wherein they serve as a means to clone and represent the complete genetic material of an organism by incorporating vast pieces of its genomic DNA. Libraries of this nature play a crucial role in the field of gene mapping, sequencing initiatives, and the identification of genes that are of particular interest. Cosmids have played a crucial role in genome research by serving as intermediaries between small plasmid vectors and bigger yeast artificial chromosomes (YACs), due to their ability to accommodate extensive DNA sequences.[24]

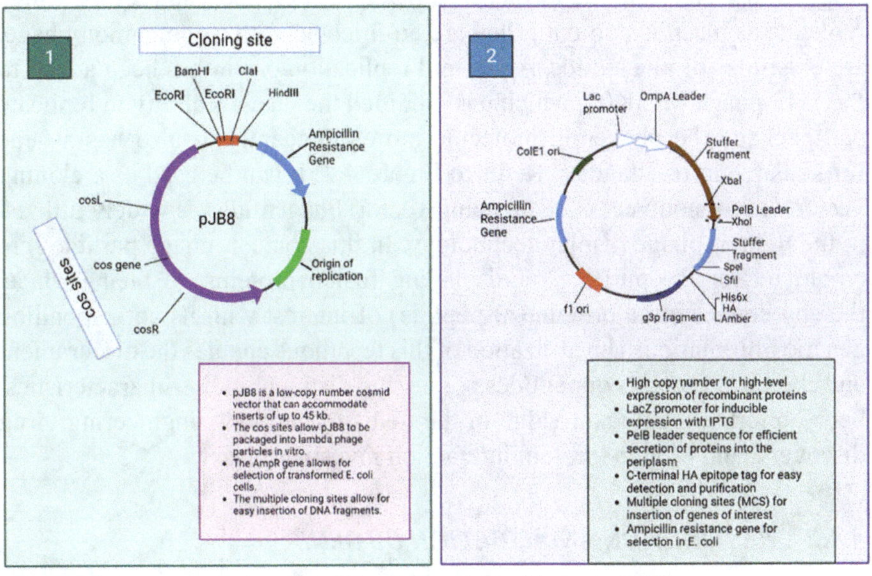

FIGURE 4.4 Vector design. (1) pJB8 vector and (2) pBluescriptIIKSÞ vector.

Source: Created with Biorender.com

4.4.1.3 PHAGEMID VECTOR

Phagemid vectors are hybrid vectors that combine features of plasmids and bacteriophages. They carry both an origin of replication for plasmid propagation in bacteria and a phage origin, enabling packaging into phage particles.[25] Although M13 vectors are extremely beneficial for producing single-stranded copies of cloned genes, they have one drawback. Restriction sites that cloned to M13 vector only possess a capacity of 1500 bp. To address this issue, a variety of new vectors that are hybrids of plasmids and M13 vectors have been developed.[26]

4.4.1.3.1 Construction of Phagemid Vector

1. Phage M13 origin of replication.
2. A portion of lac Z′ gene driven by lac promoter.
3. A MCS with lac Z′ gene.
4. Phage T7 and T3 promoter sequences flanking the MCS sequences.
5. ColE1 origin of replication.
6. Amp^r resistant gene.

Plasmids possessing both a normal dsDNA synthesis origin and the M13 replication origin can synthesize single-stranded DNA from the latter origin or dsDNA from the former. The pUC118, 119, and 120 vectors, which replicate as plasmids up until they are co-infected with a companion phage, are examples of phagemids. A plasmid replication origin has been added to the M13 phage M13K07, which has modified the phage's ability to replicate and package the phagemid molecule more abundantly. It also has a gene for kanamycin resistance.[27] Refer to Figure 4.4. It can be used as a cloning vector, expression vector, sequencing vector. Phagemids are widely utilized in the field of phage display technology. In this context, phage particles are employed for the purpose of displaying fusion proteins on their surface, thereby connecting a protein (or peptide) of interest with its corresponding genetic information. The utilization of this technique enables the discernment and choice of proteins or peptides possessing distinct binding characteristics, hence offering significant value in the fields of antibody engineering, drug discovery, and protein–protein interaction investigations.[28]

4.4.2 RESTRICTION AND LIGATION OF GENE

Following the first stage, the vector is digested by a particular restriction enzyme, producing blunt or cohesive ends. The same restriction enzyme is used for the digestion of the insert DNA gene. DNA ligase enzyme is used to bind the digested vector and insert it. It joins the two fragments together by building phosphodiester bonds between the nucleotides to form circular DNA, commonly known as plasmid.[29] Restriction enzymes, alternatively referred to as restriction endonucleases, are a class of proteins that possess the ability to identify and cleave particular DNA sequences. The finding made a significant impact on the field of molecular biology since it facilitated the manipulation of DNA, particularly in the context of cloning methodologies. These molecular entities function as enzymatic tools, precisely cleaving DNA molecules at certain sites, and facilitating the isolation, modification, or incorporation of desired DNA pieces into vectors.[24] There are different types of restriction enzymes: Type I, II, III, and IV, with Type II being the most commonly used for cloning. These enzymes recognize palindromic sequences, typically 4–8 base pairs in length, and make predictable cuts either within or a defined distance from their recognition site.[30]

Some notable examples include:

a. EcoRI, which recognizes the sequence 5′-GAATTC-3′ and cuts between the G and the A.
b. BamHI, which cleaves the sequence 5′-GGATCC-3′ between the G and the G.
c. HindIII, targeting the sequence 5′-AAGCTT-3′ and cleaving between the A's.

The overhangs produced during the procedure, commonly referred to as "sticky ends," play a crucial role in promoting the ligation of DNA fragments into vectors. This mechanism ensures the preservation of directionality and specificity, hence facilitating the cloning process. The selection of a suitable restriction enzyme is of utmost significance when assessing the intended cloning technique, taking into account many criteria such as the cloning site, enzyme efficiency, and buffer compatibility.[31]

4.4.2.1 LIGATING AGENT

The single polypeptide Bacteriophage T4 *DNA ligase* has a molecular weight of 68,000 Dalton and needs ATP to function. At pH 6.9, it operates at 40%, and pH 8.0; it operates at 65% activity. The ideal Mg^{++} ion concentration is 10 mM, and its presence is crucial. It may be inactivated by warming to 65°C for 10 min and has the peculiar ability to attach sticky and blunt-ended components.[32] *E. coli DNA ligase* cannot catalyze blunt-end ligation, but T4 *DNA ligase* can. This means that T4 DNA ligase can join two DNA molecules with blunt ends, regardless of their sequences. However, this also means that any two blunt ends can join, including those from the same DNA molecule. Blunt-end ligation is a useful way to join DNA fragments that have not been produced by the same restriction enzyme and, therefore, have incompatible sticky ends. To do this, the sticky ends are removed using the enzyme S1 nuclease, which digests single-stranded DNA.[33] Blunt-end ligation is most useful when blunt-ended fragments are joined *via* linker molecules. Linker molecules can be ligated to both ends of the foreign DNA to be cloned and then treated with a restriction enzyme to produce sticky-ended fragments. The vector DNA is then ligated to the foreign DNA via the linker molecules. These linkers also help to excise and recover the gene of interest after cloning and amplification in the host bacterium. Any two blunt DNA ends can be joined together, but it requires 10–100 times additional enzyme to achieve the same ligation efficiency. In a duplex molecule, it does not catalyze the attachment of RNA to either an RNA or DNA strand.[33]

A ligation reaction has three major components:

1. Two or more DNA segments with compatible ends.
2. A buffer containing 0.25–1 mM ATP to supply the required energy for the process.
3. T4 *DNA ligase* is the third enzyme.

4.4.3 PLASMID TRANSFORMATION OR TRANSFECTION INTO THE HOST CELL

It operates as a tiny, extrachromosomal DNA molecule within a cell that is physically distinct from chromosomal DNA and may reproduce autonomously. This ligated plasmid containing the desired gene snippet is changed into a bacterial cell.

There are several methods for transforming and transfecting the plasmid into the host cell. Plasmid DNA is introduced into bacteria, plants, and yeast through the process of transformation, and it is introduced into mammalian cells through the process of transfection.[34] Physical heat shock, electroporation, the gene gun method, calcium phosphate co-precipitation, liposomes, and microinjection are among the techniques for transforming plasmid DNA. The other method includes using chemical transformation, electroporation, and partial bombardment techniques for transfection. [35]

These procedures cause plasmid DNA to enter host cells; as a result, those cells are referred to as transformed or transfected cells. However, not all cells transformed in this procedure, thus some selection marker genes are incorporated into the vector to help identify transformed cells. Antibiotic resistance is the most common approach; since the vector carries this selection marker, only transformed cells were able to survive in the presence of antibiotics. Another technique is blue–white screening for the lacZ gene, which is a selection marker and codes for *-galactosidase*. Using the Green Fluorescent Protein (GFP) gene is another common technique.[36] After the transformation of cells, one can utilize this rDNA. rDNA technology is used in this way to create several recombinant products for use in various industrial and medical applications.

4.5 TRANSFECTION

4.5.1 *ELECTROPORATION*

Direct gene transfection may be accomplished using electroporation. High-voltage electrical pulses must be temporarily applied to a combination of cells and DNA. Plasmid DNA can enter cells through transient micropores formed by cell membranes.[37] Electroporation, a technique involving the application of high-voltage electric shocks to facilitate the introduction of DNA into cells, demonstrates compatibility with a wide range of cell types. This method exhibits a notable success rate in achieving both stable transformation and transient gene expression. Furthermore, electroporation offers the advantage of simplicity compared to alternative procedures, since it involves fewer procedural steps. Cells stopped in the metaphase of the cell cycle are particularly suited for electroporation due to the lack of the nuclear envelope and the unique permeability of the plasma membrane. Protoplasts are employed to change plant cells because dense plant cell walls prevent DNA from moving freely. Low transient transgene expression and a low frequency of regeneration are both related to protoplast transformation. Electroporation is quick, easy, and inexpensive, with little cell toxicity. The disadvantage is that it is difficult to regenerate plants from protoplasts.[38]

4.5.2 *MICROINJECTION*

The hand insertion of transgene at the microscopic or almost macroscopic scale using a fine glass micropipette is known as microinjection.[39] The transgene can be supplied *via* plasmids, cosmids, phage, YACs, or PCR products. It can be circular or linear. DNA is physically injected straight into the cytoplasm or nucleus using a glass microcapillary injection pipette. The target cell is positioned, and the nuclear envelope and cell membrane are penetrated using two micromanipulators. This process can result in stable transformants, but it takes a lot of effort and technical expertise. Microinjection has shown only sporadic success in plant transformation because of the thick cell walls and absence of a single cell-to-plant regeneration pathway.[40]

4.5.3 CHEMICAL TRANSFECTION

Because of its ease of use, low cost, and variety of available transfection chemicals, chemical transfection is a widely used approach. It is utilized for recombinant protein analysis, investigations on gene silencing, short-term gene expression, and gene expression research. Using liposomes, which are tiny molecules that may fuse with cell membranes and release their contents, the transfection technique known as lipofection can be used to deliver genetic material. The protocol must be optimized to guarantee high transfection effectiveness and that the procedure is safe for the cells being transfected. In the discovery and development of novel medications, liposome transfection techniques are particularly useful in cancer, where they may be utilized to target certain genes with small molecule-based therapies or to silence genes with RNA interference (RNAi) technology.[41]

4.5.4 PHYSICAL TRANSFECTION

Two well-known physical transfection techniques that can transport nucleic acids but are harsh on cells owing to membrane breakage and frequently result in cell death are electroporation and cell injection.[36] Physical transfection includes electroporation, microinjection, and biolistic particle delivery (Figure 4.5). An electroporation cuvette must be employed, which is dependent on the DNA content and the type of cells utilized. Using a micromanipulator and microscope, the method known as microinjection is frequently used to transfer DNA and RNA into single cells. High cell death rates are associated with biolistic particle delivery, which inserts nucleic acid-carrying microparticles into cells to transfer DNA or RNA.[42] Additionally, it is imperative to take into account the variables that could potentially impact the efficacy of transfection and the cytotoxicity toward the host cells, as well as the methods by which these parameters can be precisely and simply evaluated. Similarly, the incorporation of suitable controls in a transfection experiment holds comparable significance in facilitating a just and impartial evaluation of the experimental results.

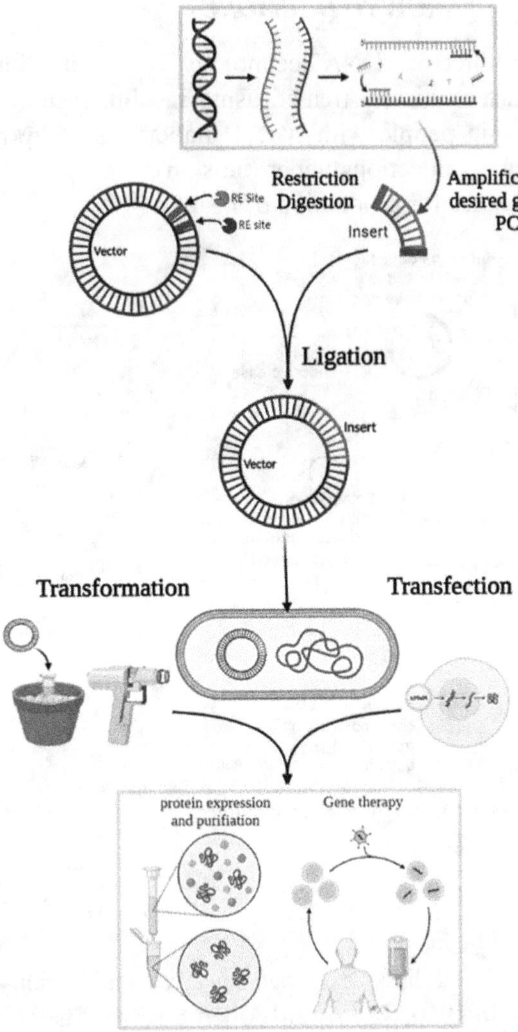

FIGURE 4.5 Preparation of rDNA molecule via cloning and its application.

Source: Created with Biorender.com

4.6 TRANSFECTION AND PROTEIN PURIFICATION IN PHARMACEUTICAL INDUSTRY

rDNA technology is used for various purposes in pharmaceutical industries. It is crucial to pinpoint the proteins that are lacking or faulty in hereditary disorders. Human proteins that can be utilized to treat disorders with a genetic component can be produced using the rDNA technology.[4]

4.6.1 *INSULIN AND GROWTH HORMONE*

The most used product of rDNA technology is insulin. Some types of diabetes mellitus are medically treated using insulin. Insulin is no longer produced internally in people with type 1 diabetes, so they must rely on external sources (usual injections under the skin).[43] Refer to Figure 4.6 for diagrammatic procedure for production of insulin.

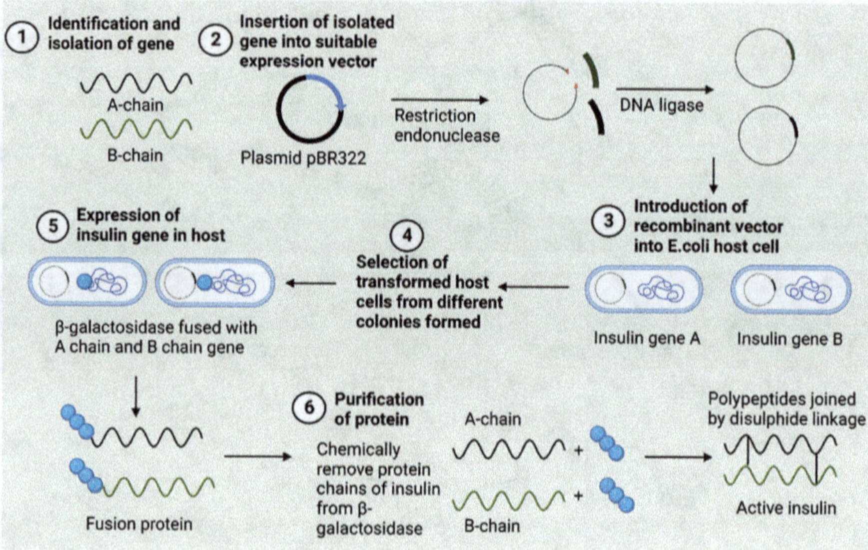

FIGURE 4.6 Formation of insulin by rDNA techniques.

Source: Created with Biorender.com

4.6.1.1 *HISTORY*

Diabetes patients did not have extended survival rates prior to the 1921 discovery of insulin. In 1910, Sir Edward Albert Sharpey-Shafer hypothesized that diabetes patients only lacked one hormone from their pancreas, and in 1921, a young surgeon by the name of Frederick Banting discovered how to extract insulin from a dog's pancreas.[43] A young surgeon by the name of Frederick Banting found how to extract insulin from a dog's pancreas in 1921; refuting Sir Edward Albert Sharpey-hypothesis Shafer's that diabetes sufferers only lacked one hormone from their pancreas in 1910. Soon after, a pharmaceutical business by the name of Eli Lilly began producing massive amounts of insulin. Several firms have developed slower-acting insulins throughout the years. Using *E. coli* bacteria, the first genetically altered synthetic "human" insulin was produced in 1978, and the FDA approved its use for human medicine in 1982.[43]

4.6.1.2 PRODUCTION OF INSULIN THROUGH P. PASTORIS

There are primarily two methods for producing human insulin. Initially, *E. coli* was used as an expression host, and then *S. cerevisiae* was used. Nevertheless, because of the large and diverse number of promoters, the methylotrophic yeast, *P. Pasotris*, is now employed as an expression host, and its capacity to produce insulin is considerably more than the prior expression hosts that were used.[44] The pPI9K-IP vector, which was electroporated in *P. Pastoris* with the BgIII restriction site, was employed here. Moreover, the converted clones were chosen with G418 antibiotic and analyzed in liquid culture using SDS-PAGE and RP-HPLC. Additionally, insulin promoter growth was permitted to occur under fermentation conditions, and the insulin promoter was cleared using a tangential flow filtering method with turbidity rate measurement. The insulin promoter was purified using cation exchange chromatography, and the concentration at which the IP was eluted was measured to be 80 mg/mL. The insulin promoter was converted into human insulin using a transpeptidation technique, which was then purified using a resin column. Finally, insulin was analyzed using mass spectrometry. As a result, all of the technologies were intended to generate human insulin on a massive scale while being cost-efficient (Figure 4.7).[44]

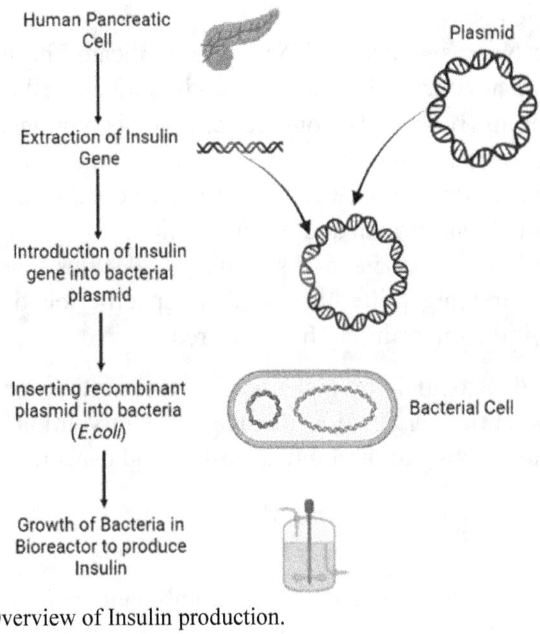

FIGURE 4.7 Overview of Insulin production.

Source: Created with Biorender.com

The pituitary gland produces growth hormones. Growth retardation, often known as pituitary dwarfism in the medical profession, affects young children who do not generate enough human growth hormone (HGH). By inserting the HGH gene into an *E. coli* plasmid, growing the cells, and extracting the HGH from the extracellular medium, the pharmaceutical sector produces hormones utilizing rDNA technology.[45]

4.6.2 INTERLEUKINS

Cytokines encompass a wide range of tiny protein molecules that are released by various cells, predominantly those within the immune system, with the primary function of facilitating intercellular communication.[46] These molecules serve crucial functions in regulating cellular proliferation, modulating immunological reactions, mediating inflammatory processes, and facilitating tissue regeneration. Cytokines can be classified into many categories based on their function and origin. There are several prominent categories that can be identified, namely:

a. Interleukins (ILs) are a class of signaling molecules that are predominantly synthesized by and exert their effects on leukocytes. They play a crucial role in facilitating intercellular communication during immune responses.
b. Tumor Necrosis Factors (TNFs) are implicated in the regulation of inflammatory responses and the modulation of cellular apoptosis.
c. Interferons (IFNs) are recognized as pivotal components in the initiation and regulation of antiviral responses and immunomodulation.
d. Growth factors are molecules that play a crucial role in the regulation of cellular growth and differentiation.
c. Chemokines are a class of signaling molecules that play a crucial role in directing white blood cells to specific locations in the body where infection or injury has occurred.

Cytokines, due to their pivotal function in immune modulation, have emerged as potential targets for therapeutic intervention in a range of diseases, encompassing autoimmune disorders and cancer.

4.6.2.1 IL-4

Manufacturing of cytokines and other recombinant biologics has recently been made possible *via* rDNA technology. Recombinant IL-4 has been linked to autoimmunity, allergies, and cancer. Human IL-4 was cloned, produced in *E. coli* using a particular promoter, and its biological effects were investigated. The correct hIL-4 gene construct was created, and it was then subcloned into the HindIII restriction enzyme site of the pcDNA3.1(+) vector. The recombinant plasmid was injected and expressed in BL21 cells. The rhIL-4 protein was assessed using SDS-PAGE and Western blotting. It was cleaned using Ni-NTA affinity chromatography.[47] The concentration and accuracy of the pure protein were evaluated using enzyme-linked immunosorbent assay (ELISA). The development of an erythroleukemic cell line was monitored using the MTT assay to see how rhIL-4 affected it. Successful cloning and insertion of the rhIL-4 gene into *E. coli* bacterium. Western blotting, therefore, showed a prominent band on the nitrocellulose and SDS-PAGE membranes. The amount of pure protein was 500 pg/mL. Cells treated to rhIL-4 multiplied in a dose-dependent manner, according to the MTT assay. T7 RNA polymerase was used to successfully transcribe the rhIL-4 gene in the prokaryotic system using a specific eukaryotic promoter.[48]

4.6.2.2 IL-2

In considering the earlier IL-2-based studies, it played a pivotal role in cancer treatment and furthermore it resulted in the earlier detection of tumor growth and led to build a platform for extensive cancer research.[49] rIL-2 products such as aldesleukin and tecelukin stimulate autocrine and paracrine changes in the body.[50] A recombinant vector (TAC promoter added from the pDR540 between HindIII and BamHI sites), DNA encoding for human interleukin-2, and a human interleukin-2 signal peptide sequence into a mammalian cell results in the replacement of the gene's 5′ noncoding sequence in both the human interleukin-2 and the human interleukin-2 signal peptide genes. IL-2 expression was studied using the pTIL2E/2TT vector, which had and was then synthesized and ligated to achieve optimum IL-2 expression. IL-2 was purified using RP-HPLC and validated using the SDS-PAGE technique.[51]

4.6.2.3 IL-12

In response to infections by bacteria, fungi, intracellular parasites, and viruses, phagocytes (monocytes/M and neutrophils) and DCs are the main physiological producers of IL-12.[52] TLRs and other receptors activate membrane-bound and soluble signals and extracellular inflammatory matrix components (for example, low molecular weight proteins) CD44 and TLRs.

With the advent of rDNA technology, it became possible to employ IL-12 clones from diverse species. As previously stated, IL-12 is a key participant in cancer treatment as well as numerous illnesses such as leishmaniasis and listeria. The production of rhIL12, which is well folded and very physiologically active, had a significant contribution. The AG181 plasmid, which had simian and hCMV promoters and was expressed in HEK293 cell lines, was optimized for the generation of rhIL12. Additionally, the plasmid's accuracy is obtained by conducting restriction digestion of the HindIII site, which is subsequently evaluated by plasmid purification. (Midi Prep).[52]

To trigger the synthesis of recombinant gene, the expression-based vector is used as a viral and bacterial promoter. The supernatant containing the transfected rhIL-12 was examined using SDS-PAGE and sandwich ELISA. For rhIL-12 protein purification, an accurate column-based approach, such as a heparin sepharose column, is used since hIL-12 has a high affinity for heparin.[53] To summarize, when the purified rhIL-12 was tested for bioactivity in the NK-92MI cell line, it was discovered that it promoted the synthesis of IFN-ɤ without the addition of any protein stabilizing agent, unlike the other commercially available samples. [54]

4.6.3 VACCINE

The field of vaccine manufacture underwent a significant transformation with the advent of rDNA technology, which involved the incorporation of genes encoding antigenic proteins from pathogens into appropriate expression systems. The administration of this substance results in the production of antigenic proteins that provoke an immune response without causing harm. In contrast to conventional approaches employing attenuated or inactivated pathogens, rDNA-derived vaccinations have improved safety, specificity, and scalability. Illustrative instances encompass the hepatitis B vaccine, which is generated through the expression of its surface antigen in yeast, and the COVID-19 mRNA vaccines, which are formulated by encoding the spike protein of the virus.[55]

4.6.3.1 HUMAN PAPILLOMAVIRUS (HPV) VACCINE (VVAX001)

Two of the most notable HPV vaccinations that have been developed utilizing rDNA technology are Gardasil and Cervarix.[56] The vaccines are formulated through the introduction of genetic material encoding the virus-like particles (VLPs) of the HPV into either yeast or insect cells.[57] Subsequently, the cells generate VLPs, which are subsequently isolated and employed as the active constituent of the vaccine. VLPs possess a structural resemblance to the virus, hence inducing an immune response; nevertheless, they do not possess the ability to cause infection.

4.6.3.1.1 Vvax001 Production (Upstream Processing)

BcuI restriction enzyme, ethanol precipitation, SP6 RNA polymerase, TURBO DNase, lithium chloride precipitation and centrifugation, and pure RNA dissolved in nuclease-free water were used to produce linearize the DNA of each plasmid. Post linearizing, all the RNA transcripts were electroporated, and Vvax001 replicons were produced in Vero cells. Hence, the transfected product were maintained at 5% FBS and recovered at 24 and 48 hours, post that supernatant was collected and centrifuged to avoid the dead population. Hence, these harvests were maintained at 60°C until the time for purification (Figure 4.8).[58]

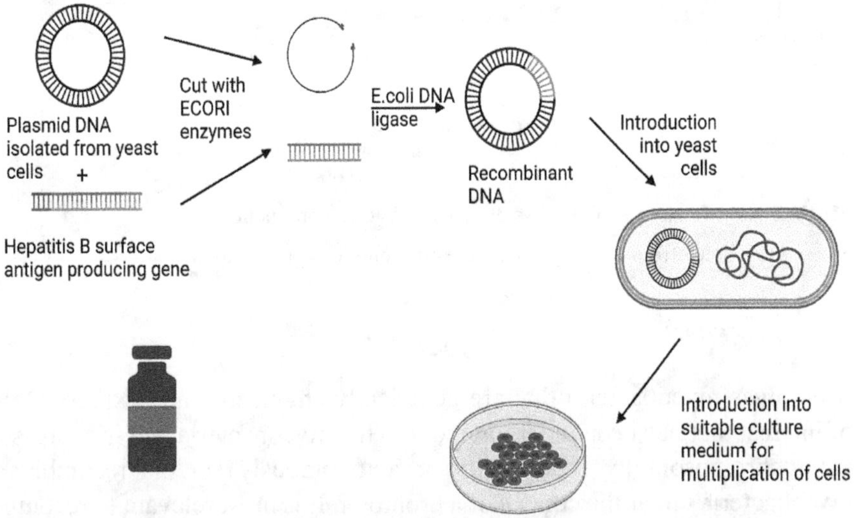

FIGURE 4.8 Hepatitis B vaccine production by rDNA techniques.
Source: Created with Biorender.com

4.6.3.1.2 Vvax001 Purification (Downstream Processing)

Pools of Vvax001 were thawed at 37°C and further pooledVvax001 was treated with benzonase in the presence of magnesium chloride for the period of 30 min at 37°C. Following, the treated substances were purified through solid-state anion exchange chromatography, hence loaded substance was washed with HEPES buffer and eluted with Nacl (Figure 4.9).[59]

Cation exchange chromatography was used to further purify the anion exchange purified fraction. Before processing to the former Drug Substance, the purity and quantity evaluation needs to be done, for that HSA (Human serum albumin) was added and kept at 60°C.[60]

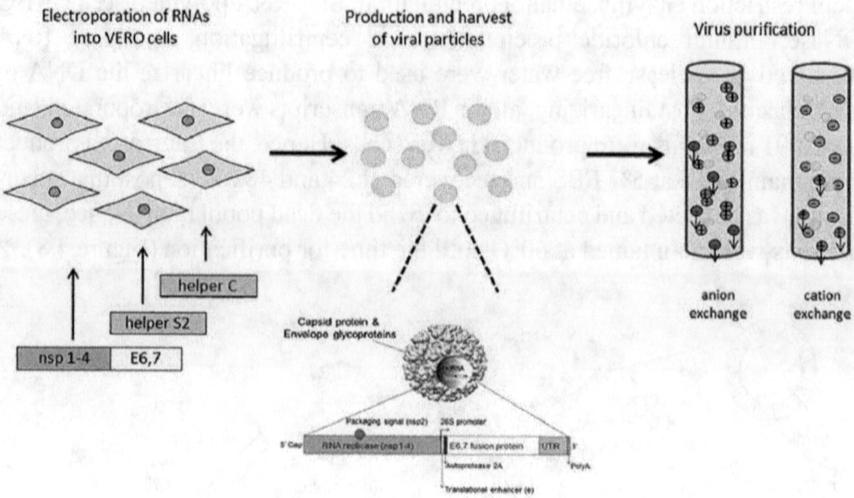

FIGURE 4.9 Downstream process of hepatitis vaccine production.

Source: Reprinted from Ref. [4]. https://creativecommons.org/licenses/by/4.0/

4.6.3.2 BCG VACCINE

It involves mycobacteria that are genetically engineered and express DNA of interest that has been incorporated into the mycobacteria and is expressed extra chromosomally (episomally or autonomously) in the recombinant mycobacteria under the control of a promoter. It is most relevant to recombinant M. bovis-BCG, in which the DNA of interest is produced extrachromosomally under the control of a mycobacterial hsp promoter, such as hsp70 or hsp60.[61] The DNA of interest is characterized as heterologous DNA (DNA from a source other than the mycobacterium into which it is introduced) and consists of all or a portion of a gene or genes producing enzymes, cytokines, lymphokines, and immunopotentiators. Taking into account the two wild strains of the wild-type BCG, conn and tice, which were placed into the recombinant plasmid (pMTB30) which had its own promoter, r30, for the construction of the rBCG30 vaccination approach, rBCG30 strains were further delivered into mice by electroporation.[62] Earlier comparisons from BCG vaccinations were conducted by analyzing the CFU parameter of the spleen and lymph node of animal models. Following that, the rBCG30 strains were expressed and purified using SDS-PAGE. To summarize, rBCG30 vaccinations stimulate T cell proliferation, but regular BCG vaccines that were live attenuated were significantly less effective than the recombinant rBCG30 vaccine.[63]

4.6.4 HEPATITIS B SUBUNIT VACCINE

The hepatitis B subunit vaccine was created using hepatitis B surface antigen (HBsAg), which was extracted from the virus' genome and cloned into a shuttle vector that produces high amounts of gene expression in yeast cells.[64] The LEU2 marker recognizes yeast cells with plasmids and the ability to thrive without leucine. Bioreactors were used to multiply certain cells on a larger scale, and the HBsAg was purified for the hepatitis B vaccine.[64]

In 1987, the recombinant vaccine for hepatitis B (HBsAg) became the first synthetic vaccine for public use. Marketed under the trade names Recombivax and Engerix-B, this vaccine is safe, effective, and does not produce allergic reactions.[65] As a result, it has been in continuous use since its introduction. Three doses of the vaccine are administered over a period of six months. Immunization against hepatitis B is strongly recommended for anyone who may come into contact with blood or body secretions.[66]

Moreover, HBsAg is a 226 Bp amino acid found on the surface of antigen. The pGPD-1 vector was used to create the vector for the yeast-based expression system, and HBs was cloned into it (HBs) this was also introduced into the RH218 plasmid to express *S. cerevisiae*.[67] Following that, fed-batch fermentation was initiated for the bulk production of yeast cells under restraint conditions, i.e., glucose concentration, while also providing aerobic conditions for growth. To determine the results of HBsAG in yeast cell lysate, sucrose content was measured, which was validated by the Auszyme II immunoassay. Additionally, the harvested cells were washed with isotonic buffer, agitated, and centrifuged, supernatant was collected and purified using ion-exchange chromatography, and HBsAg was detected using SDS-PAGE in reducing circumstances.[68] Purified HBsAg reacted similarly to plasma-derived particles with immunogenic applications during the chimp challenge test. As a result, it was determined to be a critical immunogen in the manufacturing of hepatitis B vaccine (Figure 4.10).

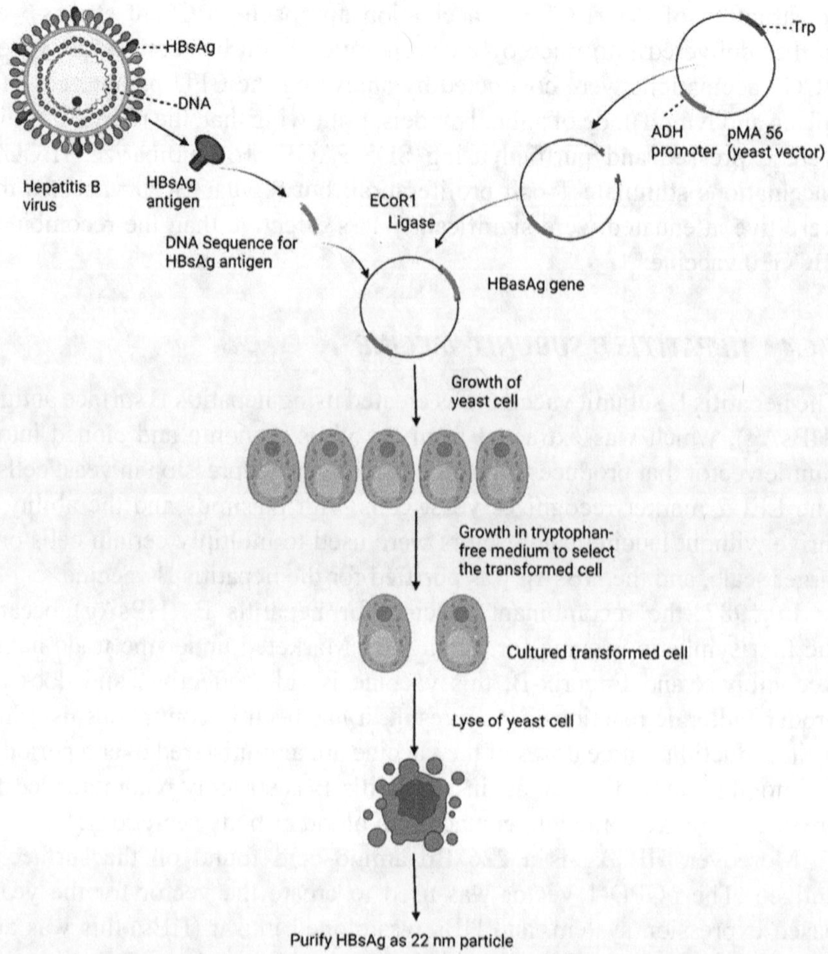

FIGURE 4.10 Formation of Hepatitis B vaccine by rDNA technology. HBsAg gene is responsible for the antigenic identification of antigen and formation of robust antibody response. The specific gene coding for the surface protein is extracted with the same restriction enzyme use for the cutting of pMA 56 (yeast cell). They are ligated in the presence of ligase and further growth is facilitated. Then, the yeasts are moved to the tryptophan-free media which allow only transformed cell to grow. On optimum growth, the cell is lysed and the antigenic portions are collected for the formation.

Source: Created with Biorender.com

4.6.5 INTERFERONS

Interferons, or INFs, are glycoproteins with a low molecular weight that are generated by certain eukaryotic cells in response to viral infections.

They increase the creation of antiviral proteins, which block viral RNA and protein synthesis, hence reducing viral infection. There are two types of IFN which are diversified based on their interaction with receptor.[69]

Type I interferons (INF and INF) are produced in response to viral or bacterial infections by diseased leukocytes and epithelial cells. They share a common ancestor gene with eukaryotes. Important pro-inflammatory cytokines including Interleukin 1 (IL-1) and Tumor Necrosis Factor- (TNF-) efficiently activate INF/ in cells, which is crucial for the innate immune system's defense against viral infections by preventing DNA replication and translation. Moreover, they promote MHC Class I molecule expression and turn on natural killer cells (Figure 4.11).[70]

Interferon (IFNγ) is a Type II interferon that is generated by T-helper type 1 cells and natural killer cells. It promotes the production of Class II MHC molecules in many cell types, which improves phagocytic characteristics and boosts the antigen-presenting capability of antigen-presenting cells. It is required for macrophages to operate properly, as they can only destroy invading germs and parasites by creating reactive oxygen molecules.[71]

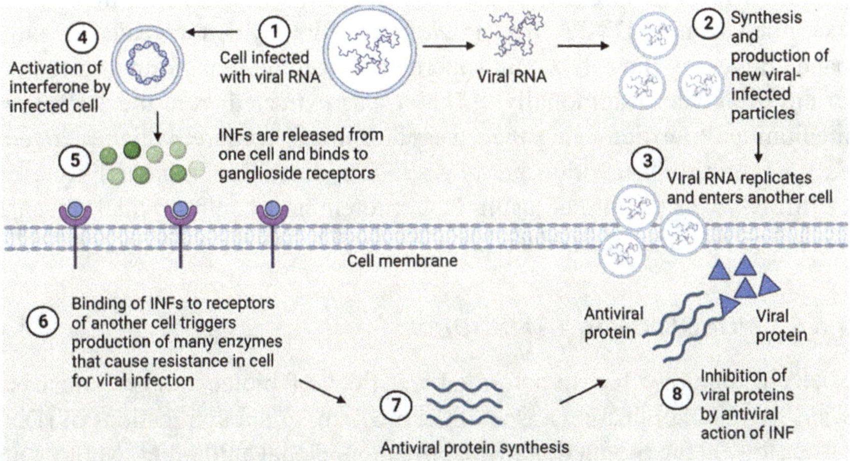

FIGURE 4.11 Production in interferon using rDNA technology.

Source: Created with Biorender.com

4.6.5.1 IFNA-2B

IFN-2b, a protein having anticancer and antiviral capabilities, can function as a key therapeutic protein for the treatment of cancer and many other diseases. During the past ten years, IFN-2b has played a critical role in more than ten diseases around the world.

E. coli strains BL21 (DE3), BL21 (DE3) pLysS, and *E. coli* DH5 were examined for IFN-2b expression. Heat shock was utilized to introduce the IFN-2b/pET28a+ gene into *E. coli* BL21 by amplifying the gene on the plasmids acquired from transformed bacterial cells.[72] The target protein can be produced with the help of IPTG during the induction, which is followed by the separation of the target protein using sodium dodecyl sulfate polyacrylamide gel electrophoresis (SDS-PAGE) of total proteins.[73] Western blotting proved that transformed bacteria successfully produced IFN-2b.[72]

4.6.5.2 *rhIFN-ɤ*

Human interferon-gamma is regarded as a key factor in immunomodulation as well as causing antiviral and anticancer characteristics. hIFN-ɤ was produced by introducing the BspT1041 and kpnI sites into the pPICZαC plasmid to create the recombinant plasmid, pNhIFN. Additionally, this plasmid was integrated into transformed *p.pastoria* colonies, and pNhIFN and pαIFN sites were identified and validated using PCR using the AOX1 gene (Universal Primer).[74,75] Following that, the expression of rhIFN-ɤ was examined using SDS-PAGE and Western Blot, which revealed a single band containing the rhIFN-ɤ in supernatant from the transformed p.pastoris of the pIFN site. Additionally, rhIFN-ɤ was extracted from the *P. Pastoris* medium and purified using the vivaspin 2 mL procedure, which recovered 85% of the rhIFN-ɤ with a purity of 95.6%. After that, a bioactivity assay of rhIFN-ɤ was performed using Novoprotein as a control, which revealed decreased amounts of rhIFN-ɤ.[76]

4.6.6 *MONOCLONAL ANTIBODIES*

rDNA technology has transformed the field of biotechnology and has a wide range of applications. One of the most important applications of rDNA technology is the production of specific monoclonal antibodies. Monoclonal antibodies are highly specific and can be used for a variety of diagnostic and therapeutic purposes.[77] The following is a description of how rDNA technology is used to create specific monoclonal antibodies:

a. **Selection of Antigen:**
The first step in generating monoclonal antibodies is to select the target antigen. This could be a protein, peptide, or any other molecule of interest. Once the antigen is identified, its DNA sequence can be cloned or synthesized using rDNA techniques.

b. **Production of Recombinant Antigen:**
In some cases, it may be necessary to produce the antigen in large quantities for immunization. rDNA technology allows for the expression of the antigen in bacterial, yeast, or mammalian cells, ensuring a consistent and abundant source of the antigen.

c. **Immunization:**
Animals (usually mice or rats) are immunized with the recombinant antigen to stimulate the production of antibodies. The immune system of these animals will recognize the antigen as foreign and produce a variety of antibodies against it.

d. **Fusion of B Cells with Myeloma Cells:**
To create hybridoma cells that can produce monoclonal antibodies, B cells from the immunized animals are fused with myeloma cells, which are immortal cancer cells. This fusion can be facilitated using polyethylene glycol (PEG) or other fusion techniques.

e. **Screening for Monoclonal Antibodies:**
The resulting hybridoma cells are screened to identify those that produce antibodies against the specific antigen of interest. This is usually done through techniques like ELISA or flow cytometry.

f. **Cloning and Expansion:**
The selected hybridoma cells are cloned to ensure monoclonality, meaning that they produce a single type of antibody. These monoclonal cell lines are then expanded to produce a renewable source of monoclonal antibodies.

g. **Production of Monoclonal Antibodies:**
The monoclonal antibody-producing cell lines can be cultured in bioreactors or other cell culture systems, allowing for the large-scale production of specific monoclonal antibodies. These antibodies can then be harvested, purified, and characterized.[78]

Specific monoclonal antibodies can be used in various applications, such as:

- Diagnosis: They can be used in diagnostic tests to detect specific antigens in clinical samples, such as blood tests for disease markers.
- Therapeutics: Monoclonal antibodies are used in immunotherapy to treat conditions like cancer and autoimmune diseases. They can target specific molecules on the surface of cells or in circulation.

- Research: Monoclonal antibodies are invaluable tools in scientific research. They can be used to study the function of specific proteins, validate experimental results, and investigate molecular pathways.

Vaccine Development: Monoclonal antibodies can be used to identify potential vaccine targets and evaluate vaccine efficacy.[79]

Several types of cancer have been treated using various methods, including gene-directed enzyme prodrugs, oncolytic virotherapy, and the insertion of tumor suppressor genes into immunotherapy. A prime example is the tumor suppressor gene p53, which is frequently inherited and is a crucial component of cancer treatment. Chemotherapy or radiation therapy may be combined with p53 gene transfer using rDNA technology in some of the procedures.[80]

Using tumor cells that have been genetically altered to express immunostimulatory molecules—for instance, a recombinant host cell that has been altered to express tumor antigens—is one of the most important aspects of vaccination. Metastatic cancer can be retreated using immunotherapy based on the adoptive transfer of gene-engineered T-cells. Targeting T-cells using gene engineering to retarget antigens produced by tumors.[81]

4.7 CELL CULTURE TECHNIQUES AND RECOMBINANT PROTEIN EXPRESSION

Mammalian proteins cannot be expressed without cell culture techniques. By using mammalian cell culture, one may overcome the restrictions of the bacterial or prokaryotic expression systems.

a. **Bacterial cell:** The chosen host for recombinant proteins is a bacterial expression host system, frequently *E. coli*, because of its low cost, well-known biochemistry and genetics, rapid growth, and high productivity. Inadequate post-translational modifications (PTMs), the production of inclusion bodies (IBs), codon bias, and endotoxin issues are some of the disadvantages of this method.[82]

b. **Mammalian cell culture system:** Of all approved recombinant protein-based biopharmaceuticals, mammalian cells outweigh all other recombinant protein expression methods. Mammalian cells have the capacity to create large and intricate recombinant proteins. In this method, the introduction of the gene and clone selection takes longer than in microbiological systems. The selection of an expression host, vectors, and transfection are the first steps in cell-line creation, which are then followed by cell-line selection.[83] The most frequently used cell lines for the expression of recombinant biopharmaceuticals are CHO, NS0, and Sp2/0. Eighty-four percent of authorized biopharmaceuticals are produced using CHO cell-based methods, according to a recent survey. Mammalian cell lines contain PTM, however, it has a different glycosylation pattern from human PTM. Also, the expression of proteins in HEK293, HKB11, PER.C6, HeLa, and CAP cells is being studied (all human cell lines). On the other hand, commercial manufacturing of these cells is still in its infancy. This method also has the disadvantage of animal virus infection. It takes a lot of work to make a culture medium for a cell line since it needs so many different things, such as growth hormones, amino acids, reducing agents, and vitamins. Four lineages of CHO cells are recognized: CHO-K1, CHO-S, CHO-DG44, and CHO-DXB11.[84]

4.7.1 HOW TO GROW THE CELL LINES

After performing the first subculture of culture, it is considered a cell line. Cell lines are of a definite period of life span and the most confluent remains the most viable with the diverse genotypic and phenotypic characteristics.

There are two types of cell lines:

1. **Finite:** After a given amount of time, a cell loses its viability and experiences senescence; these cell lines are referred to as finite.
2. **Continuous:** The term "continuous cell line" refers to a cell line that transforms via chemical or viral processes and divides unceasingly.[85]

There are two types of characteristics of cell lines according to their growth pattern.

1. **Adherent culture:** Cells that are adherent, which indicates they will stay, attached to the surface of the cell culture flask. However, for the adhering cell to separate from the surface and be utilized for further preservation or sub-culturing, trypsinization will be necessary.
2. **Suspension culture:** Suspension-grown cells develop in three dimensions and are housed in conventional, untreated culture flasks. The constant agitation of large-scale suspension cultures ensures proper gas and nutrient exchange.[86]

Components and conditions required during the continuous culture of cell line:

Tissue Culture Media: Amino acids, vitamins, minerals, and glucose are found in Eagle's MEM (minimum essential medium), also known as simple medium or basal media. Numerous formulations and variations may be purchased from various commercial sources. Depending on the cell lines for their optimum growth, a different formulation is adopted each time.[87]

4.7.2 ADDITIONAL COMPONENT

a. **Fetal calf serum** is the most widely used type of serum, while it can also be of human, horse, or calf origin. Serum supplies elements that are absent in basal tissue culture medium, such as those that encourage cell attachment and proliferation.

b. **L-glutamine** is routinely added to tissue culture media because it may provide energy through the citric acid cycle, either as a separate supplement or as part of the basal medium.

c. **Antibiotics and antimycotics** (a combination of penicillin, streptomycin, and amphotericin B) are often added to aid in preventing the growth of contaminating bacteria, fungus, and mycoplasma. Moreover, eukaryotic cells changed by genetic engineering are selected and established using specific selection antibiotics.[88]

Conditions: A 37°C temperature, a pH range of 7.2–7.4, and 5% CO_2 are required for the cell lines. When added to the medium, phenol red ranges in color from yellow at a pH of 6.5 to orange at 7.0 to red at 7.4 to pink at 7.6 to purple at 7.8. As bicarbonate is a common ingredient in tissue culture medium, a 5% carbon dioxide atmosphere is needed to maintain a pH of 7.4. HEPES, an organic buffer, is typically added to a medium to reduce its reliance on bicarbonate. Basal cell culture media with the additionally added component (serum, antibiotic, glutamine, etc.) is called the complete media.

4.7.3 CELL LINE REVIVAL, PASSAGING AND STORAGE

A well-characterized cell line that has been cryopreserved is gently transferred to the pre-warm medium in a falcon tube. The cells should be pelleted down and then re-suspended in the complete cell culture medium.

Add pelleted cells to the cell culture flask and let them grow. Observe the confluency of the expanding cells after incubation.[89] 60–80% confluent cells passage using trypsin. Most laboratories employ commercially available, purified trypsin. Contrary to popular belief, trying to speed up detachment by boosting trypsin concentration can actually speed up cell removal. Trypsin's maximum activity is reached around 37°C, hence pre-warming the enzyme may hasten cell separation. Extended incubation durations and high trypsin concentrations are detrimental and often cause cell injury, which makes the cells adhere to one another. Add complete media to the flask to stop the trypsin activity and palate down the cells. Palleted cells transfer into the new tissue culture flask or cryopreserve the cells for future use.[90]

Add freezing medium (10% DMSO in complete media) to the desired cell density for cryopreservation. This is typically 1,000,000/mL of freezing medium for mammalian cells. Cells should not spend longer than 10 min at room temperature in the frozen medium. Divide 1 mL among the cryovials. Place the cryovials into a –80°C freezer.[91]

4.7.4 GENERATE PLASMID EXPRESSING RECOMBINANT PROTEIN

The plasmid includes all of the genetic material required for recombinant protein production and cell selection to achieve the desired output. The creation of a plasmid clone carrying the appropriate gene was previously described.

4.7.5 INTRODUCE PLASMID INTO HOST CELL LINE

By using biological, chemical, or physical methods, foreign nucleic acids can be introduced into cells through the process of transfection. For the best outcomes, it should have a high transfection efficiency, low cell toxicity, and little impact on regular physiology. The biological method makes use of viral vectors for nucleic acid delivery. Retroviral transduction for DNA insertion into cells has a long history. The chemical reaction uses a number of cationic reagents, including calcium phosphate, lipids, and polymers. The earliest method was CaPO4 precipitation, but lipid-based reagents are now more widely available and more dependable, replacing it (Lipofectamine, Fugene, Transfectin). Physical approaches include micro-needle injection, magnetic nanoparticles, and electroporation (Figure 4.12).[92]

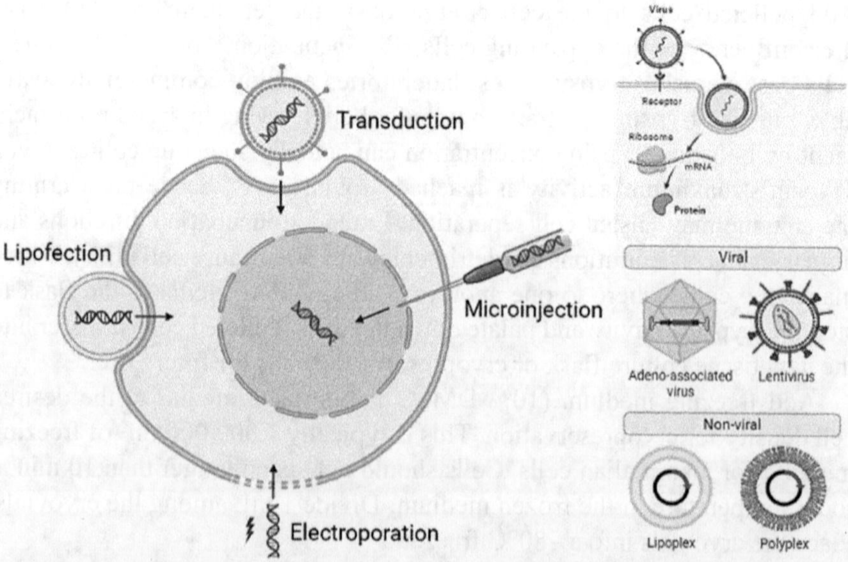

FIGURE 4.12 Method of the introduction of recombinant protein into the host cell.
Source: Created with Biorender.com

4.7.6 CLONAL CELL LINES IDENTIFICATION EXPRESSING HIGH LEVELS OF RECOMBINANT PROTEIN

(APH 3′ II) for G418 geneticin resistance, the phosphotransferase gene for hygromycin resistance, the stable gene for zeocin resistance, or puromycin N-acetyl-transferase, which encodes puromycin resistance are just a few examples of selectable markers that the plasmid may also include.[93]

4.8 LIMITATION OF RDNA TECHNOLOGY

rDNA technology, also known as genetic engineering, has revolutionized the field of biology and biotechnology. However, it is not without limitations and concerns.

- Ethical concerns: Manipulating the genetic material of organisms, particularly humans and animals, raises ethical debates about the creation of genetically modified organisms (GMOs), gene editing, and cloning.
- Environmental concerns: The release of GMOs into the environment can have unintended ecological consequences, such as outcompeting native species or disrupting ecosystems.

- Health and safety concerns: Working with rDNA in laboratories requires strict safety measures to prevent accidental exposure or the release of potentially harmful organisms.
- Unintended consequences: Genetic modifications can sometimes lead to unexpected side effects or the development of resistance to certain modifications.
- Limited understanding: Despite significant advancements in genetic engineering, our understanding of the long-term effects of genetic modifications is incomplete.
- Biodiversity loss: The widespread adoption of GMOs in agriculture can lead to a reduction in genetic diversity in the food supply.
- Regulatory and legal challenges: The regulation of rDNA technology is a complex and evolving field.
- Socioeconomic impacts: Genetic engineering can lead to disparities in access to advanced technologies, as well as issues related to intellectual property rights.
- Safety concerns for human health: There are concerns about the safety of GMOs in the human food supply. Allergenicity and potential health risks must be rigorously assessed.

Unpredictability: The behavior of GMOs in the field or the human body can be unpredictable, making it difficult to anticipate all potential consequences accurately.[94]

4.9 CONCLUSION

This technology has expanded rapidly over several decades, providing us with a platform to bring novel and complex research concepts into the field of biotechnology. Furthermore, the discovery of rDNA technology allowed us to sophisticated procedures such as vector design, transformation, transfection, restriction digestion, and purification. rDNA technology has played a critical part in delivering a wide range of applications, ranging from inserting a foreign gene into a plasmid to generating a vaccine for a wide range of diseases. As a result, it presented us with a platform to generate more hopeful prospects to mitigate with modified strains of the antigen and deliver the antidote in a highly cost-effective manner.

KEYWORDS

- **rDNA**
- **hepatitis**
- **insulin**
- **polymerase chain reaction (PCR)**

REFERENCES

1. Frier, B. M.; Sullivan, F. M.; Mair, F. S.; Koch, I. M.; Scotton, J. B. Pharmacokinetics of Human Insulin Zinc Suspension (Recombinant DNA) in Normal Man: A Comparison with Porcine Insulin Zinc Suspension. *Diabet. Med.* **1984,** *1* (3), 219–221. https://doi.org/10.1111/j.1464-5491.1984.tb01956.x.
2. Berg, P.; Mertz, J. E. Personal Reflections on the Origins and Emergence of Recombinant DNA Technology. *Genetics* **2010,** *184* (1), 9–17. https://doi.org/10.1534/genetics.109.112144.
3. Landgraf, W.; Medical Affairs Diabetes Division, Sanofi-Aventis Frankfurt, Germany; Sandow, J.; Professor, Centre of Pharmacology, Johann-Wolfgang-Goethe University, Frankfurt/Main, Germany. Recombinant Human Insulins – Clinical Efficacy and Safety in Diabetes Therapy. *Eur. Endocrinol.* **2016,** *12* (1), 12. https://doi.org/10.17925/EE.2016.12.01.12.
4. Khan, S.; Ullah, M. W.; Siddique, R.; Nabi, G.; Manan, S.; Yousaf, M.; Hou, H. Role of Recombinant DNA Technology to Improve Life. *Int. J. Genomics* **2016,** *2016*, 1–14. https://doi.org/10.1155/2016/2405954.
5. Bouchard, C.; Rankinen, T.; Timmons, J. A. Genomics and Genetics in the Biology of Adaptation to Exercise. In *Comprehensive Physiology*; Prakash, Y. S., Ed.; Wiley, 2011; pp 1603–1648. https://doi.org/10.1002/cphy.c100059.
6. Rees, H. A.; Liu, D. R. Base Editing: Precision Chemistry on the Genome and Transcriptome of Living Cells. *Nat. Rev. Genet.* **2018,** *19* (12), 770–788. https://doi.org/10.1038/s41576-018-0059-1.
7. Huang, Z.; Chavda, V. P.; Bezbaruah, R.; Uversky, V. N.; P, S.; Patel, A. B.; Chen, Z.-S. An Ayurgenomics Approach: Prakriti-Based Drug Discovery and Development for Personalized Care. *Front. Pharmacol.* **2022,** *13,* 866827–866827. https://doi.org/10.3389/fphar.2022.866827.
8. Hood, L.; Rowen, L. The Human Genome Project: Big Science Transforms Biology and Medicine. *Genome Med.* **2013,** *5* (9), 79. https://doi.org/10.1186/gm483.
9. Cohen, S. N. DNA Cloning: A Personal View after 40 Years. *Proc. Natl. Acad. Sci.* **2013,** *110* (39), 15521–15529. https://doi.org/10.1073/pnas.1313397110.
10. Hagedoorn, J.; Letterie, W.; Palm, F. The Information Value of R&D Alliances: The Preference for Local or Distant Ties. *Strateg. Organ.* **2011,** *9* (4), 283–309. https://doi.org/10.1177/1476127011421535.

11. Suzuki, K. Enzymic Diagnosis of Sphingolipidoses. *Methods Enzymol.* **1978,** *50,* 456–488. https://doi.org/10.1016/0076-6879(78)50049-9.
12. Garibyan, L.; Avashia, N. Polymerase Chain Reaction. *J. Invest. Dermatol.* **2013,** *133* (3), 1–4. https://doi.org/10.1038/jid.2013.1.
13. Lorenz, T. C. Polymerase Chain Reaction: Basic Protocol Plus Troubleshooting and Optimization Strategies. *J. Vis. Exp.* **2012,** *63,* 3998. https://doi.org/10.3791/3998.
14. Valasek, M. A.; Repa, J. J. The Power of Real-Time PCR. *Adv. Physiol. Educ.* **2005,** *29* (3), 151–159. https://doi.org/10.1152/advan.00019.2005.
15. Smith, C. J.; Osborn, A. M. Advantages and Limitations of Quantitative PCR (Q-PCR)-Based Approaches in Microbial Ecology. *FEMS Microbiol. Ecol.* **2009,** *67* (1), 6–20. https://doi.org/10.1111/j.1574-6941.2008.00629.x.
16. Dewan, I.; Uecker, H. A Mathematician's Guide to Plasmids: An Introduction to Plasmid Biology for Modellers. *Microbiology* 2023, *169*. https://doi.org/10.1099/mic.0.001362.
17. Wright, O.; Delmans, M.; Stan, G.-B.; Ellis, T. GeneGuard: A Modular Plasmid System Designed for Biosafety. *ACS Synth. Biol.* **2015,** *4* (3), 307–316. https://doi.org/10.1021/sb500234s.
18. Xavier, M. A. S.; Kipnis, A.; Torres, F. A. G.; Astofi-Filho, S. New Vectors Derives from pUC 18 for Clonig and Thermal-Induced Expression in Escherichia Coli. *Braz. J. Microbiol.* **2009,** *40*.
19. Nora, L. C.; Westmann, C. A.; Martins-Santana, L.; Alves, L. de F.; Monteiro, L. M. O.; Guazzaroni, M.-E.; Silva-Rocha, R. The Art of Vector Engineering: Towards the Construction of next-Generation Genetic Tools. *Microb. Biotechnol.* **2019,** *12* (1), 125–147. https://doi.org/10.1111/1751-7915.13318.
20. Ramos, A. E.; Muñoz, M.; Moreno-Pérez, D. A.; Patarroyo, M. A. pELMO, an Optimised in-House Cloning Vector. *AMB Express* **2017,** *7* (1), 26. https://doi.org/10.1186/s13568-017-0324-2.
21. Gomez-Escribano, J. P.; Bibb, M. J. Chapter Fourteen—Streptomyces Coelicolor as an Expression Host for Heterologous Gene Clusters. In *Methods in Enzymology*; Hopwood, D. A. Ed.; Academic Press, 2012; Vol. 517, pp 279–300. https://doi.org/10.1016/B978-0-12-404634-4.00014-0.
22. Cronan, J. E. Cosmid-Based System for Transient Expression and Absolute Off-to-On Transcriptional Control of *Escherichia Coli* Genes. *J. Bacteriol.* **2003,** *185* (22), 6522–6529. https://doi.org/10.1128/JB.185.22.6522-6529.2003.
23. Bajpai, B. High Capacity Vectors. In *Advances in Biotechnology*; Ravi, I., Baunthiyal, M., Saxena, J., Eds.; Springer India: New Delhi, 2014; pp 1–10. https://doi.org/10.1007/978-81-322-1554-7_1.
24. Bertero, A.; Brown, S.; Vallier, L. Chapter 2—Methods of Cloning. In *Basic Science Methods for Clinical Researchers*; Jalali, M., Saldanha, F. Y. L., Jalali, M., Eds.; Academic Press: Boston, 2017; pp 19–39. https://doi.org/10.1016/B978-0-12-803077-6.00002-3.
25. Lowman, H. B. Phage Display for Protein Binding. In *Encyclopedia of Biological Chemistry*, 2nd ed.; Lennarz, W. J., Lane, M. D., Eds.; Academic Press: Waltham, 2013; pp 431–436. https://doi.org/10.1016/B978-0-12-378630-2.00061-X.
26. Ray, D. S.; Hines, J. C.; Kim, M. H.; Imber, R.; Nomura, N. M13 Vectors for Selective Cloning of Sequences Specifying Initiation of DNA Synthesis on Single-Stranded Templates. *Gene* **1982,** *18* (3), 231–238. https://doi.org/10.1016/0378-1119(82)90160-3.
27. Kim, M. H.; Hines, J. C.; Ray, D. S. Viable Deletions of the M13 Complementary Strand Origin. *Proc. Natl. Acad. Sci.* **1981,** *78* (11), 6784–6788. https://doi.org/10.1073/pnas.78.11.6784.

28. Smith, G. P. Filamentous Fusion Phage: Novel Expression Vectors That Display Cloned Antigens on the Virion Surface. *Science* **1985,** *228* (4705), 1315–1317. https://doi.org/10.1126/science.4001944.
29. Gates, K. S. Covalent Modification of DNA by Natural Products. In *Comprehensive Natural Products Chemistry*; Elsevier, 1999; pp 491–552. https://doi.org/10.1016/B978-0-08-091283-7.00074-6.
30. Amarakoon, I. I.; Hamilton, C.-L.; Mitchell, S. A.; Tennant, P. F.; Roye, M. E. Chapter 28 - Biotechnology. In *Pharmacognosy*; Badal, S., Delgoda, R., Eds.; Academic Press: Boston, 2017; pp 549–563. https://doi.org/10.1016/B978-0-12-802104-0.00028-7.
31. Lessard, J. C. Chapter Seven—Molecular Cloning. In *Methods in Enzymology*; Lorsch, J., Ed.; Academic Press, 2013; Vol. 529, pp 85–98. https://doi.org/10.1016/B978-0-12-418687-3.00007-0.
32. Shi, K.; Bohl, T. E.; Park, J.; Zasada, A.; Malik, S.; Banerjee, S.; Tran, V.; Li, N.; Yin, Z.; Kurniawan, F.; et al. T4 DNA Ligase Structure Reveals a Prototypical ATP-Dependent Ligase with a Unique Mode of Sliding Clamp Interaction. *Nucleic Acids Res.* **2018,** *46* (19), 10474–10488. https://doi.org/10.1093/nar/gky776.
33. Alexander, R. C. Canonical Nucleosides Can Be Utilized by T4 DNA Ligase as Universal Template Bases at Ligation Junctions. *Nucleic Acids Res.* **2003,** *31* (12), 3208–3216. https://doi.org/10.1093/nar/gkg415.
34. Tüzmen, Ş.; Baskın, Y.; Feyda Nursal, A.; Eraslan, S.; Esemen, Y.; Çalıbaşı, G.; Banu Demir, A.; Abbasoğlu, D.; Hızel, C. Chapter 14 - Techniques for Nucleic Acid Engineering: The Foundation of Gene Manipulation. In *Omics Technologies and Bio-Engineering*; Barh, D., Azevedo, V., Eds.; Academic Press, 2018; pp 247–315. https://doi.org/10.1016/B978-0-12-804659-3.00014-2.
35. Gelvin, S. B. *Agrobacterium*—Mediated Plant Transformation: The Biology behind the "Gene-Jockeying" Tool. *Microbiol. Mol. Biol. Rev.* **2003,** *67* (1), 16–37. https://doi.org/10.1128/MMBR.67.1.16-37.2003.
36. Chong, Z. X.; Yeap, S. K.; Ho, W. Y. Transfection Types, Methods and Strategies: A Technical Review. *PeerJ*, **2021,** *9,* e11165. https://doi.org/10.7717/peerj.11165.
37. Potter, H. Transfection by Electroporation. *Curr. Protoc. Mol. Biol.* **2003,** *62* (1), 9.3.1–9.3.6. https://doi.org/10.1002/0471142727.mb0903s62.
38. Gehl, J. Electroporation: Theory and Methods, Perspectives for Drug Delivery, Gene Therapy and Research. *Acta Physiol. Scand.* **2003,** *177* (4), 437–447. https://doi.org/10.1046/j.1365-201X.2003.01093.x.
39. Zhao, Y.; Sun, H.; Sha, X.; Gu, L.; Zhan, Z.; Li, W. J. A Review of Automated Microinjection of Zebrafish Embryos. *Micromachines* **2019,** *10* (1). https://doi.org/10.3390/mi10010007.
40. *Genetic Engineering*; Setlow, J. K., Ed.; Springer US: Boston, MA, 2003. https://doi.org/10.1007/978-1-4615-0073-5.
41. Biswas, S.; Knipp, R. J.; Gordon, L. E.; Nandula, S. R.; Gorr, S.; Clark, G. J.; Nantz, M. H. Hydrophobic Oxime Ethers: A Versatile Class of pDNA and siRNA Transfection Lipids. *ChemMedChem* **2011,** *6* (11), 2063–2069. https://doi.org/10.1002/cmdc.201100259.
42. Kim, T. K.; Eberwine, J. H. Mammalian Cell Transfection: The Present and the Future. *Anal. Bioanal. Chem.* **2010,** *397* (8), 3173–3178. https://doi.org/10.1007/s00216-010-3821-6.
43. Vecchio, I.; Tornali, C.; Bragazzi, N. L.; Martini, M. The Discovery of Insulin: An Important Milestone in the History of Medicine. *Front. Endocrinol.* **2018,** *9,* 613. https://doi.org/10.3389/fendo.2018.00613.

44. Vieira Gomes, A.; Souza Carmo, T.; Silva Carvalho, L.; Mendonça Bahia, F.; Parachin, N. Comparison of Yeasts as Hosts for Recombinant Protein Production. *Microorganisms*, **2018,** *6* (2), 38. https://doi.org/10.3390/microorganisms6020038.
45. Ayuk, J.; Sheppard, M. C. Growth Hormone and Its Disorders. *Postgrad. Med. J.* **2006,** *82* (963), 24–30. https://doi.org/10.1136/pgmj.2005.036087.
46. Zhang, J.-M.; An, J. Cytokines, Inflammation, and Pain. *Int. Anesthesiol. Clin.* **2007,** *45* (2).
47. Li, H.; Yang, Y.; Hong, W.; Huang, M.; Wu, M.; Zhao, X. Applications of Genome Editing Technology in the Targeted Therapy of Human Diseases: Mechanisms, Advances and Prospects. *Signal Transduct. Target. Ther.* **2020,** *5* (1), 1. https://doi.org/10.1038/s41392-019-0089-y.
48. Ching, W. M.; Angkasekwinai, N.; Grieco, J.; Chao, C. C.; Atkins, E. H.; Romero, S. An Evaluation Study of Enzyme-Linked Immunosorbent Assay (ELISA) Using Recombinant Protein Pap31 for Detection of Antibody against Bartonella Bacilliformis Infection among the Peruvian Population. *Am. J. Trop. Med. Hyg.* **2014,** *90* (4), 690–696. https://doi.org/10.4269/ajtmh.13-0131.
49. Jiang, T.; Zhou, C.; Ren, S. Role of IL-2 in Cancer Immunotherapy. *OncoImmunology* **2016,** *5* (6), e1163462. https://doi.org/10.1080/2162402X.2016.1163462.
50. Baigent, G.; Barngrover, D. Recombinant Interleukin-2 (Aldesleukin) for Oncology and HIV Disease and Recombinant Protein Treatment (Fabrazyme) for Fabry's Disease (No. 14 in a Series of Articles to Promote a Better Understanding of the Use of Genetic Engineering). *J. Biotechnol.* **2002,** *95* (3), 277–283. https://doi.org/10.1016/S0168-1656(02)00019-6.
51. Muhammad, S.; Fan, T.; Hai, Y.; Gao, Y.; He, J. Reigniting Hope in Cancer Treatment: The Promise and Pitfalls of IL-2 and IL-2R Targeting Strategies. *Mol. Cancer* **2023,** *22* (1), 121. https://doi.org/10.1186/s12943-023-01826-7.
52. Hamza, T.; Barnett, J. B.; Li, B. Interleukin 12 a Key Immunoregulatory Cytokine in Infection Applications. *Int. J. Mol. Sci.* **2010,** *11* (3), 789–806. https://doi.org/10.3390/ijms11030789.
53. Rosano, G. L.; Ceccarelli, E. A. Recombinant Protein Expression in Escherichia Coli: Advances and Challenges. *Front. Microbiol.* **2014,** *5*. https://doi.org/10.3389/fmicb.2014.00172.
54. Jayanthi, S.; Koppolu, B. P.; Nguyen, K. G.; Smith, S. G.; Felber, B. K.; Kumar, T. K. S.; Zaharoff, D. A. Modulation of Interleukin-12 Activity in the Presence of Heparin. *Sci. Rep.* **2017,** *7* (1), 5360. https://doi.org/10.1038/s41598-017-05382-1.
55. Chavda, V. P.; Soni, S.; Vora, L. K.; Soni, S.; Khadela, A.; Ajabiya, J. mRNA-Based Vaccines and Therapeutics for COVID-19 and Future Pandemics. *Vaccines* **2022,** *10* (12). https://doi.org/10.3390/vaccines10122150.
56. Yousefi, Z.; Aria, H.; Ghaedrahmati, F.; Bakhtiari, T.; Azizi, M.; Bastan, R.; Hosseini, R.; Eskandari, N. An Update on Human Papilloma Virus Vaccines: History, Types, Protection, and Efficacy. *Front. Immunol.* **2022,** *12*.
57. Wang, J. W.; Roden, R. B. Virus-like Particles for the Prevention of Human Papillomavirus-Associated Malignancies. *Expert Rev. Vaccines* **2013,** *12* (2), 129–141. https://doi.org/10.1586/erv.12.151.
58. Kondo, T.; Mukai, M.; Kondo, Y. Rapid Isolation of Plasmid DNA by LiCl-Ethidium Bromide Treatment and Gel Filtration. *Anal. Biochem.* **1991,** *198* (1), 30–35. https://doi.org/10.1016/0003-2697(91)90501-J.

59. Ko, G. N.; Zhang, L. D.; Yan, W. W.; Zhang, M. D.; Buchner, D.; Xia, Z. Y.; Wyatt, R. J.; Jeste, D. V. The Shanghai 800: Prevalence of Tardive Dyskinesia in a Chinese Psychiatric Hospital. *Am. J. Psychiatry* **1989,** *146* (3), 387–389. https://doi.org/10.1176/ajp.146.3.387.
60. Mishra, V.; Heath, R. J. Structural and Biochemical Features of Human Serum Albumin Essential for Eukaryotic Cell Culture. *Int. J. Mol. Sci.* **2021,** *22* (16), 8411. https://doi.org/10.3390/ijms22168411.
61. Singh, V.; Kumar, B. R.; Singh, B. N. Double Recombinant Mycobacterium Bovis BCG Strain for Screening of Primary and Rationale-Based Antimycobacterial Compounds. *Antimicrob. Agents Chemother.* **2014,** *58* (3), 1389–1396. https://doi.org/10.1128/aac.01301-13.
62. Kanno, A. I.; Goulart, C.; Rofatto, H. K.; Oliveira, S. C.; Leite, L. C. C.; McFadden, J. New Recombinant Mycobacterium Bovis BCG Expression Vectors: Improving Genetic Control over Mycobacterial Promoters. *Appl. Environ. Microbiol.* **2016,** *82* (8), 2240–2246. https://doi.org/10.1128/AEM.03677-15.
63. Hoft, D. F.; Blazevic, A.; Abate, G.; Hanekom, W. A.; Kaplan, G.; Soler, J. H.; Weichold, F.; Geiter, L.; Sadoff, J. C.; Horwitz, M. A. A New Recombinant Bacille Calmette-Guérin Vaccine Safely Induces Significantly Enhanced Tuberculosis-Specific Immunity in Human Volunteers. *J. Infect. Dis.* **2008,** *198* (10), 1491–1501. https://doi.org/10.1086/592450.
64. Bitter, G. A.; Egan, K. M.; Burnette, W. N.; Samal, B.; Fieschko, J. C.; Peterson, D. L.; Downing, M. R.; Wypych, J.; Langley, K. E. Hepatitis B Vaccine Produced in Yeast. *J. Med. Virol.* **1988,** *25* (2), 123–140. https://doi.org/10.1002/jmv.1890250202.
65. Moradi Vahdat, M.; Hemmati, F.; Ghorbani, A.; Rutkowska, D.; Afsharifar, A.; Eskandari, M. H.; Rezaei, N.; Niazi, A. Hepatitis B Core-Based Virus-like Particles: A Platform for Vaccine Development in Plants. *Biotechnol. Rep.* **2021,** *29,* e00605–e00605. https://doi.org/10.1016/J.BTRE.2021.E00605.
66. Keating, G. M.; Noble, S. Recombinant Hepatitis B Vaccine (Engerix-B??): A Review of Its Immunogenicity and Protective Efficacy Against Hepatitis B. *Drugs* **2003,** *63* (10), 1021–1051. https://doi.org/10.2165/00003495-200363100-00006.
67. Decker, M. D.; Edwards, K. M.; Bogaerts, H. H. 40—Combination Vaccines. In *Vaccines*, 6th ed.; Plotkin, S. A., Orenstein, W. A., Offit, P. A., Eds.; W.B. Saunders: London, 2013; pp 981–1007. https://doi.org/10.1016/B978-1-4557-0090-5.00015-X.
68. Elghanam, M. S.; Attia, A. S.; Shoeb, H. A.; Hashem, A. E. M. Expression and Purification of Hepatitis B Surface Antigen S from Escherichia Coli; a New Simple Method. *BMC Res. Notes* **2012,** *5* (1), 125. https://doi.org/10.1186/1756-0500-5-125.
69. Fritsch, S. D.; Weichhart, T. Effects of Interferons and Viruses on Metabolism. *Front. Immunol.* **2016,** *7*. https://doi.org/10.3389/fimmu.2016.00630.
70. Boxx, G. M.; Cheng, G. The Roles of Type I Interferon in Bacterial Infection. *Cell Host Microbe* **2016,** *19* (6), 760–769. https://doi.org/10.1016/j.chom.2016.05.016.
71. Alspach, E.; Lussier, D. M.; Schreiber, R. D. Interferon γ and Its Important Roles in Promoting and Inhibiting Spontaneous and Therapeutic Cancer Immunity. *Cold Spring Harb. Perspect. Biol.* **2019,** *11* (3), a028480. https://doi.org/10.1101/cshperspect.a028480.
72. Robichon, C.; Luo, J.; Causey, T. B.; Benner, J. S.; Samuelson, J. C. Engineering Escherichia Coli BL21(DE3) Derivative Strains To Minimize E. Coli Protein

Contamination after Purification by Immobilized Metal Affinity Chromatography. *Appl. Environ. Microbiol.* **2011,** *77* (13), 4634–4646. https://doi.org/10.1128/AEM.00119-11.

73. Fazaeli, A.; Golestani, A.; Lakzaei, M.; Rasi Varaei, S. S.; Aminian, M. Expression Optimization, Purification, and Functional Characterization of Cholesterol Oxidase from Chromobacterium Sp. DS1. *PLoS ONE* **2019,** *14* (2), e0212217. https://doi.org/10.1371/journal.pone.0212217.
74. Ow D W; Ausubel F M. Conditionally Replicating Plasmid Vectors That Can Integrate into the Klebsiella Pneumoniae Chromosome via Bacteriophage P4 Site-Specific Recombination. *J. Bacteriol.* **1983,** *155* (2), 704–713. https://doi.org/10.1128/jb.155.2.704-713.1983.
75. Vogl, T.; Leigh, G.; Palfreyman, R. W.; Robert, S. Effect of Plasmid Design and Type of Integration Event on Recombinant Protein Expression in Pichia Pastoris. *Appl. Environ. Microbiol.* **2018,** *84* (6), e02712–e02717. https://doi.org/10.1128/AEM.02712-17.
76. Gomi, K.; Morimoto, M.; Nakamizo, N. Characteristics of Antiviral and Anticellular Activities of Human Recombinant Interferon-Gamma. *Jpn. J. Cancer Res.* **1985,** *76* (3), 224–234.
77. Lu, R.-M.; Hwang, Y.-C.; Liu, I.-J.; Lee, C.-C.; Tsai, H.-Z.; Li, H.-J.; Wu, H.-C. Development of Therapeutic Antibodies for the Treatment of Diseases. *J. Biomed. Sci.* **2020,** *27* (1), 1. https://doi.org/10.1186/s12929-019-0592-z.
78. Mitra, S.; Tomar, P. C. Hybridoma Technology; Advancements, Clinical Significance, and Future Aspects. *J. Genet. Eng. Biotechnol.* **2021,** *19* (1), 159. https://doi.org/10.1186/s43141-021-00264-6.
79. Berger, M.; Shankar, V.; Vafai, A. Therapeutic Applications of Monoclonal Antibodies. *Am. J. Med. Sci.* **2002,** *324* (1), 14–30. https://doi.org/10.1097/00000441-200207000-00004.
80. Belete, T. M. The Current Status of Gene Therapy for the Treatment of Cancer. *Biol. Targets Ther.* **2021,** *15,* 67–77. https://doi.org/10.2147/BTT.S302095.
81. Zhang, X.; Cui, H.; Zhang, W.; Li, Z.; Gao, J. Engineered Tumor Cell-Derived Vaccines against Cancer: The Art of Combating Poison with Poison. *Bioact. Mater.* **2023,** *22,* 491–517. https://doi.org/10.1016/j.bioactmat.2022.10.016.
82. Bhatwa, A.; Wang, W.; Hassan, Y. I.; Abraham, N.; Li, X.-Z.; Zhou, T. Challenges Associated With the Formation of Recombinant Protein Inclusion Bodies in Escherichia Coli and Strategies to Address Them for Industrial Applications. *Front. Bioeng. Biotechnol.* **2021,** *9,* 630551. https://doi.org/10.3389/fbioe.2021.630551.
83. Tripathi, N. K.; Shrivastava, A. Recent Developments in Bioprocessing of Recombinant Proteins: Expression Hosts and Process Development. *Front. Bioeng. Biotechnol.* **2019,** *7,* 420. https://doi.org/10.3389/fbioe.2019.00420.
84. Dumont, J.; Euwart, D.; Mei, B.; Estes, S.; Kshirsagar, R. Human Cell Lines for Biopharmaceutical Manufacturing: History, Status, and Future Perspectives. *Crit. Rev. Biotechnol.* **2016,** *36* (6), 1110–1122. https://doi.org/10.3109/07388551.2015.1084266.
85. Weiskirchen, S.; Schröder, S. K.; Buhl, E. M.; Weiskirchen, R. A Beginner's Guide to Cell Culture: Practical Advice for Preventing Needless Problems. *Cells*, **2023,** *12* (5), 682. https://doi.org/10.3390/cells12050682.
86. Noel, V.; Berry, M. D. Culture of Adherent Cancer Cell Lines. In *Cancer Cell Biology*; Christian, S. L., Ed.; Methods in Molecular Biology; Springer US: New York, NY, 2022; Vol. 2508, pp 19–29. https://doi.org/10.1007/978-1-0716-2376-3_3.

87. Verma, A.; Verma, M.; Singh, A. Animal Tissue Culture Principles and Applications. In *Animal Biotechnology*; Elsevier, 2020; pp 269–293. https://doi.org/10.1016/B978-0-12-811710-1.00012-4.
88. Pilgrim, C. R.; McCahill, K. A.; Rops, J. G.; Dufour, J. M.; Russell, K. A.; Koch, T. G. A Review of Fetal Bovine Serum in the Culture of Mesenchymal Stromal Cells and Potential Alternatives for Veterinary Medicine. *Front. Vet. Sci.* **2022,** *9,* 859025. https://doi.org/10.3389/fvets.2022.859025.
89. Luhur, A.; Klueg, K. M.; Roberts, J.; Zelhof, A. C. Thawing, Culturing, and Cryopreserving Drosophila Cell Lines. *J. Vis. Exp.* **2019,** *146,* 59459. https://doi.org/10.3791/59459.
90. Koshikawa, N.; Hasegawa, S.; Nagashima, Y.; Mitsuhashi, K.; Tsubota, Y.; Miyata, S.; Miyagi, Y.; Yasumitsu, H.; Miyazaki, K. Expression of Trypsin by Epithelial Cells of Various Tissues, Leukocytes, and Neurons in Human and Mouse. *Am. J. Pathol.* **1998,** *153* (3), 937–944. https://doi.org/10.1016/S0002-9440(10)65635-0.
91. Whaley, D.; Damyar, K.; Witek, R. P.; Mendoza, A.; Alexander, M.; Lakey, J. R. Cryopreservation: An Overview of Principles and Cell-Specific Considerations. *Cell Transplant.* **2021,** *30,* 096368972199961. https://doi.org/10.1177/0963689721999617.
92. Fus-Kujawa, A.; Prus, P.; Bajdak-Rusinek, K.; Teper, P.; Gawron, K.; Kowalczuk, A.; Sieron, A. L. An Overview of Methods and Tools for Transfection of Eukaryotic Cells in Vitro. *Front. Bioeng. Biotechnol.* **2021,** *9,* 701031. https://doi.org/10.3389/fbioe.2021.701031.
93. Zhu, M. M.; Mollet, M.; Hubert, R. S.; Kyung, Y. S.; Zhang, G. G. Industrial Production of Therapeutic Proteins: Cell Lines, Cell Culture, and Purification. In *Handbook of Industrial Chemistry and Biotechnology*; Kent, J. A., Bommaraju, T. V., Barnicki, S. D., Eds.; Springer International Publishing: Cham, 2017; pp 1639–1669. https://doi.org/10.1007/978-3-319-52287-6_29.
94. Krimsky, S.; Ozonoff, D. Recombinant DNA Research: The Scope and Limits of Regulation. *Am. J. Public Health* **1979,** *69* (12), 1252–1259. https://doi.org/10.2105/AJPH.69.12.1252.

CHAPTER 5

Microbial Biotransformation: From Conventional to Advance Technological Insights

VIVEK P. CHAVDA[1], RAJASHRI BEZBARUAH[2,3*], NASIMA AHMED[2], SHAHNAZ ALOM[2,4], FARAK ALI[2,5], VISHWA JYOTI BARUAH[6], BHASWATI SARMAH[7,8], RUPSHIKHA MALAKAR[4], DONG-HUA YANG[9], and QIUXU TENG[10]

[1]*Department of Pharmaceutics and Pharmaceutical Technology, L. M. College of Pharmacy, Ahmedabad, Gujarat, India*

[2]*Department of Pharmaceutical Sciences, Faculty of Science and Engineering, Dibrugarh University, Dibrugarh, India*

[3]*Institute of Pharmacy, Assam Medical College and Hospital, Dibrugarh, Assam, India*

[4]*Department of Pharmacology, Girijananda Chowdhury Institute of Pharmaceutical Science-Tezpur, Sonitpur, Assam, India*

[5]*Department of Pharmaceutical Chemistry, Girijananda Chowdhury Institute of Pharmaceutical Science-Tezpur, Sonitpur, Assam, India*

[6]*Centre for Biotechnology and Bioinformatics, Faculty of Biological Sciences, Dibrugarh University, Dibrugarh, Assam, India*

[7]*Department of Plant Breeding and Genetics, Assam Agricultural University, Jorhat, Assam, India*

[8]*Plant Biology Section, School of Integrative Plant Science, Cornell University, Ithaca, NY, USA*

[9]*New York College of Traditional Chinese Medicine, Mineola, NY, USA*

[10]*College of Pharmacy and Health Sciences, St. John's University, New York, USA*

**Corresponding author.*

Advances in Pharmaceutical Biotechnology, Volume 1: Exploring Cutting-Edge Innovations and Applications.
Vivek P. Chavda and Zhe-Sheng (Jason) Chen (Eds.)

DOI: 10.1201/9781774919798-5

ABSTRACT

Metabolism or biotransformation is the enzymatic alteration of natural and chemically synthesized products, into a substance having a specifically modified structure. The catalytic potential of a major range of microorganisms namely bacteria and fungi can be used as a channel to break down compounds with perplexing structures and compositions to derive profitable products via biosynthesis. Microbial biotransformation also known as microbial biotechnology is acquiring emphasis and broadly exercised to produce/synthesize metabolites in large portions with exceeding specificity. To carry out a broad range of chemical reactions, which include hydrolysis, oxidation, condensation, reduction, isomerization, synthesis of chiral compounds, construction of new C–C bonds, and reversal of hydrolytic reactions, microorganisms have been utilized in diverse biotransformation. In biotransformation, the conversion reactions have to be carried out in a sterile environment; and the reactions are to be monitored chromatographically or spectroscopically to prevent the creation of defective conversion products. Microbial biotransformation has been designed and regarded as a convincing means for the bioconversion of steroids, non-steroidal compounds, and pesticides for the bioremediation of an extensive collection of chemicals, including pharmaceuticals, radionuclides, hydrocarbons (such as oil), and polychlorinated biphenyls. Upgrading the genetic and metabolic engineering techniques will eventually heighten the research in the field of microbial transformations. This chapter expounds on the history, basic concept of bioconversion reactions, practical aspects, methodologies, and applications of microbial biotransformation.

5.1 INTRODUCTION

The term "biotransformation" refers to the biological processes or reagents used to change organic material (such as plant or animal waste) into valuable products or energy (such as certain microorganisms).[1–4] Before being transported to various areas of the body, substances in the body are biotransformed by several enzymes in the liver, stomach, kidney, skin, and placenta. Although some compounds are bio-transformed into more harmful forms, biotransformation generally makes substances easier to excrete in urine or bile. Therefore, the biotransformation of chemicals is not always beneficial. Microbes have evolved this mechanism to adapt to environmental changes and contribute to various biotechnological processes. The foremost and vital feature of biotransformation is to derive a product that maintains the original/primary carbon skeleton. There are two types of biotransformation: enzymatic and non-enzymatic.[5] Furthermore, enzymes are branched into microsomal and non-microsomal. Enzymatic elimination is biotransformation that occurs in the body due to various enzymes. Enzymes cause microsomal biotransformation in the smooth lipophilic membrane of the endoplasmic reticulum. Enzymes found in mitochondria are involved in non-microsomal biotransformation.[6] The area of microbial transformation that is studied by biocatalysis is expanding dramatically from a niche field of chemistry to one that is currently quite active and includes the creation of therapeutic items.[7] Biotransformation as it may be is best interpreted as the precise conversion of a given compound into a given product with structural similarity through the aid of biocatalysts, including microorganisms such as fungi.[8] A biocatalyst may be specified as an enzyme or an entire inactivated microorganism comprising one or more enzymes produced therein. For this study, in particular, microbial transformation is yet another term related to biotransformation.[9] There is only a small difference between biotransformation and bioconversions. Utilizing the catalytic abilities to live things, biotransformation can involve several different chemical reaction processes. Since a living microorganism constantly produces enzymes, bioconversions frequently use enzymes that are quite unstable for the substrates that are being used. The phrases "biotransformation" and "bioconversion" are sometimes used interchangeably due to their comparable features.[10,11]

Biotransformation is used to produce products with specific chemical structures certainly related to the substrates or starting materials of the reaction merely through a few chemical changes, which in many cases result from the action of a single enzyme.[12] The development of biotransformation processes as part of a synthetic process to manufacture L-ascorbic acid (Vitamin C) and ephedrine is documented in the 19th century chemical literature. Alcohol is oxidized to acetic acid by the bacteria *Acetobacter xylinum*, however, the oxidation of glucose to gluconic acid and sorbitol to sorbose by *Acetobacter xylinum* has not been adequately utilized.[13] Biotransformation was undoubtedly achieved in the early 20th century when D-sorbitol was converted to L-sorbose by *Acetobacter suboxydans* and benzaldehyde was converted to phenyl (milk-based methanol) by yeast. The earliest evidence of yeast oxidizing steroid nuclear hydroxyl groups and yeast reducing steroid nuclear double bonds was reported by Mamoli and Vercellone in 1936.[14] While Kramli and Horvath (1949) were able to lower cholesterol by *Pencillium roseum* and *Azatobacter sp.*[15], Hench (1952) demonstrated that the efficacy of cortisone in rheumatoid arthritis can only be achieved by introducing O_2 at the C-11 position with the assistance of *Rhizopus*[16], which is chemically very difficult. Subsequently, several people were able to solve the task with different fungi, actinomycetes, and bacteria. Various types of biocatalysts are currently used for biotransformation reactions.

5.2 MICROBIAL GENETICS

In molecular biology, the insertion of external/foreign DNA or RNA into eukaryotic or bacterial cells is a typical approach. To fully benefit from the abundance of knowledge produced by genome sequencing efforts and consequent global gene and protein expression assessments, genetic tools are needed.[17] Escherichia coli is where bacterial genetics was first developed and refined.[18] Foreign DNA can enter cells in diversified ways, which comprise transduction, transformation, conjugation, and transfection. Horizontal gene transfer occurs naturally through transformation, transduction, and conjugation, but transfection differs from these processes in how it transfers genes.[19]

5.2.1 TRANSFORMATION

Bacterial cells pick up genetic material from their environment using transformation. In the world of nature, the frequent origin of this genetic material is from neighboring lysed bacteria and may comprise plasmid DNA or DNA fragments dispersed into the environment.[20] In different bacteria, spontaneous transformation is driven by a bunch of conditions, namely the cell cycle or the presence of particular chemicals.[21] Although not all bacteria are capable of absorbing DNA by nature, this ability can be developed in the lab through chemical manipulation.[22] For a better visual understanding, refer to Figure 5.1.

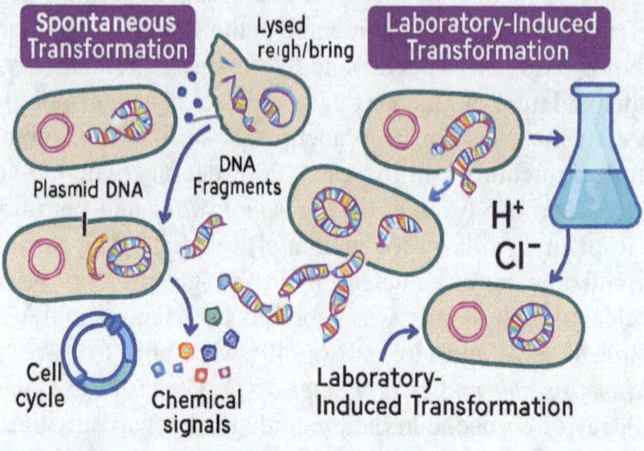

FIGURE 5.1 Transformation process involving the entry of naked and lysed DNA in the host cell.

This is frequently accomplished using calcium chloride, which makes the cell membrane permeable so that the bacteria can quickly pick up your target plasmid.[24] Researchers have been employing the electroporation approach, which involves applying an electrical charge to cells to intensify the penetrable property of cell membranes and, consequently, the effectiveness of transformation.[25]

5.2.2 TRANSDUCTION

The method by which a virus spreads its genetic material from one bacterium to another is called transduction.[26] Bacterial cells can be infected by viruses known as bacteriophages, which then use those cells as hosts to produce new viruses.[27] Once these viruses have multiplied, they assemble and occasionally take some of the bacterial DNA from the host cells. This portion of DNA might eventually be incorporated into the genome of the new host when one of these bacteriophages infects a different host cell.[28] Refer to Figure 5.2.

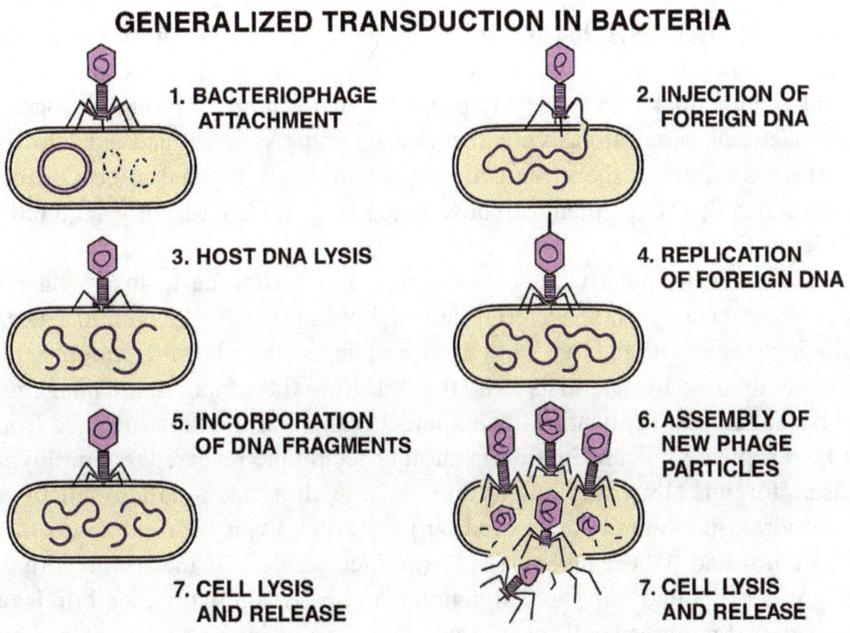

FIGURE 5.2 Generalized transduction process. It involves phage injecting the foreign DNA in the cell which leads to lysis of host DNA(Bacterial). The foreign DNA replicates in the hist and gets incorporated with the fragments of bacterial DNA.

Transduction comes in two categories: generalized and specialized. Bacteriophages can pick up any part of the host's genome during generalized transduction. The bacteriophage can pick up a particular section of the host's DNA if it is specialized. Using viruses, scientists have used the transduction process to transfer desired genes into a variety of host cells.[29] There are two possible cycles for generalized transduction: lytic and lysogenic.[30] Here, by coincidence, the viral genome is packed into bacteriophages together with any arbitrary DNA fragment. The lytic stage of the phage life cycle is when it happens. When bacterial DNA from a virus infects another cell, it may be integrated into the host genome or, if the DNA originally belonged to a plasmid, it may rebuild the plasmid. Linkage data, gene mapping, comparing the genomes of two different bacteria, mutagenesis, and other topics are studied using generalized transduction. The P1 phage's transduction of E. coli is an illustration of generalized transduction.[28]

Only the lysogenic cycle, or by a temperate phage, is capable of specialized transduction.[31] In this case, the virus only contains a specific portion of the bacterial DNA. In the lysogenic cycle, it happens when the prophage, or viral DNA, that is inserted into the bacterial genome, excises. A portion of the bacterial DNA that is flanked on both sides of the prophage is also excised when it excises from bacterial DNA. Here, both a bacterial and a viral genome make up the newly packed phage genome. Through lysogeny, the bacterial gene, along with the viral genome is later inserted into the host genome, when the virus with the recombinant genome infects a fresh bacterial cell. The recipient cell now displays the traits it recently acquired.[32] Refer to Figure 5.3.

Scientists frequently use transduction as a technique to insert various DNA sequences into a bacterial cell or host genome.[33] Phagemids, DNA cloning vectors that have both bacteriophage and plasmid features, are frequently used by scientists to do this.[34] Before transduction, the phagemid is packaged into replication-incompetent phage parts with assistance from a helper phage.[35] Transduction is another technique researchers employ to insert foreign DNA into eukaryotic cells, such as mammalian cell lines. Lentiviral and Adeno Associated Viruses (AVV) can be used to do this. Lentiviral and AVV can be used to produce stable cell lines with foreign DNA incorporated into the cell genome and carried down via cell division as well as temporary cell lines in which a gene of interest is expressed but not integrated into the genome.[36]

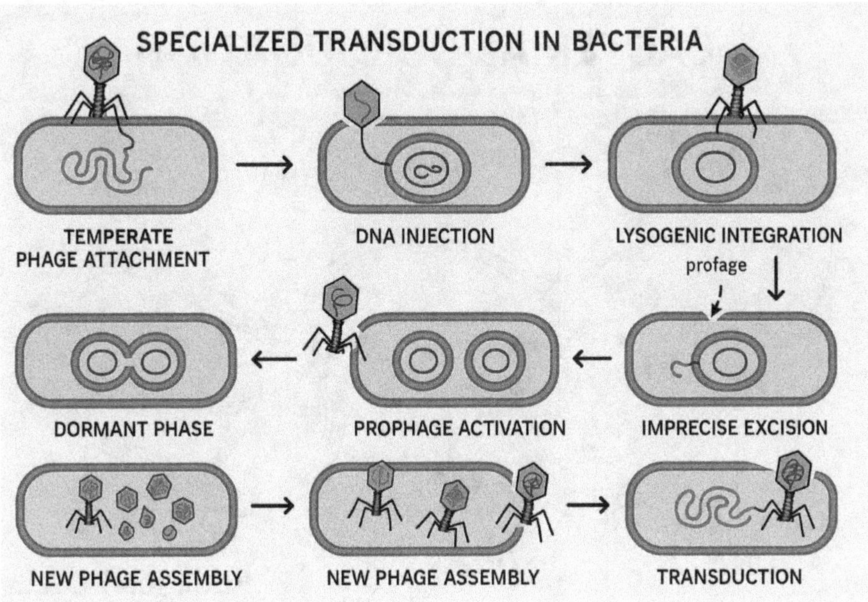

FIGURE 5.3 Specialized Transduction Process. It involves the addition of a specific gene into the genome of the host which can lead to a permanent reservoir of desired product or characteristics.

5.2.3 CONJUGATION

One bacterium can exchange genetic material with another directly through the process of conjugation.[37] One bacterium acts as the genetic material giver during conjugation, and another bacterium acts as the recipient. The fertility factor, or F-factor, is a DNA sequence that is carried by the donor bacterium.[38] The donor can communicate with the recipient by using a thin, tube-like structure called a pilus, which is made possible by the F-factor.[39] The donor bacterium transfers genetic material to the recipient bacterium when the pilus draws the two bacteria together. The genetic material typically takes the shape of a plasmid, which is a little circular piece of DNA.[40] The receiving bacterium frequently gains some type of genetic advantage thanks to the genetic material transferred during conjugation. For instance, conjugation frequently facilitates the transfer of plasmids containing genes for antibiotic resistance.[41] Do refer to Figure 5.4.

The primary utility of microbial conjugation is the transmission of genetic information via the transfer of a DNA segment that may code for antibiotic resistance, xenobiotic resistance, and the capacity to utilize or synthesize novel metabolites.[42]

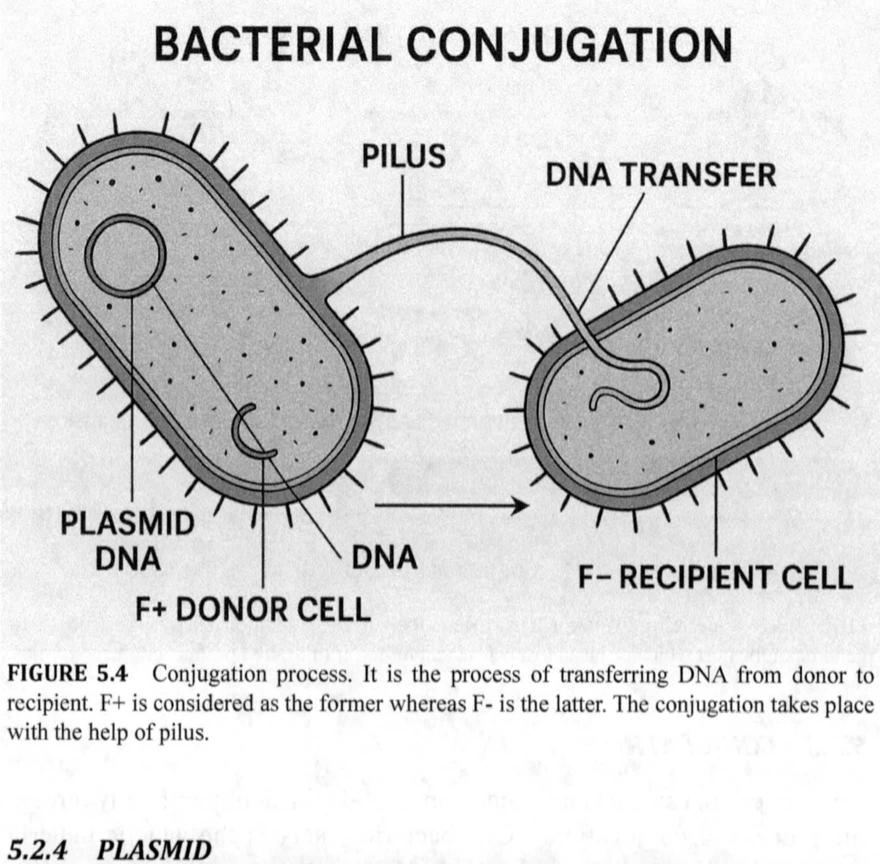

FIGURE 5.4 Conjugation process. It is the process of transferring DNA from donor to recipient. F+ is considered as the former whereas F- is the latter. The conjugation takes place with the help of pilus.

5.2.4 PLASMID

Circular DNA molecules called plasmids are found in the cytoplasm of bacteria and can replicate themselves on their own.[43] It functions as a vector in genetic engineering and is involved in gene transfer from one cell to another.[44] c. Plasmids can be transformed or conjugated to transfer between species, but they typically have a small range of hosts.[45] Do refer to Figure 5.5 for the diagrammatic representation of the plasmid.

Plasmids can deliver an array of functions. Organisms might continue to live longer and better as a result of the genes in the plasmid.[46] Plasmids may either eradicate the rival organisms or protect the host cell by building poisons. Additionally, bacterial reproduction is favored by plasmids.[47] Owing to their minuscule structure, plasmids generally have a few genes with precise functions (as opposed to a large amount of non-coding DNA). Numerous plasmids with diversified functions can dwell side by side in the same cell.[48]

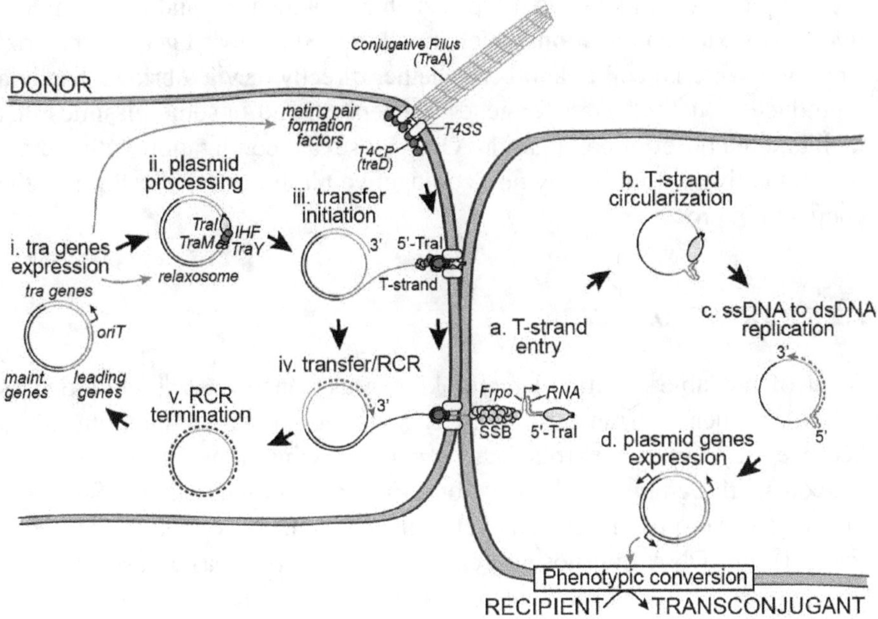

FIGURE 5.5 Schematic diagram of the life cycle of the F plasmid during conjugational transfer from the donor to the recipient cell. This F plasmid backbone is composed of the tra regions encoding all genes involved in conjugational transfer (light blue); the origin of transfer oriT (red); the leading region (green), which is the first to be transferred into the recipient cell; and the maintenance region (dark blue) involved in plasmid replication and partition. (i) The initiation of conjugation requires the expression of the tra genes. Some of the produced Tra proteins form the T4SS and the conjugative pilus that will recruit the recipient cell and mediate mating pair stabilization. (ii) Other Tra proteins constitute the relaxosome (TraI, TraM, and TraY), which, in combination with the integration host factor (IHF), bind to the oriT and prepare the plasmid for transfer by inducing the nicking reaction by the TraI relaxase. (iii) Interaction between the relaxosome and the Type IV Coupling Protein (T4CP) initiates the transfer of the T-strand by the T4SS. (iv, v) Transfer of the TraI-bound T-strand in the recipient is concomitant with the conversion of the ssDNA into dsDNA by Rolling Circle Replication (RCR) in the donor. (a) Upon entry into the recipient, the ssDNA T-strand is coated by the host chromosomal SSB, and the single-stranded promotor Frpo adopts a stem-loop structure recognized by the host RNA polymerase to initiate the synthesis of RNA primers. (b) TraI performs the circularization of the fully internalized T-strand. (c) The RNA–DNA duplex is recognized by the host DNA polymerase to initiate the complementary strand synthesis reaction. (d) Once the conversion of the ssDNA plasmid into dsDNA is completed, plasmid gene expression results in the phenotypic conversion of the recipient cell into a transconjugant cell.

Source: Reprinted from Ref. [49]. https://creativecommons.org/licenses/by/4.0/

The plasmid can be of two types such as conjugative and non-conjugative.[50] By sexual conjugation, which involves the transfer of genetic material from one bacterial cell to another, whether directly or via a bridge, bacteria reproduce sexually.[51] Transfer genes are genes found in some plasmids that facilitate the onset of conjugation. Only by sexual conjugation with the aid of conjugative plasmids may non-conjugative plasmids transfer through the conjugation process.[52]

5.2.5 TRANSPOSONS

A set of movable genetic elements known as transposons is classified as a DNA sequence. Transposons are sometimes known as jumping genes because they can move around the genome.[53] Some transposons, however, are constantly retained at the location of insertion in the genome. Since the majority of transposons are inactivated, they are immobile. Retrotransposons (class I) and DNA transposons (class II) are the two basic categories for transposons.[54] Eukaryotes frequently include retrotransposons. Eukaryotes and prokaryotes both include DNA transposons.[52,55] The DNA transposons and Tn family, which are typically the carriers of extra genes for antibiotic resistance, include bacterial transposons. Antibiotic resistance genes in bacteria are spread by transposons, which can move from one plasmid to another or from a DNA chromosome to a plasmid and vice versa. Due to widespread antibiotic resistance, which is partly brought on by transposons, treating bacterial infectious illnesses is challenging. The global mortality rate is rising, and the cause is bacterial infectious illnesses.[56] Refer to Figure 5.6 for transposable element.

It is very likely that methods for high-throughput gene functional analysis, including transposon insertion sequencing analyses, will be used extensively to identify important elements of plant-bacterial interactions.[57] Transposon insertion sequencing is a technique that combines high-throughput sequencing with saturation transposon mutagenesis to simultaneously examine the function of every non-essential gene in a bacterial genome. This method can be applied to both in vitro and in vivo studies to find the genes responsible for pathogen virulence, microbe-plant interactions, and stress tolerance.[58] Transposable elements are ubiquitous and serve as non-viral gene delivery systems. Transposon-based vectors can integrate the genome steadily and express transgenic constructs in cells for an extended period.[57]

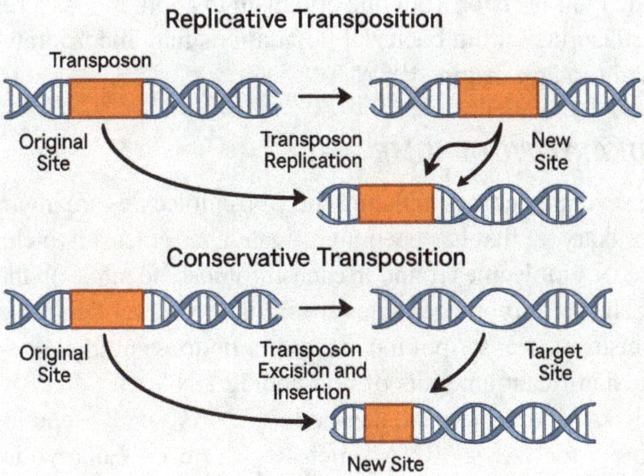

FIGURE 5.6 Process of transposition. It can be either replicative or conservative.

5.3 GENETIC ORGANIZATION OF PROKARYOTIC AND EUKARYOTIC

To understand molecular evolution, one must be familiar with both prokaryotic and eukaryotic organisms. Their shared lineage requires that many aspects of adaptation and limitation be shared by their molecular activities, but their variations in size, cellularity, structural complexity, and genetic organization also result in diverse evolutionary alternatives.[59]

5.3.1 PROKARYOTIC GENOME

They contain 580 to more than 4600 kbp of single circular DNA molecule. Extrachromosomal bacteria or plasmids also contain such compounds as genetic bacteriophages.[60] Numerous bacteria also carry plasmids, ranging in size from a few to 100 kbp. Small genetic components called genes on plasmids replicate separately from the bacterial chromosome.[61] Bacteria are known as haploid because they typically contain only one copy of each of their chromosomes because of this, a gene's mutation will have a more noticeable impact on the cell.[62] Replicons are DNA rings that contain the genetic data required for their replication. Also, genetic elements called transposons have many kbp of DNA, along with the information needed for them to move from one genetic locus to another. Numerous plasmids contain genes involved in the acquisition or rearrangement of DNA as well as those that mediate their transfer from one organism to another.[63] As a result, widely dispersed plasmids within bacterial populations may incorporate genes with distinct evolutionary origins.[64,65]

5.3.2 *EUKARYOTIC GENOME*

They have several genomes in them. The DNA molecules are invariably linear, and every eukaryote that has been investigated has at least two chromosomes. The linear DNA molecules found in each chromosome make up the eukaryotic genome.[66] The prokaryotic genomes are similar to those of eukaryotic plastids. The monocistronic transcription units, discontinuous coding areas like introns and exons, significant amounts of non-coding DNA, and diversity of coding (mRNA, rRNA, and tRNA) and non-coding RNAs are all found in the eukaryotic genome (snRNA, snoRNA, microRNA, etc.).[67] Eukaryotic cells have many compartments for transcription and translation, sexual reproduction, and the common recombination mechanism known as meiosis.[64,65]

Refer to Table 5.1 to evaluate the differences between eukaryotic and prokaryotic.

TABLE 5.1 Difference Between the Genetic Organization of Prokaryotic and Eukaryotic.[64,65,68–70]

Sl No.	Genetic organization of prokaryotic	Genetic organization of eukaryotic
1.	Contains a small number of genes.	Contains a large number of genes.
2.	DNA is less than 0.1 pg in size.	DNA is more than 1 pg in size.
3.	Contains one copy of the genome and haploid chromosome.	Contains more than one copy of the genome and diploid or more chromosomes.
4.	Homologous chromosome is absent.	Homologous chromosome is present.
5.	Operons are the units that organize prokaryotes.	Eukaryotic chromosomes are arranged as distinct entities.
6.	Chromosome is devoid of centromeres and telomeres.	Chromosomes contain centromeres and telomeres.
7.	DNA is not present inside organelles.	Inside chloroplast and mitochondria, some DNA is found.
8.	Genome is not packed with histones.	Genome is packed with histones.
9.	Naked DNA with a circular shape.	Protein-bound DNA with two codes and is linear in shape.
10.	Introns are not found in their structure.	Introns are found in their structure.
11.	Replication of DNA occurs in the cytoplasm.	Replication of DNA occurs in the nucleus.
12.	Replication of DNA is fast.	Replication of DNA is slow.
13.	At the beginning of the cell, division replication occurs.	During the S phase of cell division, replication occurs.
14.	Contain transposons in their structure.	Does not contain transposons in its structure.

TABLE 5.1 *(Continued)*

Sl No.	Genetic organization of prokaryotic	Genetic organization of eukaryotic
15.	Nucleus is not found in their structure.	Nucleus is present in their structure.
16.	70S ribosomes are present.	80S ribosomes are present.
17.	Organelles are present without membranes.	Membrane-bound organelles are present.
18.	Reproduction occurs in an asexual manner such as binary fission.	Reproduction occurs through mitosis and meiosis.
19.	More frequent HGT (horizontal gene transfer) is present.	Less frequent HGT is present.
20.	They are evolving both in sequence and structure quite quickly.	The evolution process in eukaryotic is more conservative.

5.4 MICROBIAL BIOTRANSFORMATION

Microbial transformation is a scope of biocatalysts, swiftly capturing interest in diverse areas. Microbial biotransformation exploits biological catalysts that are microorganisms for the sake of modifying definite compounds to distinct products with structural similarity. Many different chemicals, including metals, hydrocarbons, pharmaceuticals, and other contaminants are transformed using microbial biotransformation.[3] Microbiological biotransformation has long been recognized as a vital method to curtail the manufacture of numerous chemicals employed in the pharmaceutical, food, agrochemical, and other sectors.[71] Animal models have been used extensively in biotranslational research in the field of drug discovery to explore the metabolism of substances.[72] The phenomena of microbial biotransformation is thus frequently used to contrast drug metabolism pathways and improve interesting metabolites recognized in animal models for additional pharmacological and toxicological assessments.[71,73] White biotechnology uses microbial biotransformation to create desired products from bacteria, plants, mammals, algae, yeast, actinomycetes, and other living cells.[74]

5.4.1 ADVANTAGES OF MICROBIAL TRANSFORMATION

Microbial translational research can result in a variety of advantages. The processes required for microbial transformation are likely to take place at near-neutral pH, ambient temperature, and atmospheric pressure.[75] On the contrary, chemistry frequently calls for these conditions to be taken to their logical extremities, which is not only unfavorable to the environment but also undesirable from an industrial standpoint. Additionally, extreme pH, temperature, and pressure can adversely affect the manpower performing the drastic procedures, as well as the surrounding communities in the area. Above all other considerations, as biocatalysts are highly reactive, enantiomeric, and regiospecific, tons of microorganisms that are versatile biocatalysts perhaps be availed to execute immensely specific conversions using inexpensive substrates.[75] This directly and primarily relates to the chemical structures of compounds that one may particularly wish to obtain.

Microorganisms do have an enormous capacity to induce multiple alternative innovative and facile enzymatic systems capable of metabolizing unknown substrates. Henceforth, many larger-scale studies can be performed on different endophyte species in terms of chemical changes in molecules and compounds of interest.[76] The genome of a new species of thermophilic fungi can be probed to provide gene sequences encoding thermostable enzymes that are extra resistant to fluctuations in reaction temperature.[77]

Microorganisms, although incredibly small, much smaller than their name would imply, and invisible to the human eye, can play a crucial part in the pharmaceutical industry today. Microorganisms have an innate capacity to produce heat, alkali, and acid-resistant unique enzymes. One such study was performed on the thermophilic archaeon *Pyrobaculum calidifontis* VA1, which produces a thermostable esterase.[78,79]

Comparing them to other creatures, they have by far the highest surface area to volume ratio due to their small size. Because of the extensive exchange of chemicals and metabolites on their surfaces, this enables them to maximize metabolic rates. Under suitable culture conditions, microorganisms grow exponentially.[80] Due to their natural ability to reproduce, microorganisms can produce large amounts of enzymes in a short period. Additionally, it is possible to acquire and grow microorganisms that can endure harsh conditions like extremely high or low temperatures, as well as acidic or alkaline environments. Microbial transformation can enable feasible reactions that are impossible with conventional synthetic methods. Furthermore, endophytes can produce natural, biodegradable compounds.[81]

5.5 TYPES OF MICROBIAL TRANSFORMATION/BIOCONVERSION

Microorganisms have been exploited in a variety of biotransformation to carry out a wide range of chemical reactions. These chemical reactions include but are not limited to, oxidation, reduction, hydrolysis, condensation, isomerization, formation of new C–C bonds, synthesis of chiral compounds, and reversal of hydrolytic reactions.[82] Several of these microbe-assisted transformations/bioconversion can exhibit several types of chemical reactions simultaneously or sequentially.[3,83–86]

Microbial transformation/bioconversion can also be categorized based on the type of reactions that occur on the reactant/substrate. Bioconversion can be categorized based on the resultant biochemical pathways that follow and are divided into phase I, phase II, and phase III.[83–86]

Phase I: Phase I biotransformation produces a polar, water-soluble, and frequently active metabolite. Many of the products from this phase can also serve as phase II substrates.

- Oxidation with cytochrome P450 (most common)
- Reduction
- Hydrolysis

Phase II: Phase II biotransformation reaction combines the metabolite with natural hydrophilic groups to produce a significant polar metabolite that is eliminated by the body as water-soluble inactive molecules.

- Methylation
- Glucuronidation (most common)
- Acetylation
- Sulfation
- Conjugation with glutathione
- Conjugation with amino acids (glycine, taurine, and glutamic acid)

Phase III: Phase III biotransformation reaction occurs after phase II when a drug can continue to be metabolized and excreted. The superfamilies into which it is divided are as follows:

- ATP-binding cassette (ABC)
- Solute carrier (SLC) transporters

Among these, the following top four types of biotransformation reactions are generally aided by microorganisms or their enzymes and cover almost all sorts of chemical reactions:

i. Oxidation: Hydroxylation, epoxidation, dehydrogenation of C–C bonds, oxidation of alcohol and aldehydes, oxidative degradation of alkyl, carboxyalkyl, or ketoalkyl chains, oxidative removal of substituents, oxidative deamination, oxidation of heterofunctions, and oxidative ring fission.
ii. Reductions: Reduction of organic acids, aldehydes, ketones, and hydrogenation of C–C bonds, reduction of heterofunctions, dehydroxylation, and reductive elimination of substituents.

iii. Hydrolysis: Hydrolysis of esters, amines, amides, lactones, ethers, lactams, etc.
iv. Condensation: Dehydration, O- and N-acylation, glycosidation, esterification, lactomization, and amination.
v. Isomerization: Migration of double bonds or oxygen functions, racemization, rearrangements, formation of C–C bonds, or hetero atom bonds.
iv. Mixed Reactions: Hydroxylation with reduction; hydroxylation with oxidation; hydroxylation with side chain degradation; rupture of C–C linkages with oxidation of side chain.

Various microorganisms employed to carry out the reactions are given in Table 5.2.

TABLE 5.2 Microorganisms Exploited for Oxidative and Hydrolytic Biotransformation.

Reaction	Microorganisms used
Oxidation	Engineered baker's yeast with Cyclohexanone monooxygenase from *Acinetobacter* sp. NCIB 9871[87,88]
	Cunninghamella echinulate, Cunninghamella blakesleeana[89,90]
	Beauveria bassiana, Rhodicoccus erythropolis and *Agrobacterium sp.*[91,92]
	E. coli JM109(pDTG601) strain[93]
	Acetobacter acetii DSMZ3508, *A. pasteurianus* DSMZ2324 *and Gluconobacter oxidan* DSMZ44615[94]
	Microbacterium campoquemadoensis MB5614 and *Mycobacterium* M156[95,96]
Hydrolysis	*Trichosporon, Rhodotorula*, and *Rhodosporidium* sp.[97,98]
	Agrobacterium radiobacter AD1[99]
	Rhodotorula glutinis[100]
	Nocardia EH1[101]
	Rhodococcus rhodochrous IFO 15564[102]

Source: Adapted Using CC BY4.0 license from Ref. [42].

5.6 PRACTICAL ASPECTS OF BIOTRANSFORMATION

Conversion reactions are executed in sterile, agitated, aerated fermenters in case of large-scale equipment with the transformation process being monitored chromatographically or spectroscopically.[103] When a maximum titter is attained, the process is over. Because contamination can result in the creation of defective conversion products as well as complete breakdown, maintaining a sterile environment is essential. Since the buffer solution is where all the biotransformation reaction primarily takes place, therefore, recovering the product is not too difficult. Extracellularly, transformation processes by-products are present in either dissolved or suspended form. The reaction product that could be adsorbed to cell material is then separated from the cell material by repeatedly washing it with water or organic solvents.[104] It is feasible to use a polyphase system for the biotransformation of lipophilic compounds. The substrate containing a water-immiscible fluid phase has a deposition of enzyme or cell material containing an aqueous phase on its top. As the transformation reaction progresses, the product slowly goes back into the solvent phase after first entering the aqueous phase as the substrate. At the boundary between the aqueous and solvent phases, the actual transition can occasionally take place[105]. The recovery methods vary depending on the solubility of the product and include adsorption to ion exchangers, extraction using the right solvents, precipitation as the calcium salt, and straight distillation from the medium for volatile compounds.[106] Additionally, biological reactions are carried out in ambient, aqueous settings as opposed to other types of reactions, necessitating fewer solvents and less severe conditions, leading to processes that are more ecologically friendly.[107]

5.6.1 EQUIPMENT USED

In many cases, typical stirred tanks used for large-scale chemical synthesis with temperature and pH control are sufficient for biocatalysts and no extra equipment is needed. The process of biotransformation on a large scale has been carried out commonly inside the bioreactors. If the enzymes are not commercially available, they are produced using fermentors.[108] They are vessels with stirring technology that enables sterile (monoseptic) operations. This indicates that the only organism that can be utilized for biocatalysis in the bioreactor is the one carrying the appropriate enzyme. Normal stirred-tank reactors, as employed in chemistry, can be used in many situations, though.[109] The most common bioreactors used in the biotransformation process are the electrokinetic bioreactor and glass cell bioreactor.[107]

5.6.1.1 ELECTROKINETIC BIOREACTOR

The electrokinetic bioreactor included the following components: a rack, a polyvinylchloride support block, a square platinized titanium anode (0.01

m^2), a cation exchange membrane, an anolyte compartment, electrolyze project, an anion exchange membrane, a compartment for recovering benzoic acid or any solvent, a catholyte compartment, a bipolar membrane, a square stainless-steel cathode (0.01 m^2).[110] The electrolyte and electrode frames were machined from PVC, while the middle compartment was made from translucent Perspex.[107]

5.6.1.2 GLASS CELL BIOREACTOR

It is a spherical reaction vessel with a capacity of 250ml, a flange bore of 75 mm in diameter, and a height of 90 mm made up of the glass cell reactor. The anode and cathode electrode housings were constructed to order.[111] They had an opening at the top that was 30 mm wide, 183 mm in height, sealed with a rubber septum, and equipped with a glass tube that was 3 mm broad and 10 mm long for the electrode connections. Each electrode compartment has a fine glass frit with a 20 mm diameter.[112] A metal clip was used to fasten a flange top with a sample opening to the bottom of the central compartment. The 3 cm^2 platinized titanium paddle electrodes were put into their housings in line with the frits, capped, and sealed with heat shrink tubing. The glass cell bioreactor was supported on a shaking plate in a water bath that was thermostatically controlled and connected to a power supply via crocodile clips to provide a continuous current. A cell polarity reverser is used to automatically reverse the current direction every three minutes.[107]

5.6.2 METHOD OF SELECTION OF MICROORGANISMS

The choice of strains, either from their natural sources or from the numerous available cultures, that are entirely in charge of catalyzing the desired biotransformation reaction is not only crucial but also very significant. Numerous bacteria have been shown to frequently perform the needed bioconversions with the aid of a related chemical entity. Microorganisms should meet two conditions[113]:

1. To produce a certain product, the bacterium must possess the enzyme that causes the required transformation.
2. The substrate used should be the initial material. In the medium, organisms that can utilize the substrate flourish.

5.6.3 *PREPARATION OF MICROBIAL CULTURES*

5.6.3.1 *RANDOM SCREENING*

The procedure involves adding the substrate to a culture of many microorganisms and then testing the medium to see if the product is present after a predetermined amount of time. If a specific organism is responsible for the change, it is chosen for additional research.[114]

5.6.3.2 *COMPARATIVE SYSTEM*

Searching the literature for comparable types of conversion and testing the organism or nearby ones in the system under study will increase the likelihood of finding a suitable organism.[114]

5.6.3.3 *MIXED CULTURE*

In some circumstances, it may be useful to combine two or more microorganisms to convert a specific substrate.[109]

5.6.3.4 *ENRICHMENT METHOD*

The soil samples are enhanced by adding a substantial amount of substrate, water, and additional nutrients. In the medium, organisms that can utilize a substrate grow; these organisms are then isolated and employed for transformation.[115]

5.6.3.5 *MODIFIED ENRICHMENT METHOD*

The substrate dissimilation mechanism-blocked mutants are always isolated using the modified enrichment approach. In this specific case, a steroid substrate is typically included as the sole C source exclusively in a minimum medium seeded sufficiently with the soil dilutions. The cells that eventually cause the substrate's degradation will develop, and they are then moved to a medium that is specifically enriched with another C source, such as glucose, to continue their growth. It has been widely established and documented that a sizable number of microbial strains, including yeasts, eubacteria, and molds, may be kept strictly following the advised "standard methods," such as agar slant, soil culture, frozen culture, and lyophilized culture preserved at temperatures ranging from − 20 to −170°C.[114]

5.6.3.6 *FILTRATION ENRICHMENT METHOD*

In this method, mutagenesis is used to induce the development of filamentous organism spores, such as actinomycetes and fungi, in a liquid minimum media. Filtration is used to carefully separate the microcolonies of prototrophs that have so formed, with the filtrate containing the auxotroph spores that were unable to proliferate. After plating the filtrate produced in this way, the colonies are thoroughly examined for auxotrophic features.[116]

5.6.3.7 *PENICILLIN SELECTION PROCEDURE*

In the penicillin selection process, the antibiotic treatment selectively kills the prevailing developing cells, enriching the auxotrophs that are unable to thrive on the minimum media.[117] Therefore, a variety of inhibitors other than penicillin may also be used successfully in this approach based only on their mode of action, such as dihydrostreptomycin for *Pseudomonas aeruginosa.*[118]

5.6.3.8 *AGAR PLATE METHOD*

The agar plug method, one of the most dependable and accurate techniques, involves transferring agar cylinders with single colonies to test plates after enough incubation, preferably in a wet environment.[114]

5.6.3.9 *OTHERS*

Some other methods of culture preparation are spraying with a reagent or adding indicator dyes, inhibition of assay organism, and sodium pentachlorophenolate method.[114]

5.6.4 STERILIZATION

Sterility is necessary because contamination could stop the desired reaction from happening, lead to the production of conversion products with low quality, or entirely damage the substrate. By using steam, autoclave, or boiling techniques, fermentation media are sterilized. Sterilization times are shortened with synthetic media. Longer sterilizing times are required for media with higher viscosities.[109]

5.6.5 AERATION AND STIRRING IN THE BIOTRANSFORMATION TECHNIQUE

Oxygen should be brought into close contact with the cellular structure for effective microbial transformation. Sterile air is created by forcing air through a sterile filter.[119] Impellers are employed to help with aeration since filamentous organism growth results in non-Newtonian broth, which is a constant concern. Airlift fermenters and surface-attached microorganisms (biofilms) in fixed fluidized bed reactors are examples of contemporary approaches.[109]

5.6.6 PRODUCT RECOVERY FROM BIOREACTORS

There are several methods for in situ product recovery from bioreactors. The desired end products in the majority of biotransformation processes are extracellular. The product may be suspended or in a soluble condition.[120] When using complete cells, they must be separated and repeatedly cleaned (as needed with water or an organic solvent). The regularly employed methods, such as adsorption to ion exchangers, solvent extraction, precipitation by salts, etc. can be utilized to recover the extracted product.[121] Direct distillation from the medium can be used to recover the volatile compounds. Electrodialysis, which involves passing a direct electric current through a membrane to remove organic acids from fermentation broths, has been used after the first filtration step.[122] The advantage of a continuous process is that it only requires one step. Additionally, product removal may also release end product inhibition, which is a frequent barrier to the effectiveness of bioprocesses. With lactic acid fermentation, which accomplished constant pH control, the promise of this method was shown.[123] The viability, physiology, and metabolism of bacterial cells may be impacted by the introduction of an electric current. To sterilize solutions, for instance, pulsed high voltages are utilized in a variety of industries. However, high voltages can also be used to permeabilize cells so that foreign genetic material can be introduced.[124]

5.6.7 FACTORS INFLUENCING BIOTRANSFORMATION

Biotransformation is heavily influenced by several variables, including cell cultures' physiological, biochemical, and environmental features. The source of plant tissue, culture media composition, carbon sources, plant growth regulators, and environmental factors like CO_2 and oxygen in the environmental, all have an impact on the production of secondary metabolites.[125]

5.6.7.1 PHYSICOCHEMICAL PROPERTIES OF A DRUG

Physicochemical properties such as acidity, basicity, molecular weight and size, pKa values, interaction with metabolizing enzymes, and lipophilicity can also affect the process of biotransformation.[125]

5.6.7.2 CHEMICAL PROPERTIES OF A DRUG

5.6.7.2.1 Induction of Enzyme

When exposed to certain substances or medication, certain enzymes are stimulated or activated, these enzymes are known as enzyme inducers. These can include a variety of medications, insecticides, and even smoked cigarettes.[113,126]

5.6.7.2.2 Enzyme Inhibitors

A characteristic that impacts the biotransformation process is the inhibition of enzyme activity brought on by exposure to a particular medication. Either a direct or indirect mechanism may be to blame for the inhibition.[127] Three additional mechanisms such as non-competitive inhibition, competitive inhibition, and product inhibition are used to further achieve direct inhibition. As an illustration, allopurinol reduces the activity of xanthine oxidase, while indirect inhibition is accomplished through two different physiological changes, that is, changed physiology and repression.[113,126]

5.6.7.3 *ENVIRONMENTAL FACTORS*

In any biotransformation process, the result can be affected by the existence of heavy metals, polycyclic aromatic hydrocarbons, organophosphate insecticides, and halogenated pesticides in the environment.[113,126]

5.6.7.4 *BIOLOGICAL FACTORS*

5.6.7.4.1 Species Variations

Different species show different biotransformation reactions due to their different genetic build-up. The gene-controlled drug metabolizing enzymes vary from species to species as a result of genetic differences. For example, glucuronosyltransferase deficiency affects cats, while acetyltransferase deficiency affects dogs.[113,126]

5.6.7.4.2 Sex Variations

Another crucial factor in the biotransformation process is gender. As drug metabolism is regulated by sex hormones, it was discovered in a population under study that there were differences in the biotransformation process between the sexes.[126]

5.6.7.4.3 Strain Variations

Pharmacogenetics refers to the variations in how members of the same species respond to the same medication. Strain difference refers to the investigation of deviations in medication response occurring between subjects. It is possible to control strain variation either monogenetically or polygenetically. In humans, fraternal twins conceived from various eggs show polygenic control. However, deviations might also result from cultural and ethnic differences.[113]

5.6.7.4.4 Age Variations

Since the body's growth condition varies depending on age in each of these categories, the process of biotransformation is distinct in neonates, children, adults, and geriatrics. As previously indicated, because their microsomal enzyme systems are undeveloped, neonates and infants are unable to metabolize routinely used medications that include alcohol.[128] On the other hand, some medicines can be metabolized by children more quickly than by adults. However, the biotransformation process is further impacted in the elderly by the liver and lower hepatic blood flow.[126]

5.6.7.4.5 *Effect of Nutrients*

The type of food consumed is a crucial aspect that influences the body's enzyme system and in turn the entire process. Considerations for drug metabolism include total protein intake, a fat-free diet, protein-to-carbohydrate ratio, dietary deficiencies in minerals and vitamins, hunger, and alcohol consumption.[126]

5.6.7.4.6 *Physiological Factors*

The biotransformation process of various medications is greatly impacted by physiological factors including illness, pregnancy, and hormone abnormalities. Due to the high levels of steroid hormones, these states can decrease a person's capacity to metabolize medications, lengthen the half-life of practically all pharmaceuticals, or the activity of some enzymes is repressed while inducing that of others.[126]

5.6.7.4.7 *Origin of Plant Tissue*

Explant's genetic makeup controls the production of secondary products in cultures. Scientists identified high fielding cultivars produced high nicotine content in their cultures, while low-fielding cultivars produced cultures with less ability to produce nicotine.[125]

5.6.7.4.8 *Growth Control Agents*

Growth regulators have an impact on the synthesis of secondary metabolites and cell growth in culture. Auxins and cytokinins can sometimes work together synergistically while other combinations can inhibit the synthesis of steroidal hormones. For example, gibberellic acid was added to the solid medium to considerably enhance the taxol-producing plant, *Taxus cuspidate.*[125]

5.6.7.4.9 Culture Media

The production of biomass and the synthesis of secondary metabolites are both influenced by the chemical composition of the culture media used to create callus or suspension cultures. By controlling an ideal balance between biomass production and secondary metabolite production, perfect culture conditions are to be maintained to produce plant products in culture.[125]

5.6.7.4.10 Temperature

Plants can frequently survive in a wide range of temperatures. In general, greater temperatures tend to boost the generation of volatile oils.[125]

5.6.7.4.11 Carbon Source in the Medium

Sucrose is the most often utilized carbon source in culture media for tissue growth and has a significant impact on the formation of secondary metabolites. Galactose, fructose, and many other carbon molecules affect how products accumulate in cultures. Sucrose and glucose produced the most encouraging results in terms of biomass generation of product field out of all the carbon smoke currently employed in cultures.[125]

5.6.7.4.12 Intensity of Light

Light stimulates the function of enzymes *in vitro*. High light levels reduced *Nicotiana tabacum's* ability to produce nicotine, whereas ongoing darkness improved nicotine synthesis and its storage. Light plays a role in determining and producing several alkaloids and glycosides in a plant, for example, stramonium, belladonna, and cinchona are grown in full sunlight as compared to shade.[125]

5.6.7.4.13 Culture with High Cell Density

To boost the productivity of secondary metabolites, high cell density culture has been examined. For example, Japonica cells were grown up to a cell mass of 759 L using a nearly formulated fermented and optimized culture media.[125]

5.6.7.4.14 Effect of pH

Before autoclaving, a medium pH is typically adjusted between 5 and 6, extremes of pH are avoided since they can damage the cell culture.[125]

5.7 METHODOLOGIES FOR BIOTRANSFORMATION

Chemical changes to substances including nutrients, toxins, amino acids, and medicaments in the body are referred to as biotransformation. For nonpolar substances to be excreted rather than reabsorbed in renal tubules, it is also necessary to convert them to polar substances.[129] The hazardous kinetics of xenobiotics may be dominated by biotransformation and the metabolites may accumulate in organisms at higher concentrations than their parent substances. It especially relates to pharmaceuticals and other xenobiotics.[130] Changes in a molecule's pharmacological activity, which can occur in the form of an increase, decrease, or little change, are usually linked to the conversion of molecules from one form to another within an organism. The living plant can be viewed as a biosynthetic laboratory for secondary plant products like glycosides in addition to the primary metabolites like amino acids and sugar.[125] Utilizing enzyme-catalyzed reactions with living cells, microbial biotransformation often takes advantage of simple chemical processes including oxidation, reduction hydrolysis, and degradation, leading to the creation of C–C or C-hetero atom bonds.[131]

5.7.1 BIOTRANSFORMATION BY IMMOBILIZED CELL CULTURE

A variety of transformation procedures involves immobilized cells, which has the advantage of allowing the process to continue continuously and allowing the cells to be recycled several times.[131] Under the same conversion conditions, it has been noted that immobilized plant cells may produce at a higher rate than freely suspended cells. This method of culture, which enables the trapping of cells in a gel of calcium alginate, and polyvinyl alcohol resin on stable support of cloth, foam, and hollow fibers was crucial in the immobilization of plant cells. It gives cells a situation that resembles membership in the tissue of an entire plant. In this method, it is anticipated that cells will expand, contract, and gather the metabolites.[125] With the aid of a particular set of enzymes and immobilized living cells, biotransformation can also be performed. The same cells can be utilized repeatedly. A single or multistage reaction may be used as the transformation mechanism.[132]

Example: L-alanine and malic acid synthesis for commercial purposes. Another example is the production of lipase on the surface of microbial cells and the usage of lipases to immobilize fungal mycelium on biomass support particles.[133]

5.7.2 BIOTRANSFORMATION BY FREE CELL CULTURE

The free cell culture method is also effective in the process of biotransformation. Free cells, as well as immobilized cell system both, have their uses in the biotransformation process.[134] Hydroxylation, oxidation, acylation, and methylation are a few examples of several types of biotransformation reactions. The following factors seem crucial in developing ways for cultured cells to release secondary metabolites as much as possible.[125,135]

i. Highly effective clone selection
ii. Selected metabolite removal from the medium
iii. Cell permeabilization
iv. Media optimization for maximum production and excretion.

Free cells developed in a variety of cultures, the most common two cultures are given below:

5.7.2.1 CALLUS CULTURE

A callus culture is a mass of cells or tissue that develops after the undifferentiated parenchyma cells of the parent tissue begin and continue to proliferate on a clearly defined semi-solid medium.[136] When a developed tissue explant is grown in a medium, to reach a meristematic state, the quiescent cells go through modifications. Differentiation is the process by which mature cells return to their meristematic stage, causing the creation of a callus. Additionally, a process known as redifferentiation allows the cells from a callus to develop into an entire plant. The endogenous auxin and cytokinin regulate callus development. By adjusting the auxin: cytokinin ratio in the callus culture, organogenesis can be induced and controlled.[135,137]

5.7.2.2 SUSPENSION CULTURE

In liquid nutrient media, the suspension culture generally consists of homogenous individual plant cells. The developed callus tissue is transferred to an agitated liquid nutrient media in Erlenmeyer culture flasks to begin the suspension cultures.[138] Except for the addition of agar, the composition of the medium for the development of suspension cultures might be the same as that specified for callus cultures. In suspension culture, the soft callus typically forms without much trouble. Typically, the suspension culture is kept at 250°C in complete darkness or under low-intensity fluorescent lighting. Usually, a cell suspension forms in 4 to 6 weeks. There is no proof that cells of root or shoot origin have a different metabolism from those developed in culture because the cells are meristematic and typically undifferentiated.[139] The suspension culture is a subculture by transferring aliquots of the suspension to a fresh medium at regular intervals that have not been treated or fractionated. There are several types of suspension cultures that are frequently used such as permanent cultures and semi-permanent cultures.[125,135]

5.7.3 BIOTRANSFORMATION BY HAIRY ROOT CULTURE

Roots are excellent sources of a range of natural compounds, including hyoscyamine, Catharanthus alkaloid, atropine, and propane. In recent years, there has been a lot of focus on using hairy root culture to obtain secondary plant products. The hairy roots are primarily subcultured on a solid medium and later on a rotary shaker at 1000 rpm at 25°C in a dark liquid medium. The generation of all root-derived metabolites from dicotyledonous plants may be possible via hairy root culture.[140] Numerous hairy root cultures of Asteraceae family members have been reported to be capable of bio-transforming BHT into stilbenequinone. It has been observed that *Panax ginseng* hairy root culture was capable of bio-transforming 18-glycyrrhetinic acid and glycosylating digitoxigenin, a precursor to cardiac glycosides.[125,135]

5.7.4 BIOTRANSFORMATION BY IMMOBILIZED ENZYME CULTURE

An immobilized enzyme system is made up of enzymes that have been confined to polymeric matrices or other carriers physically or chemically. Because of the following advantages, these cell-free enzyme systems are extensively used in biotransformations even though enzymes eventually lose function after being immobilized.[113]

i. The targeted items do not deteriorate and are stable.
ii. The end product is typically the only desired product due to unfavorable side effects.

iii. The product is simpler and easier to isolate and recover.
iv. Because it is a cell-free system, neither the substrate nor the product faces a transport barrier like a cell membrane.
v. The enzyme and system both are stable.

5.7.5 BIOTRANSFORMATION BY WHOLE-CELL CULTURE

Immobilized enzyme systems that are isolated are readily accessible commercially and they have some advantages of their own. The process is made more challenging by the fact that a cofactor is necessary for the specific reaction. The disadvantage of co-factors is offset by the extensive enzyme machinery present in all cell systems.[141] In comparison to isolated enzyme systems, this system is also more relatively affordable and environmentally beneficial. However, the system's main limitations are the time-consuming setup process necessitated by the high volume, the low productivity brought on by the lower tolerance for concentration, the formation of multiple products as a result of side reactions, and the cell membrane acting as a barrier to the transportation of biomolecules.[113] Example, the conversion of 1-phenylalanine (1-phe) in yeast to the aromatic alcohol 2-phenylethanol (2-PE).

5.7.6 BIOTRANSFORMATION BY GROWING CELL CULTURE

The most popular kind of manufacturing system is one that grows complete cell systems. Here, the necessary cells are grown in an appropriate medium. Only after a particular amount of cell growth in the media, a saturated substrate is supplied to the culture media. As in the case of steroid biotransformation, emulsifiers are occasionally added to solubilize substrate and products.[142] Spectroscopic or chromatographic methods can be used to track the conversion of the substrate to the products. When product creation is at its best, biotransformation can be stopped.[113] When cultivating growing cultures, the chosen strain is grown in an appropriate medium, and when the culture has developed as intended, a concentrated substrate solution is applied (6–24 h). Using a very big inoculum and adding the concentrated substrate right away without giving it time to grow is one variation of this process.[143] To aid in solubilizing poorly soluble chemicals, emulsifiers like tween or solvents such as acetone, ethanol, dimethyl sulfoxide, and dimethyl formamide may be utilized.[131]

5.7.7 BIOTRANSFORMATION BY NON-GROWING CELL CULTURE

The biotransformation approach for non-growing cells separates the production phase from cell growth or the enzyme manufacturing phase. Cells that are dormant or not actively developing transform substrates into the necessary products.[113] If enzyme induction by the addition of substrate is not required, resting cells may be employed. The elimination of substrate-induced growth inhibition is a significant benefit of this culture method. The following are some reasons why it is the favored approach for biotransformation reactions[113,144]:

i. As the cell development ends at this concentration, a very high substrate concentration can be employed.
ii. Product recovery and isolation are simple.
iii. Low risk of contamination because cells can be cleaned and reused.
iv. The substrate-to-product conversion efficiency is good.

Example: Mycobacterium species NRRL B-3805 free resting cells utilized to cleave the side chains of β-sitosterol in biphasic systems containing bis(2-ethylhexyl) phthalate (BEHP).[113]

5.7.8 TOOLS FOR ENHANCING SECONDARY METABOLITE PRODUCTION

5.7.8.1 ELICITORS

Elicitors are biologically derived substances that are involved in the communication between plants and microbes. Elicitors are regarded as mediator substances that cause the synthesis of secondary metabolites in cell cultures, from the accumulation product of cell cultures. Secondary metabolites have been produced using a variety of elicitors.[145] Alginates polysaccharides are known to operate as an inducer of the formation of shikkonin in *Lithospermum erythtrorhizen* cells and echinatin in *Glycyrrhiza echinata* cells. Abiotic elicitors have also been used in addition to elicitors of biotic origin. They either have a physical or chemical origin. Elicitation is the process of inducing certain reactions in cell cultures using microbial pathogens.[125,137]

Elicitation enhances secondary product accumulation effectiveness in plant cell culture.[125]

i. Cutting down on times.
ii. Preventing changes in media.
iii. Causing metabolites to be excreted into the media.
iv. The activation of enzymes that are part of the biosynthetic pathway.

5.7.8.2 PRECURSOR FEEDING

It is another interesting but contradicting method for boosting the production of secondary metabolites. Even precursors boost the production of several phytochemicals in vitro. There are two different ways to increase the precursor supply inside the cell. The first is by adding more medium, provided the uptake mechanism is not restrictive.[146] Second, the intracellular level may be altered, for example, by choosing resistance to precursor analogs.[147] When L -phenylalanin is fed to Lithospeimum species cultivated cells, shikonin synthesis increases threefold. Similar to this, when hydroquinone is added in trace amounts to Datura species cell suspension cultures, the production of arbutin increases significantly.[125]

5.8 VARIOUS APPLICATIONS OF MICROBIAL BIOTRANSFORMATION

Microbial biotransformation is a process where various microorganisms such as enzymes, and bacteria are used to obtain certain xenobiotics through chemical reactions such as oxidation or hydroxylation because of their high stereoselectivity, regio-, chemo- and enantioselectivity.[148] They are considered to be the most significant tools for the modification of complex structural compounds such as steroids which are very difficult to obtain through general reaction procedures.[149] Mostly used biologically and structurally interesting metabolites are the result of microbial biotransformation. In recent times, biologically important chemicals and medicines are obtained by biotransformation such as antibiotics (Penicillin, Erythromycin, Azithromycin, etc.), vaccines such as BCG, Polio vaccines, etc. steroidal compounds such as androstenedione, food product such as alcohol, jellies, fermented juices, beverages, etc.[106]

5.8.1 BIOTRANSFORMATION OF STEROIDS

Steroids are terpenoid lipid compounds that possess a special structure having 17 carbon atoms called perhydrocyclopentanophenanthrene ring which includes four cycloalkane ring systems: three are cyclohexane and one is a cyclopentane ring.[150] So, many cellular functions of our body are controlled by steroidal compounds such as cholesterol, bile acids, antibiotics, sex hormones, vitamin D, cardiac glycosides, corticoid hormones, etc.[151] Microbial biotransformation is the ultimate tool to produce these kinds of steroidal compounds which are not easy to produce by conventional methods. Steroids are being extensively used in some treatments as anti-inflammatory, diuretics, pregestational, anabolic, coronary heart disease, and as antifungal and contraceptive agents. Additionally, they are applied to the management of prostate or breast cancer.[150,152]

5.8.1.1 MICROBIAL TRANSFORMATION OF STEROLS

A combination of phytosterol and diosgenin is used in the pharmaceutical industry in the manufacturing of steroidal hormones. For example, testosterone (2), a male sex hormone can be prepared from cholesterol (1) as starting material by the action of *Mycobacterium* sp. NRRL B-3805.[153] In recent years, cholesterol was incubated with free cells of *Arthrobacter simplex* for the production of cholest-4-en-3-one (3) in a system of aqueous organic solvent and steroidal compounds such as androst-1, 4-dien-3-17-dione (ADD) (4) and androstenedione (5) prepared from cholesterol under the action of *Mycobacterium* sp. in a system containing Trition X-100 and Triton X-114 as non-ionic surfactant.[154] Yan et al. described the alternate route for the preparation of 6-methylsteroids from 3-hydroxy-5,6-cyclopropanocholestanes. Three important steroidal metabolites namely 5,6a-cyclopropano-5a-androsta-3,17-dione (7), 3b-hydroxy-5, 6a-cyclopropano-5a-androsta-17-one (8) and androsta-4-ene-3,17-dione (5) were being prepared through microbial biotransformation of 3β-hydroxy-5, 6a-cyclopropano-5a-cholestane (6) in the presence of *Mycobacterium* sp. NRRL B-3805. When 3b-hydroxy-5, 6b-cyclopropano-5-β-cholestane (9) has been fermented in the presence of *Mycobacterium* sp., it yields three important steroids, that is, (5), 5, 6β-cyclopropano-5β-androsta-3,17-dione (10) and 3a-hydroxy-5, 6β-cyclopropano-5β-androsta-17-one (11) as shown in Figure 5.2.[155] Ambrus and his team investigated the microbial conversion in the chemistry of C-24 and C-26 of ergosterol into various steroidal derivatives. They found that when ergosterol (12) was incubated with *Mycobacterium* sp. BCS 396, generates three important metabolites such as 3-oxo-4, 22-ergostadien-26-oic acid methyl ester (13), 3-oxo-1, 4, 22-ergostatrien-26-oic acid methyl ester (14), and 3-oxo-1, 4, 22-ergostatrien-26-oic acid (15).[156] When β-sitosterol (16) undergoes a microbial transformation in the presence of a genetically modified strain of *Mycobacterium* sp.

BCS 396, produces 4-stigmasten 3-one (17), 26-hydroxy-4-stigmasten-3-one (18), and 3-oxo-4-stigmasten-26-oic acid (19) and under same condition, stigmasterol (20) are converted into 4,22 stigmastadien-3-one (21), 6β-hydroxy-4, 22 stigmastadien-3-one (22), 26-hydroxy-4, 22 stigmastadien-3-one (23), 3-oxo-4, 22 stigmastadien-26-oic acid methyl ester (24) and 3-oxo-1, 4, 22 stigmastatrien-26-oic acid methyl ester (25). Androstenedione (5) can also be prepared by modified *Mycobacterium* sp. VKM Ac-1815D[157] as shown in Figures 5.7 and 5.8.

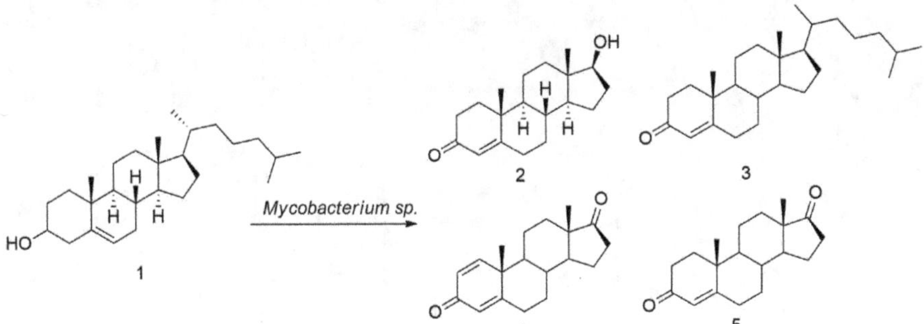

FIGURE 5.7 Microbial transformation of cholesterol.

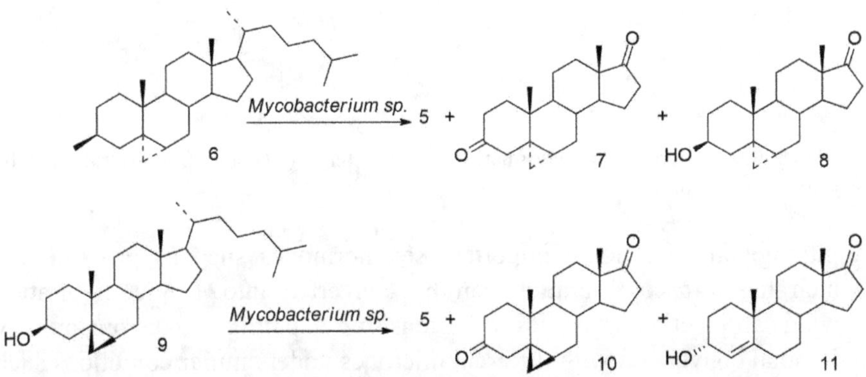

FIGURE 5.8 Microbial transformation of 3-hydroxy-5, 6-cyclopropanocholestanes.

5.8.1.2 MICROBIAL TRANSFORMATION OF STEROIDAL HORMONES

Cardiac glycosides which are having strong interaction with cardiac cell belongs to C-3 glycoside derivatives called 14-hydroxy-5β, 14β-pregnane. Generally, these kinds of compounds are obtained by various chemical reactions on a compound having 5β, 14β configuration viz. digitoxin,

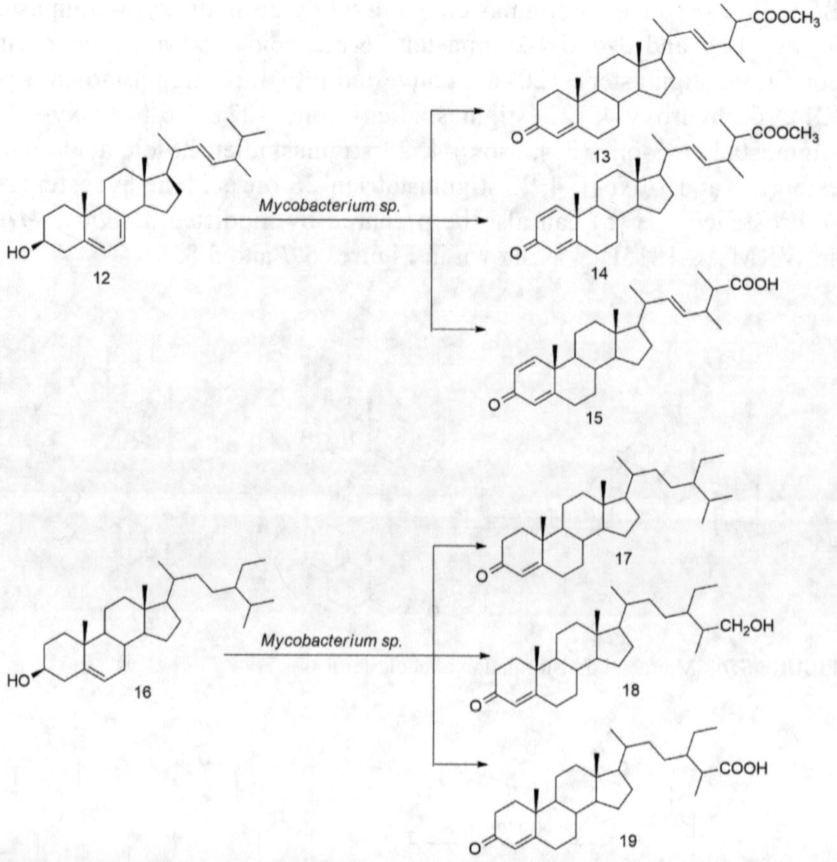

FIGURE 5.9 Microbial transformation of ergosterol (12) and β-sitosterol (16) by *Mycobacterium* sp.

digitoxigenin, etc. Some important sex hormones such as progesterone which has 14α-configuration can be converted into 14β configuration having cardioactive properties.[158] Various steroidal drugs can be prepared by microbial conversion with different microbes under similar conditions such as progesterone (26) converted into 14α- hydroxyprogesterone (27) and 9α -hydroxyprogesterone (28) under *Thamnostylum piriforme,* whereas under *M. griseocyanus* yield 6β, 14α- dihydroxy progesterone (29) and 7α, 14α -dihydroxyprogesterone (30) and efficient hydroxylation has been carried out by *Asparagillus fumigatous* to yield 11α-hydroxyprogesterone (31), 15β-hydroxyprogesterone (32), 7β -hydroxyprogesterone (33), 7β–15β-dihydroxyprogesterone (34) and 11α–15β-dihydroxyprogesterone (35) as shown in Fig. 5.11, under action of *Fusarium culmorum* progesterone (26)

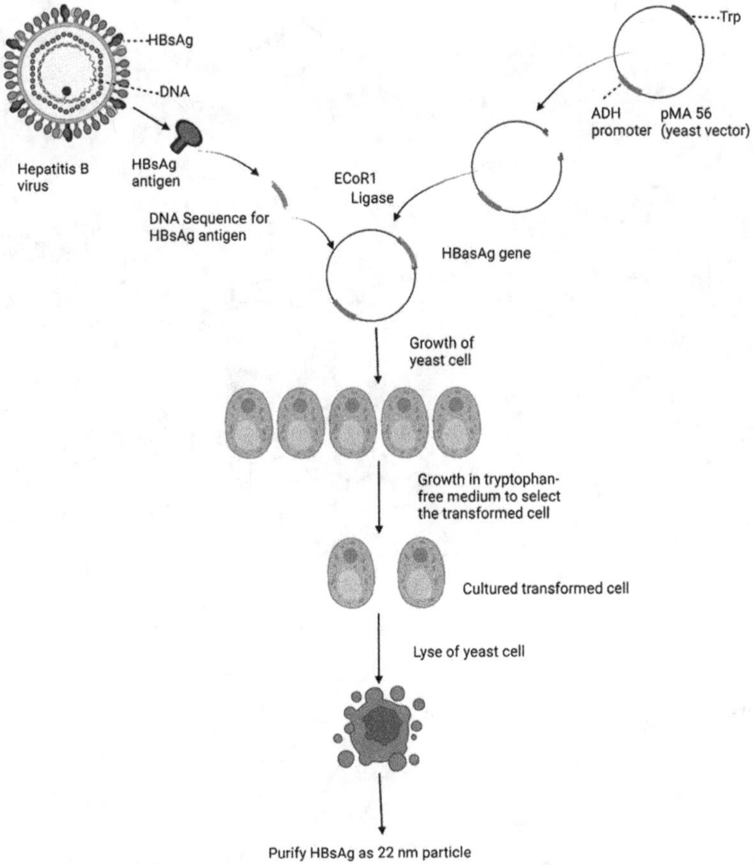

FIGURE 5.10 Microbial transformation of stigmasterol.[20]

converted into 15α-hydroxyprogesterone (36) and 12β–15α-dihydroxy progesterone (37) but when progesterone treated under *Stereothermophillus* which is a thermophilic bacterium, get converted into 20α-hydroxyprogesterone (38), 6β-hydroxyprogesterone (39), 6α-hydroxyprogesterone (40) and 9, 10-seco-4-pregnene-3, 9, 20-trione (41), Figure 5.12.[150] Madyasth and Joseph reported that 17α-hydroxyprogesterone (42) can be converted into 17a, 20a-dihydroxypregn-4-en-3-one (43), 7α, 17α -dihydroxypregn-4-en-3, 20-dione (44), 6β, 17α, 20α-trihydroxypregn-4-en-3-one (45) and 11α, 17α, 20α-trihydroxypregn-4-en-3-one (46) whereas in presence of *F. culmorum* it gives 6β, 17α-dihydroxy progesterone (47) and 15α, 17α-dihydroxy progesterone (48) as shown in Figure 5.13.[159]

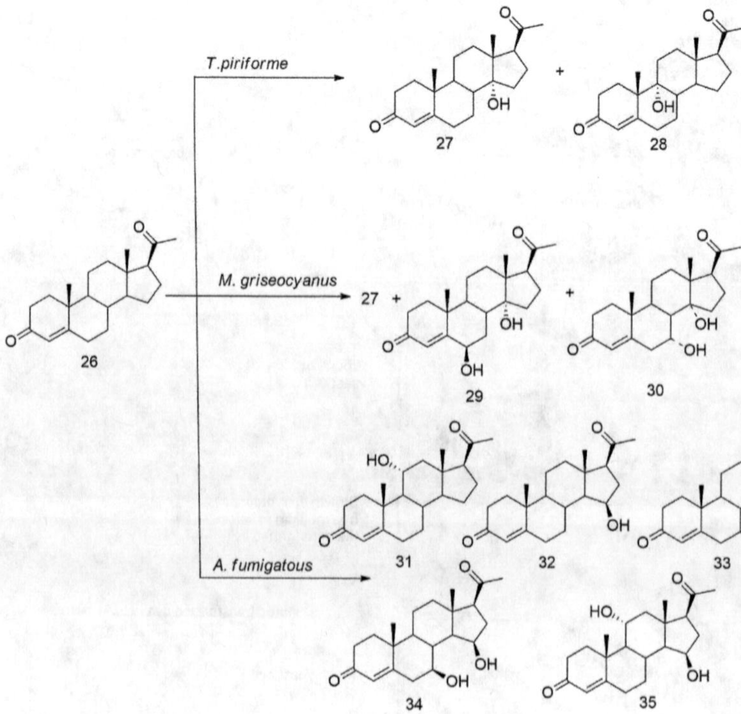

FIGURE 5.11 Microbial conversion of progesterone (40).

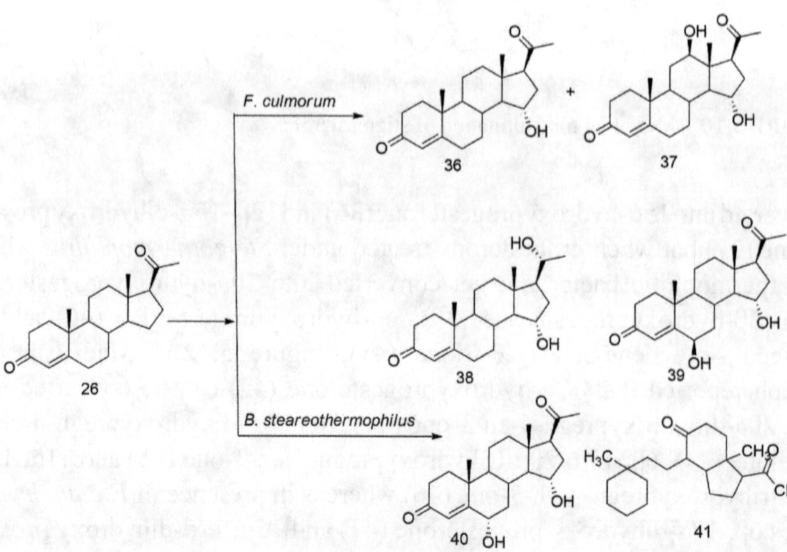

FIGURE 5.12 Microbial conversion of progesterone (40).

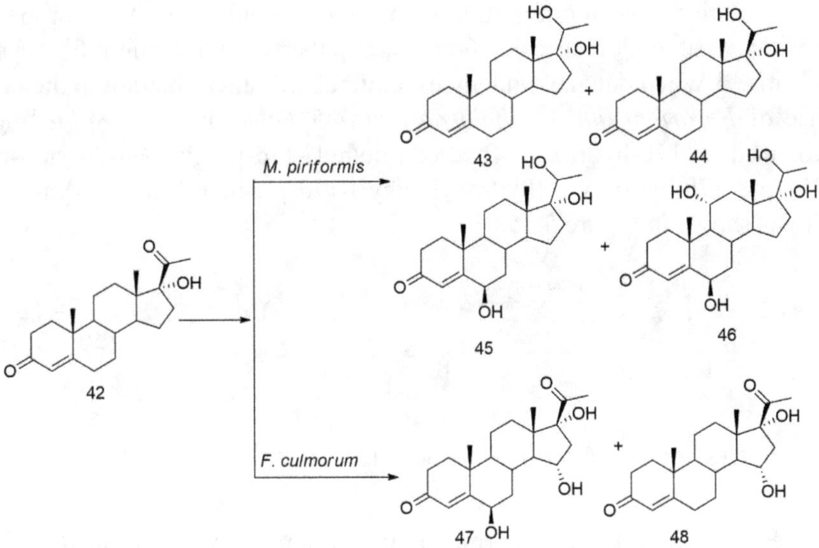

FIGURE 5.13 Microbial transformation of 17α-hydroxyprogesterone (42).

Pregnenolone (49) is another important hormone that is mainly involved in repairing damaged nerves. When pregnenolone undergoes microbial biotransformation in the presence of *Mucor piriformis* it generates two steroidal metabolites such as 3β, 7α-dihydroxypregn-5-en-20-one (50), and 3β, 7α, 11α-trihydroxypregn-5-en-20-one (51) and another two essential oxidized metabolites namely 3β, 11α, 16b-trihydroxypregn-5-en-20-one (52) and 11α, 16β-dihydroxypregn-4-en-3, 20-dione (53) can be obtained from pregnenolone by using fungus called *Botrytis cinereal*[160] and reaction scheme have been shown in Figure 5.14.

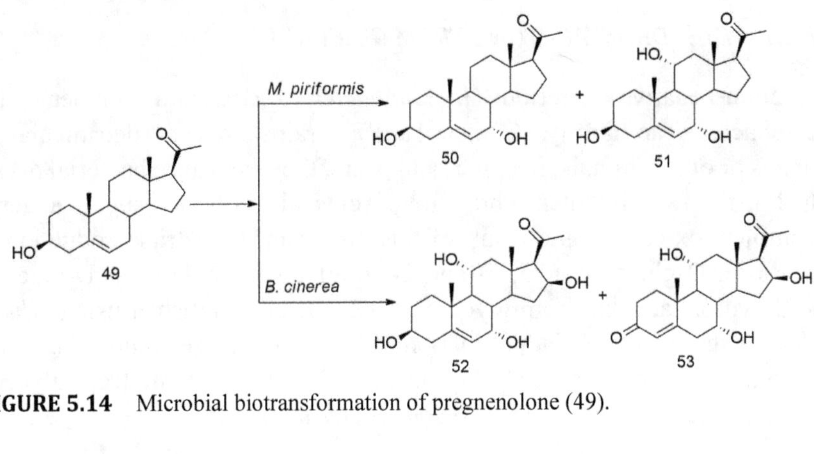

FIGURE 5.14 Microbial biotransformation of pregnenolone (49).

Danazol (54) is an oral pituitary gonadotropin inhibitor that is primarily used to treat endometriosis, premature puberty, and benign fibrocystic mastitis.[161] When danazol undergoes a microbial transformation in the presence of *Fusarium lini, Cephalosporium aphidicola* and *Aspergillus niger* produced 17β-hydroxy-2-(hydroxymethyl)-17α-pregn-4-en-20-yn-3-one (55) and 17β-hydroxy-2-(hydroxymethyl)-17α-pregn-1,4-dien-20-yn-3-one (56) as shown in Figure 5.15.[162]

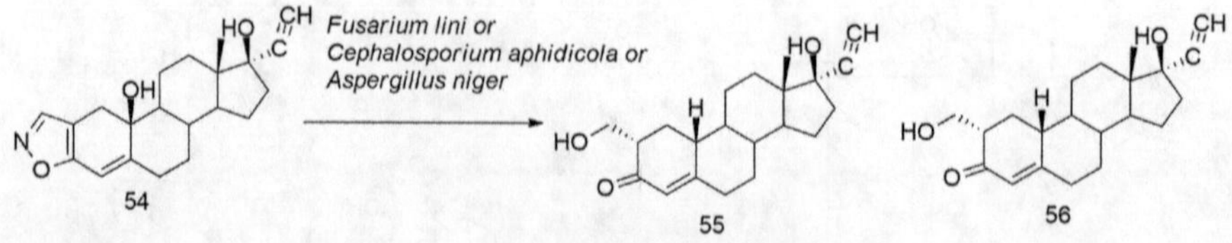

FIGURE 5.15 Microbial biotransformation of danazole (54).

Testosterone (2), a male hormone can be microbially transformed to yield various metabolites. For example, under the action of *M. griseocyanus* major metabolites 14a, 17b-dihydroxyandrost-4-en-3-one (57) and 14a-hydroxy-androst-4-en-3, 17-dione (58) are obtained while under *T. piriforme* a very minor quantity of metabolites are produced such as 14a-hydroxylated derivatives and yielded (57), 6b, 17b-dihydroxyandrost-4-en-3-one (59), 9a, 17b-dihydroxyandrost-4-en-3-one (60) and 6bhydroxyandrost-4-en-3, 17-dione (61) and two essential compounds such as 1-enetestosterone (62) and 11a-hydroxy-1-ene-testosterone (63) are being produced under the action of *Nectria haematococca* as shown in Figure 5.16.[163,164]

5.8.2 NON-STEROID COMPOUNDS

5.8.2.1 DIHYDROXYACETONE FROM GLYCEROL

The gram-negative bacterium *Gluconobacter oxydans* is a member of the Acetobacteraceae family. *Gluconobacter* strains are not detrimental to humans or other animals, in apples and pears, they can cause bacterial decay, which is followed by intense browning. Acids, alcohols, and sugars undergo incomplete oxidation as a result of this organism. Industrial production of L-sorbose, D-gluconic acid, 5- and 2-ketogluconic acids from D-glucose, and dihydroxyacetone from glycerol can all be accomplished using *Gluconobacter* strains. Industrial production of L-sorbose, D-gluconic acid, 5- and 2-ketogluconic acids from D-glucose, and dihydroxyacetone from glycerol can all be accomplished using *Gluconobacter* strains.[165]

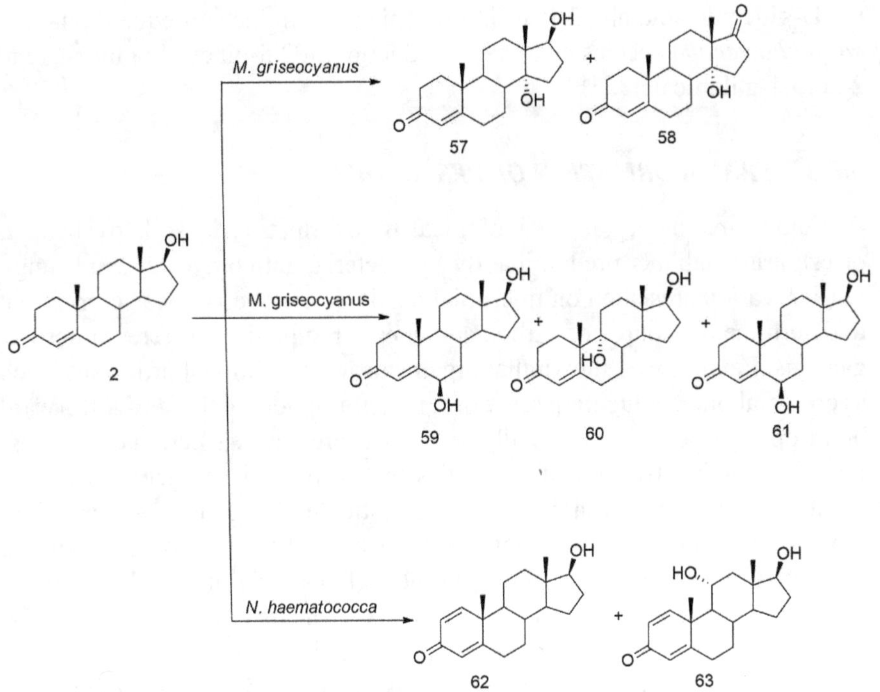

FIGURE 5.16 Microbial biotransformation of testosterone (2).

5.8.2.2 PROSTAGLANDINS

Prostaglandins are hormones made of C-20 unsaturated fatty acids. These have a variety of medical applications, including the use of PGE-2 as a contraceptive, PEG-2 to minimize childbirth pain, PEG-1 to treat congenital heart failure, and PEG-2 to treat digestive disorders. These prostaglandins can be synthesized by microbial transformation using pathogenic fungi like Cryptococcus neoformans from unsaturated fatty acids.[166]

5.8.2.3 L-ASCORBIC ACID (VITAMIN C)

Reichstein-Grussner synthesis is being used to make L-ascorbic acid. L-ascorbic acid is produced by a multi-step microbial conversion process that is used to provide vitamins or as an antioxidant in food production. *Acetobacter suboxydans* perform the oxidation step from D-sorbitol to L-sorbose in a submerged process at 30–35° C with vigorous stirring and aeration. The first stage of this continuous two-stage process comprises the oxidation of glucose to 2,5-diketo-D-gluconic acid by an Erwinia species via D-gluconic acid and 2-keto-D-gluconic acid. In the subsequent phase, a *corynebacterium* species catalyzes a reduction of 2,5-diketo-D-gluconate to 2-keto-1-gulonic acid.[3,167]

5.8.3 TRANSFORMATION OF PESTICIDES

Pesticides are undergoing a biological transformation in soil, freshwater, or estuarine habitats predominantly by bacteria, actinomycetes, and fungi. Earlier, various research on microbial pesticide metabolism was carried out to determine the source of carbon and energy required by different microorganisms. Later, it was shown that organisms used additional processes, such as co-metabolic conjugation reactions, transformation, or the straightforward build-up of a pesticide internally in a microorganism. Microbial biotransformation or inactivation of pesticides by various microorganisms can be carried out by four mechanisms such as pesticides as a source of energy and growth, growing energy may not be obtained but they can bio-transform the pesticides, it is possible to conjugate the full pesticidal molecule or one of its intermediates with naturally occurring chemicals and pesticide may be absorbed by the organism and builds up there. For example, the ring and side chain of insecticide carbaryl were hydroxylated by species of *Penicillium,* ethyl group of herbicide Atrazine was dealkylated by *Fusarium moniliforme,* dicofol-like compound, and TDE is the metabolic product of DDT and these are metabolized by *Trichoderma viride,* etc.[168,169]

5.8.3.1 PESTICIDES AS A NUTRIENT SOURCE

The disintegration of an organic molecule by microbe(s) to its inorganic components is a convenient process if one wants to avoid the persistence of potentially hazardous material in the environment. Plenty of organic pesticides may act as the only source of carbon or energy for certain microbes' development and reproduction. A pesticide will be broken down and split into parts that can be fed into well-known oxidative cycles like the Krebs cycle if it can be used in this way, providing the organism with all the energy it requires. Pesticides can be tested using the enrichment culture method to see if they serve as a source of carbon and growth. Experimentally, it was found that only a few specific microbes are able to metabolize the pesticides. For example, soil microflora and organophosphate pesticides like "Diazinon"1-3 interact in complex ways. Utilization of the chemical by an *Arthrobacter* species and a *Streptomyces* species has been proven to change the balance of the current microbial community in the case of "Diazinon" by encouraging a selective enrichment of these species.[170]

5.8.3.2 COMETABOLISM

The discovery and relevance of the fact that a bacterium manages to alter a chemical without obtaining energy to maintain its growth are tied to the widespread use of xenobiotic substances in contemporary society. Alexander (1967) was the first to recognize the potential significance of cometabolism for the transformation of pesticides; today, the results of several experiments can be explained by this mechanism (Table 5.3). Cometabolism does not completely deal with the removal of pesticide degradation but it has been found that it can be carried out to eliminate, reduce, or increase the toxicity of the environment.

TABLE 5.3 Microbial Transformation of Pesticides Along with Their Product.

Substrate	Product	Microorganisms
2,3,6-Trichlorobenzoic acid (TBA)	3,5-Dichlorocatechol	*Brevibacterium* sp.[171]
2,4,5-Trichlorophenoxyacetate	3,4-Dichlorocatechol	*Brevibacterium* sp.[171]
3-Nitropheol	Nitro hydroquinone	*Flavobacterium* sp.[171]
DDT (Dichlorodiphenyltrichloroethane)	*p,p*-Dichloro diphenylmethane	*Aerobacter aerogenes*[171]
p,p-Dichloro diphenylmethane (metabolite of DDT)	*p'*-Chlorophenyl acetic acid	*Hydrogemonas*[171]
Chloroneb	2,5-Dichloro-4-methoxyphenol	*Fusarium sp.*[171]
Cholorobenzilate	4,4'-Dicholorobenzo pheonone	*Rhodotorula gracilis*[171]

Since actinomycetes, fungi, and bacteria all have an impact on the cometabolism process, it is reasonable to believe that cometabolism occurs frequently in natural ecosystems. For a comprehensive knowledge of its true source, cometabolism, particularly playing a role in the microbial transformation of pesticides, needs to be further thoroughly studied.

5.8.3.3 MICROBIAL AGGREGATION OF PESTICIDES

The feasibility of pesticides being assimilated into microorganisms by a passive or an active accumulation process raises specific concerns since, unlike a conjugation reaction, microbial intervention only temporarily eliminates a dangerous ingredient. The chlorinated hydrocarbons DDT, dieldrin, aldrin, and heptachlor accounted for the majority of findings of pesticide build-up within the cells. When put in soil containing pentachloronitrobenzene, dieldrin, and DDT, mycelia of actinomycetes and fungi collected these chemicals to levels above ambient values.[172] A study found that pesticide uptake is found to be higher in the autoclaved cell of microbes than in normal live cells such as *Aerobacter aerogens* where uptake of methoxychlor is found to be double which indicates that polarity and lipid solubility influence the uptake process. Similarly, *Saccharomyces cerevisiae* demonstrated that the adsorption ability for lindane and dieldrin was intensified following the organism being boiled. The accumulation of pesticides in plankton in freshwater and marine habitats and aquatic microorganisms can pose a threat to fish and higher vertebrates since they are a significant source of nutrients for a variety of aquatic filter-feeding creatures. Consequently, the discovery of widespread biomagnification by these species must inspire serious worry. A sea diatom called *Cylindrotheca closterium* can absorb and concentrated DDT from its growth medium around 0.1 ppm of the insecticide which is nearly about 200-fold. Similarly, it was observed that cultures of the ciliates *Paramecium bursaria* and *P. multimicronucleatum*, as well as the blue-green alga *Anacystis nidulans,* and the green alga *Scenedesmus obliquus*, concentrated DDT and parathion ensuing exposure for 7 days at rates of 100 to 964 and 50 to 116 times, respectively.

5.8.3.4 ENZYMATIC BIOTRANSFORMATION IN PESTICIDES

Most pesticides undergo microbial or enzymatic biotransformation by opting to follow chemical reactions.

5.8.3.4.1 Hydroxylation

A common fundamental alteration of a pesticide is hydroxylation which adds a hydroxyl group and creates a more biologically reactive product that is frequently more polar, and consequently more water soluble. Various enzymes like monooxygenase, hydroxylases, or mixed-function oxidases are involved in the hydroxylation process. When aromatic pesticides undergo hydroxylation, the polar group is included in the cleaving ring system. For example, *Pseudomonas* carried out hydroxylation of 2,4-dichlorophenoxy-acetic acid, it gives 2,4-dichloro-6-hydroxyphenoxyacetic acid while *Aspergillus niger* gives 2,4-dichloro-5-hydroxyphenoxyacetic acid and minor product 2,5-dichloro-4-hydroxyphenoxyacetic acid derivatives as shown in Figure 5.17. The latter product denotes a chlorine atom shift together with the substitution of a hydroxyl group, which will be further treated as an NIH-shift under epoxidation processes.[171]

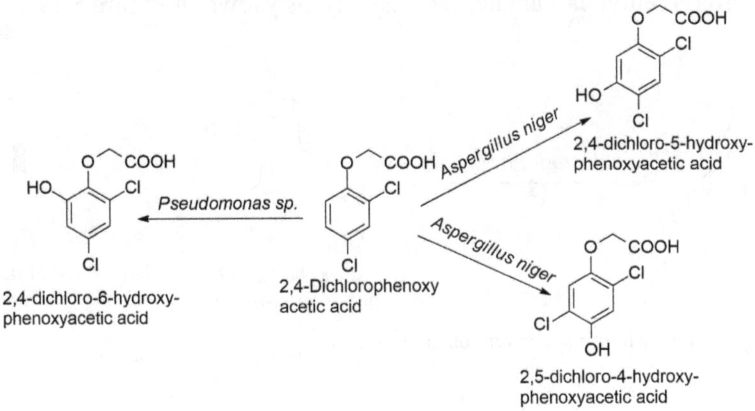

FIGURE 5.17 Microbial conversion 2,4-dichlorophenoxyacetic acid.

Similarly, 2-methyl-4-chlorophenoxyacetic acid (MCPA) is one of the powerful, selective highly used herbicides which can be bio-transform to 5-hydroxy and 6-hydroxy derivatives with *Pseudomonas sp.* and *Aspergillus niger,* respectively. *p*-Chloroaniline which is mostly present in different herbicides can be oxidized by *Fusarium oxysporum* as shown in Figure 5.18.

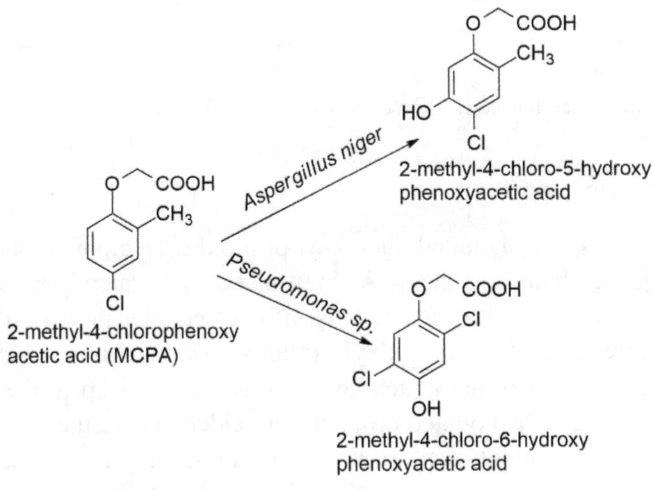

FIGURE 5.18 Microbial conversion 2-methyl-4-chlorophenoxyacetic acid.

Carbaryl is another important irreversible cholinesterase inhibitor used as a pesticide that can be transformed by side chain and ring hydroxylation into different positions by a fungus called *Gliocladium roseum* to give 4-hydroxy and 5-hydroxy-1-naphthyl methylcarbamate as well as 1-naphtyl *N*-hydroxymethyl carbamate, respectively, as shown in Figure 5.19.

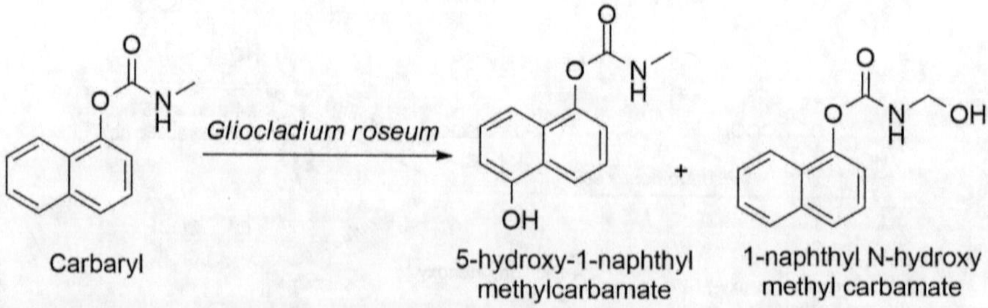

FIGURE 5.19 Microbial conversion carbaryl.

The systemic metabolism of fungicide 2,5-dimethyl-3-furancarboxylic acid anilide was investigated by Wallofer et al. 1972 and he stated that mentioned fungicide can be hydroxylated to 2-hydroxymethyl-5-methyl-3-furancarboxylic acid anilide and 2-methyl-5-hydroxymethyl-3-furancarboxylic acid anilide, respectively, by *Rhizopus japonicus* as shown in Figure 5.20.

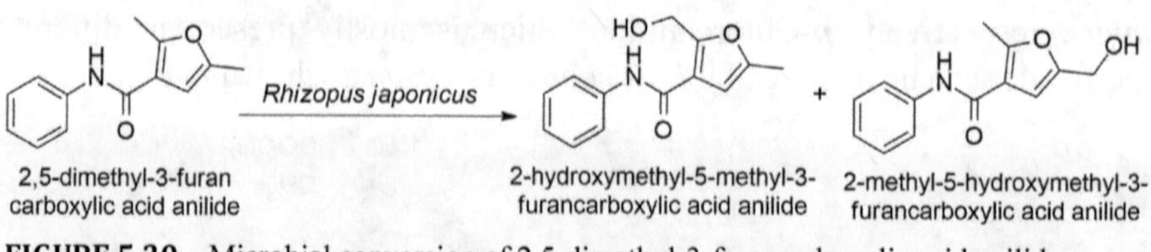

FIGURE 5.20 Microbial conversion of 2,5-dimethyl-3-furancarboxylic acid anilide.

5.8.3.4.2 Dealkylation

Alkyl moieties are included in many pesticides, including phenylureas, acylanilides, carbamates, and s-triazines, and they frequently present active groups that produce the intended harmful effect. Dealkylation reactions are, therefore, crucial because alkyl groups on side chains are typically the initial targets of microbial attack and they are a first step in the detoxication of pesticides. Dealkylated products and aldehyde are the product of the dealkylation reaction. Further, dealkylation can be classified based on group modification attachment such as *N*-dealkylation and *O*-dealkylation which are catalyzed by the mixed-function oxidase.[171]

N-dealkylation: Although the proper mechanism of *n-dealkylation* is not clear but here the alkyl group used to be removed from the terminal N-atom in a side chain by microbial biotransformation.

For example, the fungus *Aspergillus terreus* converts carbaryl to l-naphthyl N-hydroxymethyl carbamate and then to l-naphthyl carbamate which is a dealkylated product.[173] *Trichoderma viride* and *Aspergillus candidus* gradually dealkylated the herbicide diphenamid, a carbamate-related amide, to N-methyl 2,2-diphenylacetamide and then to 2,2-diphenylacetamide respectively as shown in Figure 5.21.[174]

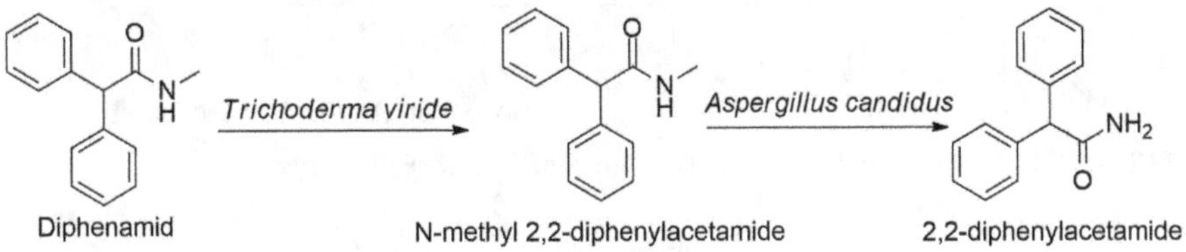

FIGURE 5.21 Microbial conversion of diphenamide.

Chloroxuron is an herbicide belonging to a class of phenylureas derivatives that can be dealkylated by a liquid mixture of soil bacteria. Stepwise dealkylation of diuron which is a very effective herbicide, results in the formation of 3-(3,4-dichlorophenyl)-1-methylurea, monolinuron, and linuron and 3,4-dichlorophenylurea in soil by *Aspergillus niger* and *Talaromyces wortmanii* converts metobromuron to -(*p*-bromophenyl) -3-methoxyurea and 1-(*p*-bromophenyl) - 3-methylurea as shown in Figure 5.22.[171]

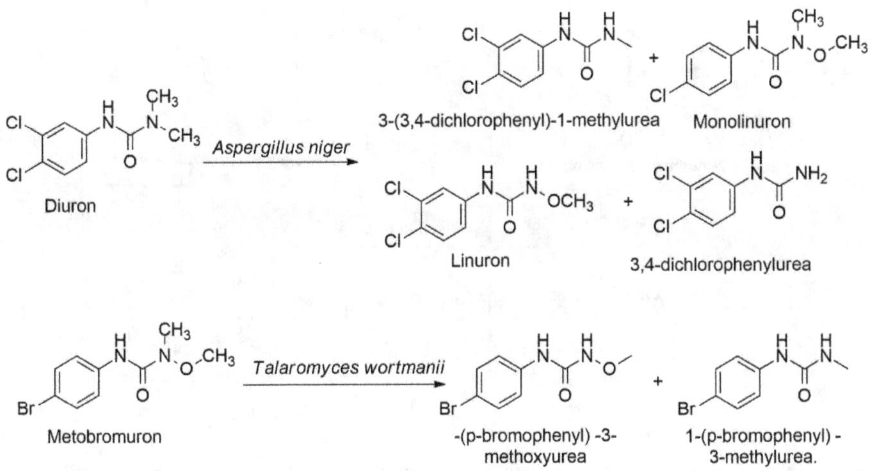

FIGURE 5.22 Microbial biotransformation of Diuron and Metobromuron.

s-Triazine derived herbicide simazine can be dealkylated through *Aspergillus fumigatus* to 2 chloro-4-ethylamino-6-amino-*s*-triazine. Different microbes have their specific ability to bio-transformed different side chains of herbicides or pesticides. Ethyl moiety and isopropyl moiety can be removed by *Aspergillus fumigatus* and *Rhizopus stolonifera,* respectively. It is found that after transformation, there is no reduction of the herbicidal or phytotoxic properties of *s*-triazine herbicide as shown in Figure 5.23.[171]

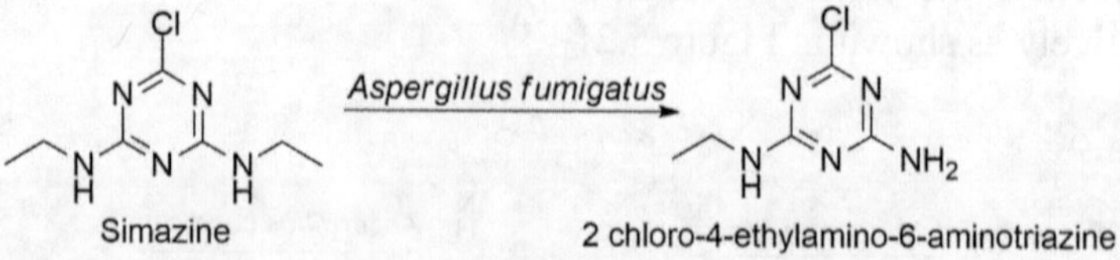

FIGURE 5.23 Microbial conversion of Simazine by *Aspergillus niger.*

O-dealkylation

A methyl or other alkyl group that served as a link between an oxygen atom and the other chemical component is taken off during an ether cleavage or O-dealkylation process. Soil fungus can demethylate methylated aromatic compounds. For example, species of *Hormodendrum, Haplographium,* and *Penicillium* can convert veratric acid to vanillic acid and *o*-, *m*-, and *p*-methoxybenzoic acids to their corresponding hydroxybenzoic acids as shown in Figure 5.24.

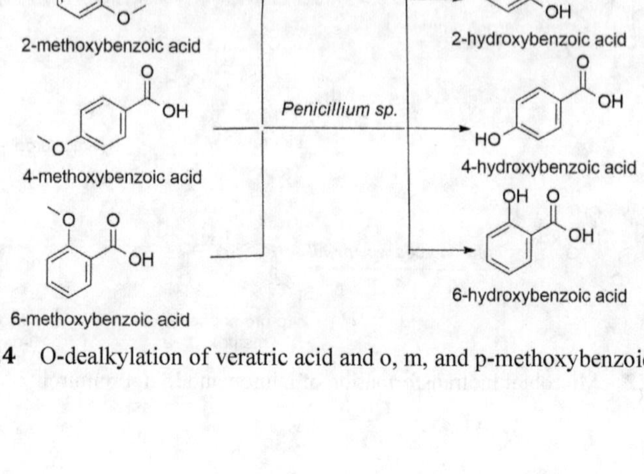

FIGURE 5.24 O-dealkylation of veratric acid and o, m, and p-methoxybenzoic acid.

Researchers mentioned that the fungus *Alternaria tenius* and *Trichoderma lignorum* oxidized the pesticide bromophos by a double-dealkylation to the bis-demethylated bromophos. Anisole, whose chlorinated derivatives were occasionally discovered as intermediates during the breakdown of chlorinated phenoxy acetic acids, was also shown to undergo *O*-dealkylation; replacement cultures of *Aspergillus niger* oxidized anisole to phenol. The fungicide griseofulvin is demethylated by various fungi at various methoxy groups linked to the molecule, as per Boothroyd et al. 1961. The 2-demethylgriseofulvin was generated by *Botrytis allii*, the 6-demethylgriseofulvin by *Cercospora rnelonis*, and the 4-demethylgriseofulvin by *Microsporum canis* as shown in Figure 5.25.

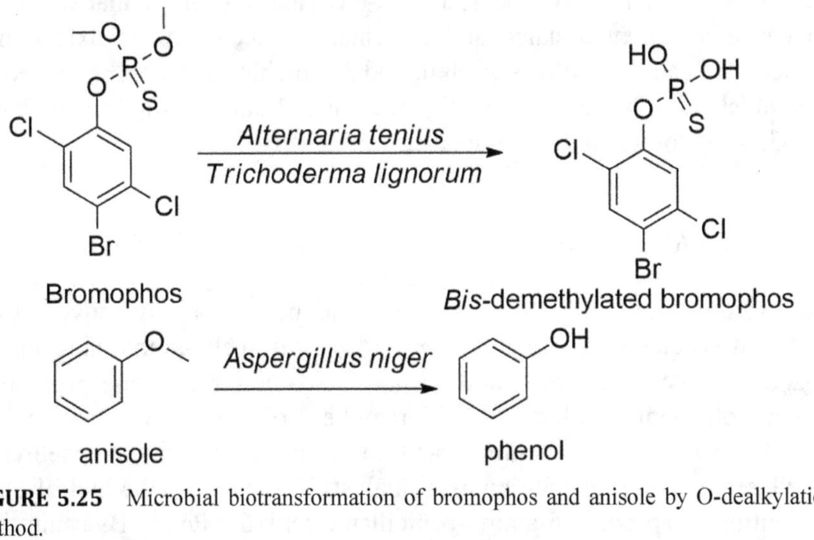

FIGURE 5.25 Microbial biotransformation of bromophos and anisole by O-dealkylation method.

5.8.3.4.3 *Cleavage of Ether Linkage*

It appears that the ether cleavage enzymes under investigation are highly flexible in the range of substrates they can oxidize, as described under *O*-dealkylation reactions, which can also be thought of as cleavage of an ether bond. Given that all thorough studies have shown that reduced pyridine nucleotides and molecular oxygen are necessary, the breaking of an ether bond is thought to be a reaction brought on by a mixed-function oxidase.

Both a reductive reaction with higher phenoxy acetic acids and an oxidative reaction with 2,4-D and MCPA have been shown as mechanisms involved in the breaking of ether bonds. According to MacRae and Alexander, a *Flavobacterium sp.* was responsible for cleaving the ether linkage of 2,4-dichlorophenoxyalkyl carboxylic acids from the propionic homolog to the undecanoic homolog. According to the cleavage, phenol and the fatty acid corresponding to the aliphatic moiety were produced.[175,176]

5.8.3.4.4 Oxidation of Aromatic Ring

Since many pesticides are cyclic molecules, their full biodegradation can only occur after the ring has been broken. Numerous microbes are capable of oxidizing aromatic chemicals and using the aliphatic molecules that arise as substrates in the intermediate metabolism. For example, *Achromobacter sp.* has catechol 1,6-oxygenase, a meta-cleaving oxygenase that was found to be active on methylated and chlorinated catechols. The oxidation of 3-methylcatechol, 4-chlorocatechol, and 3,5-dichlorocatechol produced the semialdehydes 2-hydroxy-3-methylmuconic, 4-chloro-2-hydroxymuconic, and 3,5-dichloro-2-hydroxymuconic, respectively.[171]

5.8.4 MICROBIAL BIOTRANSFORMATION OF ANTIBIOTICS

For example, antibiotic chloramphenicol can be microbially converted into different species by using *Streptomyces griseus* such as chloramphenicol-1-acetate, chloramphenicol-3-acetate, chloramphenicol-3-propionate, chloramphenicol-3-isobutyrate, chloramphenicol-3-butyrate, chloramphenicol-3-isovalerate.[177] Penicillin G acylase was used in an aqueous-methylisobutyl ketone (MIBK) solution as a catalyst for the enzymatic hydrolysis of penicillin G to produce 6-amino-penicillanic acid (6-APA).[178] By using silica gel and alumina chromatography, maridomycin (MDM) III transformation products by *Streptomyces lavendulae* were identified by using Rf values on a thin-layer chromatogram. The different spot was analyzed and it was found that the microbial biotranformed product is 18-dihydro-maridomycin III, 4”-depropionyl- maridomycin III, and 18-dihydro- depropionyl-maridomycin III which were later confirmed by different spectral analysis such as thin-layer chromatography, mass- and NMR spectrometry. In veterinary medicine, flumequine (FLU), a fluoroquinolone antibiotic belongs to the family of broad-spectrum category particularly used against gram-negative bacteria. It can kill bacteria by stopping its growth by interfering with the DNA-gyrase enzyme and, as a result, stopping normal DNA synthesis. Flumequine was subjected to microbial biotransformation to generate potent and therapeutically useful metabolites. Around eight important metabolites were obtained by reacting with various ligninolytic fungal strains. For example, FLU undergoes metabolism with all kinds of fungi to generate F1 and F4 metabolite, with *D. squalene* generating F7, with *I. lacteus, P. tigrinus* and *D. squalene* generating F2, whereas F1 undergoes react with *I. lacteus* to generate F3, F7 undergo react with *D. squalene* to generate F6 which further react with *D. squalene* to generate F8 and F5 [179] which is shown in Figure 5.26.

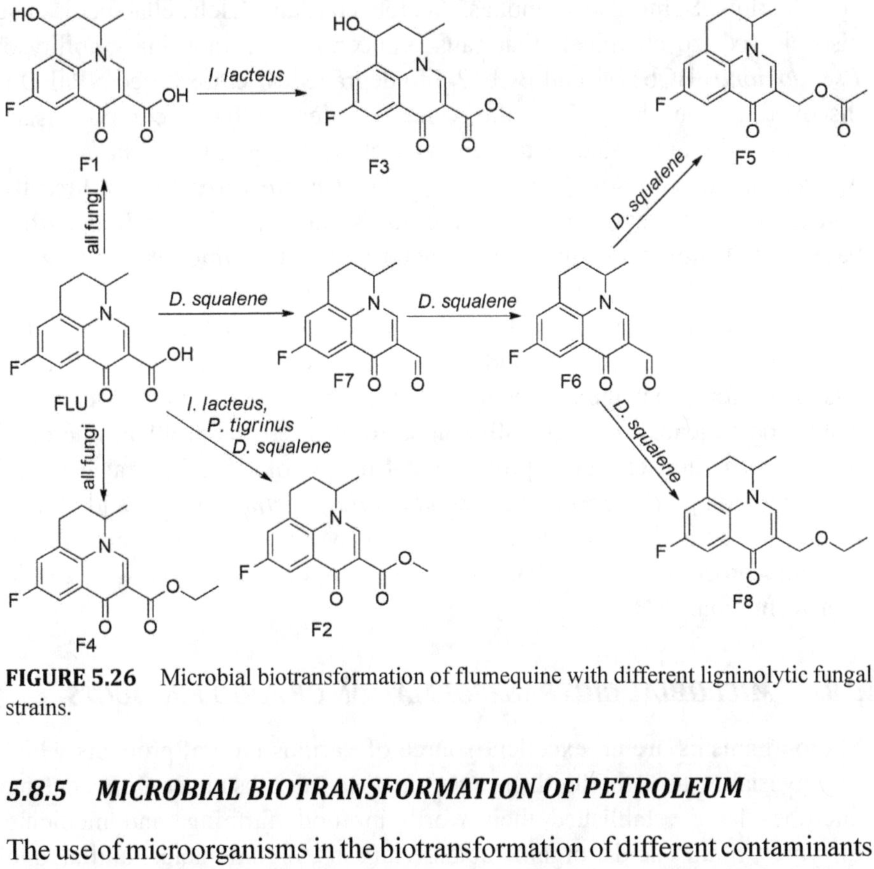

FIGURE 5.26 Microbial biotransformation of flumequine with different ligninolytic fungal strains.

5.8.5 MICROBIAL BIOTRANSFORMATION OF PETROLEUM

The use of microorganisms in the biotransformation of different contaminants has attracted a lot of attention recently to purify the environment. For the bioremediation of a wide variety of chemicals, including pharmaceuticals, radionuclides, hydrocarbons (such as oil), and polychlorinated biphenyls, the catabolic diversity of bacteria has been used (PCBs). The primary worrying trend is the public health risk that these contaminants entail. To accelerate the transformation of xenobiotic substances, significant advancements in molecular methods, bioinformatics, genomics, and metagenomics are being applied.[180,181] A huge range of xenobiotics are being bio-transformed by both aerobic and anaerobic bacteria such as *Escherichia, Bacillus, Pseudomonas, Rhodococcus, Gordonia, Methanospirillum, Pelatomaculum, Syntrophus, Methanosaeta,* etc.[182] Acetone, cyclohexane, styrene, benzene, ethylbenzene, propylbenzene, dioxane, and 1,2-dichloroethylene can all be extensively metabolized by *Mycobacterium vaccae*. PCBs (Polychlorinated Biphenyls) are known to be very effectively degraded by Pseudomonas and Bacillus. Some Pseudomonas, Acetobacter, and Klebsiella species can also bio-fix azo chemicals that cause cancer. The extraordinary ability of *Pseudomonas* BCb12/1 and BCb12/3 to degrade low ethoxylated NPnEO is discovered (Non-phenol Polyethoxylates) whereas phthalate compounds are breakdown by anaerobic methanogens such as *Methanospirillum hungatei, Methanosaeta concilii,* and *Syntrophobacter fumaroxidens*. Currently, potential contaminants with the pentafluorosulfanyl (SF5-) functional group have been demonstrated to be bio-transformed by *Cunninghamella elegans, Pseudomonas knackmussii,* and *P. pseudoalcaligenes* KF707.[183–186]

Biotransformation of petroleum products is very much essential because they are extensively harmful and carcinogenic for marine and aquatic habitats. Therefore, leakage and accidental release of petroleum products into water bodies leads to water pollution so to counter this problem microbial biotransformation can be a proper and fruitful solution. Several bacterial taxa, including *Rhodococcus, Pseudomonas, Arthrobacter*, and *Mycobacterium*, actively break down petroleum hydrocarbons through an alkyl aromatic process. Bioremediation can be fully achieved by this microbial transformation.[187,188]

5.8.6 MICROBIAL BIOTRANSFORMATION OF FOOD PRODUCTS

Microorganisms are an excellent source of various natural products which are invariably used in food production. Natural substances derived from microbes have established their worth in food, farming, and medicine. Enzymes, amino acids, organic acids, vitamins, organic acids, and alcohol are examples of primary metabolites which are used both as nutritional supplements and in the biotransformation of raw materials to create industrial goods. While secondary metabolites are organic substances that are mostly extracted from tissues or plants. Microbes are employed to prepare a range of foods as well as to provide food and feed supplements. For instance, *propionibacterium* and *pseudomonas* provide vitamins; organic acids from *Aspergillus*, *lactobacillus,* and *Rhizopus*; enzymes are obtained from *aspergillus*, *bacillus*, *corynebacterium*, *brevibacterium,* and *Escherichia coli* generate amino acid.[189–192]

Different enzymes are involved in industrial food production which are mostly obtained from microorganisms. These microbial enzymes act as catalysts in the fermentation process because of their higher stabilities, consistency, and cost-effectiveness. For example, amylolytic enzymes are mostly used in food, paper, detergent, glucose syrup, crystalline glucose, maltose syrups, and textile industries.

α-Amylase is a starch degrading enzyme that is obtained from *Bacillus stereothermophilus, Bacillus amloliquefaciencs,* and *Bacillus licheniformis* are mostly used in the production of rice cakes, powdery foods, clarification of fruit juices, etc.[193,194] Glucoamylases is an exo-acting hydrolytic enzyme obtained from *Aspergillus niger* and *Aspergillus awamori* which are mainly used in the production of high glucose syrup, high-fructose syrups, soya sauce as well as light beer.[195,196] Protease is another important enzyme mainly used in the production of dairy products, baked foods, bread, cracks, and waffles and it can easily be obtained from *Aspergillus usamii.*[197] Lactase also known as β-galactosidase carries out lactose hydrolysis. It can be obtained from yeast *Kluyveromyces fragilis* and *Kluyveromyces lactis.* In order to reduce lactose intolerance in humans, lactase is utilized in the manufacturing of milk and dairy products. It can also be used in the hydrolysis of lactose in whey.[198] Various applications of microbial enzymes in food production are mentioned in Table 5.4.

TABLE 5.4 Application of Microbial Enzyme in Food Production Technology.

Enzyme	Microbial source of enzyme	Applications
α-amylase	*Bacillus stereothermophilus, Bacillus amloliquefaciencs Bacillus licheniformis*	Production of rice cakes, powdery foods, clarification of fruit juices.[193,194]
Glucoamylases	*Aspergillus niger Aspergillus awamori*	Production of high glucose syrup, high-fructose syrups, soya sauce as well as light beer.[195]
Protease	*Aspergillus usamii*	Production of dairy products, baked foods, breads, cracks, and waffles.[197]
Lactase	*Kluyveromyces fragilis Kluyveromyces lactis*	Production of milk and dairy products to reduce lactose intolerance.[198]
Lipase	*Penicillium camemberti Aspergillus niger Aspergillus oryzae*	Production of food, beverage, beer and wine, fruit juice, ice cream, biofuel, detergents, and animal feed, flavor developing agent.[197]

TABLE 5.4 *(Continued)*

Enzyme	Microbial source of enzyme	Applications
Phospholipase	*Fusarium oxysporum*	Production of oils, dairy industries, bakery items, stability of dough for bread, preparing cheese, butter, and ice cream.[197]
Esterase	*Bacillus licheniformis*	Production of scents and flavors for food items.[197]
Lipoxygenase	*Nostoc* sp. *Anabaena* sp. *Lasiodiplodia theobromae*	Production of dough, aroma generation for food products, and making bread.[197]
Cellulase	*Aspergillus* *Trichoderma* *Bacillus* *Paenibacillus*	Food production, clarification, and stabilization of juices, purification of fruits, and wine production.[197]
Xylanase	*Streptomyces* sp., *Bacillus* sp. *Pseudomonas* sp. *Aspergillus* sp., *Fusarium* sp. *Penicillium* sp	To enhance the texture, tastiness, and palatability of biscuits. Preparing bread, juices, Reduction of beer viscosity.[197]
Glucose oxidase	*Aspergillus niger* *Penicillium glaucum*	Pharmaceutical and food industries. As aroma, flavor, and stability enhancers of foods as well to improve color, texture, and shelf life.[197]

5.8.7 USE OF MICROBES IN THE PHARMACEUTICAL INDUSTRY

As microbes are employed for the synthesis of substances in various sectors and the production of other valuable goods, they can be found everywhere and can be both helpful and detrimental to humans. On the other hand, they are also contributing to human disease at varying rates. For the production of chemical substances and pharmaceutical research, they are utilized in the pharmaceutical industry. Chemical firms now synthesize industrial goods utilizing microorganisms. As a result of their significant influence on the production of numerous valuable chemicals, they are advantageous from an industrial standpoint.[199]

Numerous antibiotics, including streptomycin from the bacteria streptococcus, are produced by microorganisms. The use of microorganisms for medically significant investigations, such as bacteriorhodopsin, plays a significant role in pharmaceuticals. In the pharmaceutical industry, microbes are also utilized to synthesize chemical compounds, chemical medicines, and other substances. Pharmacists can find antimicrobial medications that would stop an increasing number of communicable diseases due to the discovery of cellular mechanisms. Additionally, it guarantees that the pharmacological treatments target opportunistic microorganisms without damaging their human hosts.[199] Additionally, bacteria play a significant part in the production of various hormones, which help to regulate deadly diseases. The use of genetic engineering is the most crucial element in hormone production. Bacterial cells are modified and exploited in the manufacture of goods of significant economic importance. A couple of examples are the creation of human insulin, which is used to treat diabetes, and somatotrophin, also known as human growth hormone, which is used to treat pituitary dwarfism.[200]

The manipulation of genes has revolutionized genetic engineering, which incorporates biological sciences such as biochemistry, microbiology, and biotechnology. The term "recombinant DNA technology" is sometimes used. One of the initial methods used in genetic engineering was the use of a virus vector to transfer DNA (genes) into a host. Along with the rest of the host's DNA, the foreign DNA replicates and is passed on to daughter cells, becoming a permanent component of the host. This theory has been applied to the identification of new genes, the transfer of DNA into various species, and the creation of novel hormones. *Actinomyces naeslundii* and *streptococcus gordonii* have agonistic and antagonistic characteristics. These two microbes work together to create biofilms.[199,201]

Riboflavin and vitamin K are prepared commercially using *Escherichia coli* bacteria. These vitamins are blended into nutritional supplements and utilized to cure the deficiency of various ailments either owing to vitamins and minerals that are essential for optimal growth in tiny quantities and regulate bodily functions. When this bacterium, which also resides in human intestines, is active, it harms the digestive system and produces diarrhea and other issues related to the digestive system. On one hand, *E. coli* is helpful in the production of chemical compounds, but it can also be problematic.[199] Numerous nutritional supplements are made with the help of bacteria. Additionally, D-amino acids like D-p-hydroxyphenylglycine, a crucial intermediary in the manufacture of the antibiotic amoxicillin, which can be used to treat infections of the skin, ear, and urinary tract, are produced by *E. coli*. Amoxicillin works as an antibiotic to reduce the frequency of infections, but excessive doses can cause cellular toxicity, which disrupts cell membranes because of the build-up of toxins that these infections release. The liver, a significant human organ, was also harmed by the toxins that were created.[199]

The majority of human history has been dominated by bacterial diseases. The two most prevalent bacterial illnesses that impact human populations worldwide are cholera and plague. Some microorganisms are employed in the pharmaceutical industry as a potential supply of biological targets for the creation of medicines. To treat bacterial infections, antibiotics are employed. However, because of the environment, many bacterial species have become resistant to antibiotics. The prevalence of bacterial disease has decreased due to advances in plumbing and hygiene, the creation of bacterial vaccinations, and the discovery of antibacterial medicines.[199] Various microorganisms are employed to clean the environment. Some of them are utilized at an industrial level with high production rates and wasted costs. Microbes have a wide variety of metabolic capabilities, and some can detoxify or degrade pollutants like crude oil or pesticides. They can also be used in bioremediation, which is the most important method for cleaning up polluted sources and the environment, and some can even break down plastics.[202]

With the development of molecular biology and recombinant DNA, human proteins may be produced using *Escherichia coli* and other bacteria through a process known as heterologous expression. Human insulin, which is used to treat diabetes mellitus types I and II, is a prime example. Initially, pancreatic extracts from cows and pigs were used to purify insulin. However, the procedure was costly, and numerous patient cases of immunological reactions triggered by animal insulin were documented. After isolating the human insulin gene, *E. coli* was used for heterologous expression to produce the human protein.[203]

A novel and workable method for generating anti-cancer medicine is the isolation and identification of endophytic fungi that produce taxol. Worldwide development and application of taxol-producing fungi have advanced significantly. *Fusarium oxysporum* was used to make taxol when it was cultured in potato dextrose broth. Furthermore, it was discovered that the filamentous fungus *Aspergillus niger* isolated from *Taxus cuspidate* produces taxol.[204] Researchers have also looked into the significance of extracellular enzymes made by filamentous fungi. The enzyme that catalyzes the breakdown of lactose into glucose and galactose is known as β-d-galactosidase. Due to its significance in the treatment of lactose intolerance, the lactase market has been growing rapidly on a global scale. The asparaginase enzyme is an additional biological substance that is significant in fungi. Acute lymphoblastic leukemia and non-Hodgkin lymphoma are two examples of hematologic illnesses that can be treated with this enzyme.[203] A strain of *Penicillium nalgiovense* obtained from Antarctica was used by Svahn et al. to generate and isolate amphotericin B.[205] Amphotericin B was recovered from a different organism for the first time; ordinarily, it is isolated from *Streptomyces nodosus*. *A. Niger* LCF9 created an alkaline protease with collagenolytic activity. With no release of amino acids, the enzyme degraded several types of collagens and released low molecular weight peptides with potential therapeutic applications.[203] Table 5.5 enumerates the biopharmaceutical products which incorporate the use of microbes.

TABLE 5.5 Microbes and Their Biopharmaceutical Products.

Sl No.	Microbes	Biopharmaceuticals	Clinical use
1.	*Taxomyces andrenae*	Taxol	Anticancer[203,206,207]
2.	Aspergillus sp. Penicillium sp.	Proteases	Dermal ulcers, anticancer, inflammation, cardiac and digestive disorders.[203,206,207]
3.	*Penicillium nalgiovense*	Amphotericin B	Antibiotic[203,206,207]
4.	*foetidus*	Beta-galactosidase	Lactose intolerance[203,206,207]
5.	*Monascus ruber* *A. terreus*	Lovastatin	Antilipidemic[203,206,207]
6.	*A. terreus* *E. coli*	L-asparaginase	Acute lymphocytic leukemia (ALL)[203,206,207]
7.	*Claviceps purpurea*	Ergot alkaloids	Migraine, headache[203,206,207]
8.	*P. griseofulvum*	Griseofulvin	Antibiotic[203,206,207]
9.	*Clostridium botulinum*	Botulinum Toxin type A & B	Dystonia[203,206,207]
10.	*Clostridium histolyticum*	Collagenase	In burn areas and chronic dermal ulcers[203,206,207]
11.	*E. coli*	PEG-L-asparaginase	Acute lymphocytic leukemia (ALL)[203,206,207]
12.	*Borrelia burgdorferi*	OspA (Outer surface protein A)	Lyme disease vaccine[203,206,207]

5.9 ENZYME AND ITS INDUSTRIAL APPLICATION

5.9.1 AMYLASE

Amylase is one of the most important enzymes involved in the catalysis of α-1,4-glycosidic linkage in polysaccharides like starch to a product like glucose, maltose, and maltotriose, etc.[208] Amylases may be utilized in miscellaneous industrial activities, inclusive of those in the pharmaceutical sectors, fermentation, and food. Plants, animals, and microbes all produce α-amylases.[209] The microbes' amylases are given their wide range of industrial uses because of their higher stability as compared to α-amylases obtained from animal and plant sources.[210] The main benefits of employing microorganisms for the synthesis of α-amylases are their affordable bulk production capability and their ease of manipulation to produce desired enzymes. Several fungi, yeasts, and bacteria have been used to produce α-amylase. Nonetheless, utilization and demand for enzymes from fungal and bacterial sources have been prevailing in the industrial sectors.[208] α-Amylases may be used in a variety of industrial processes, including those in the food, fermentation, textile, paper, detergent, and pharmaceutical sectors. However, because of advancements in biotechnology, the use of amylase has been broadened in a variety of industries, including the textile, food, brewing, and distilling sectors as well as domains like clinical, medicinal, and analytical chemistry.[211]

5.9.1.1 INDUSTRIAL APPLICATION

5.9.1.1.1 Conversion of Starch

The starch industry uses -amylases most frequently to hydrolyze starch during the starch liquefaction process, which turns starch into fructose and glucose syrups.[212]

5.9.1.1.2 Detergent Industry

It is been ascertained that the detergent's capability to erase rough stains is especially due to enzymes in detergent formulas and this in turn ensures that the detergent is not hazardous to the environment. 90% of all liquid detergents contain amylases, the second type of enzyme employed in the creation of enzymatic detergent.[208] Since starch can collect many forms of particle dirt, its removal from surfaces is crucial in achieving the benefit of whiteness. Examples of amylases from Bacillus or Aspergillus are utilized in the detergent business.[213]

5.9.1.1.3 Textile Industry

Amylases are employed in the de-sizing process in the textile industry. Starch and other sizing agents are applied to the yarn before the cloth manufacturer to guarantee a quick and safe weaving procedure. Starch is a very appealing size because it is affordable and widely accessible. It is fairly simple to get rid of in the majority of areas on the planet. Later, in a wet process, the starch is taken out of the woven cloth by the textile finishing industry. De-sizing is the process of taking the starch out of the fabric, which acts as a strengthening agent to stop the warp thread from breaking during the weaving process. The -amylases remove the size alone, avoiding the fibers, and in a selective manner.[214]

5.9.1.1.4 Food Industry

In the processed food sector, amylases are widely used in processes including baking, brewing, making digestion aids, making cakes, fruit juices, and making starch syrups. The baking industry has made extensive use of amylases. These enzymes can be used in the bread dough to break down the starch in the flour into smaller dextrins, which the yeast then ferments. When amylase is added to the dough, the rate of fermentation is accelerated and the dough's viscosity is decreased, which improves the product's volume and texture.[209]

5.9.1.1.5 Fuel Alcohol Industry

Ethanol is the uttermost prevalent liquid biofuel. Starch being the economic and effortlessly accessible raw material around the world, it is more often utilized as a substrate for ethanol synthesis. In this process, starch must first be solubilized before going through two enzymatic stages to produce fermentable sugars.[215] Liquification and saccharification, in which starch is converted into sugar using an amylolytic microorganism or enzymes like -amylase, are the two steps in the bioconversion of starch into ethanol. Fermentation, on the other hand, involves the conversion of sugar into ethanol using an ethanol-fermenting microorganism like the yeast *Saccharomyces cerevisiae*.[216]

5.9.1.1.6 Paper Industry

The purpose of using -amylases in the pulp and paper industry is to modify coated paper's starch to create low-viscosity, high molecular weight starch. The coating process improves the paper's ability to write by making the surface sufficiently robust and smooth. In addition to being a suitable covering for the paper, starch is a good sizing agent for finishing paper, enhancing its quality and erasability. The size improves the paper's strength and stiffness.[209]

5.9.2 CATALASE

Catalase is an oxidoreductase enzyme which involved in the quenching of reactive oxygen species obtained as a byproduct of aerobic reactions like hydrogen peroxide. As a result, it functions as an antioxidant and defends the cell from oxidative stress.[217] The fundamental process by which this enzyme functions is the breakdown and subsequent hydrogen peroxide (H_2O_2), a reactive oxygen species, breaks decomposed into oxygen and the following reaction, water, therefore, reducing the oxidative stress brought on by this substrate.[218]

5.9.2.1 APPLICATIONS

Medical application: Catalase has been used in the medical field for clinical diagnosis, reverting malignant features, and treatment of various diseases caused by oxidative stress.[219]

Food industry: Catalase has got so many applications in food industries for the production of cheese, the antioxidant effect of the catalase-glycose oxidase system, determining the quality of milk, and wrappers in food packaging.[220]

Bioremediation: A waste management strategy called bioremediation employs living organisms to eliminate or remove contaminants from contaminated sites. It is a method of treatment that transforms toxic compounds into less toxic or non-toxic ones using naturally occurring organisms. Catalase has been used as an indicator in the degradation of hydrocarbons, and as an oxygen provider during aerobic bioremediation and bleached effluents treatment in textile industries.[219]

5.9.3 PEROXIDASE

Peroxidase is a heme-containing enzyme that facilitates a variety of compounds' oxidation utilizing hydrogen peroxide as an electron acceptor on several organic and inorganic substrates.[221] An enzyme derived from horseradish roots called horseradish peroxidase catalyzes the treatment of wastewater that is released from industrial sources. Using H_2O_2 or other peroxides, peroxidases catalyze the oxidation of a wide range of substrates. Different intermediary enzyme forms are involved in the peroxidase catalytic cycle.[222]

Application: Peroxidase has a wide range of industrial applications such as the decolorization of synthetic dyes, synthesis of organic as well as other polymers like cardanol, and polyaniline; phenolic compound bioremediation, delignification of paper pulp in the paper industry, for labeling of antibody for enzyme immunoassays and diagnostic as well as an analysis tool for diagnosis of different diseases.[223]

5.9.4 LIPASE

Lipase is a hydrolytic enzyme belonging to the hydrolase family of enzymes involved in the hydrolysis of carboxylic ester bonds. Hydrolysis of triglycerides into monoglyceride, diacylglyceride, glycerol, and fatty acid is done by lipase. These enzymes are widely distributed in the realms of plants and animals as well as in bacteria, molds, and fungi.[224]

Lipase is used for the production of an infant's supplements which mimic the mother's milk and as well as in cheese production. It has also been used to improve the texture and color of dough, and improvement in quality of noodles and pasta production. In pharmaceutical industries, it can be utilized for the synthesis of chiral compounds, and the interesterification of vegetable oil. In the leather industry, it can also be used for the dispersion of fats and liming of leather. Lipase is commonly used in the cosmetic industry for the synthesis of cosmetic ingredients like isopropyl myristate.[224]

5.9.5 PROTEASE

Protease is a proteolytic enzyme that is commonly called peptidases, proteinases, or proteas due to its involvement in the breakdown of peptide bonds found in the polypeptide chain of amino acids.[225] Serine proteases, aspartic proteases, threonine proteases, cysteine proteases, glutamic proteases, asparagine proteases, metalloproteases, and proteases with mixed or unknown catalytic mechanisms are the different categories of proteases based on the catalytic mechanism and the presence of amino acid residue(s) at the active site.[226]

Proteases are being extensively used in food production industries for the preparation of soy sauce, protein hydrolysate, clearing food juice and beer, cake preparation, tenderization of meat, and improvement of the texture of flour paste. In both the institutional and industrial sectors, proteases have been employed as dishwashing, cleaning, and laundry detergents. Alkaline protease is used more frequently in developing leather industries due to its elastolytic and keratinolytic properties. Alkaline protease's influencing qualities are highly useful in the leather processing industries.[225,227] Proteases are sometimes used in the photographic industry to recover silver from X-ray films. In the field of medicine, alkaline protease has been used in the composition of non-woven tissue, gauze, and ointments and the preparation of elastoterase immobilized on bandage is used to treat a variety of disorders, including burns, carbuncles, furuncles, and wounds.[225]

5.9.6 PENICILLINASE

Penicillinase is an enzyme responsible for breaking off a β-lactam ring of penicillin and cephalosporins as well as its derivatives. Penicillinase hydrolyzes penicillin to produce penicilloic acid. Penicillin acylase or amidase hydrolyzes the amide bond of the basic penicillin nucleus for the production of 6-amino penicillanic acid.[228]

5.9.6.1 APPLICATIONS

Industrially, penicillinase as well as penicillin acylase have been used for bulk production of penicillin derivatives like penicillanic acid, penicilloic acid, etc. Moreover, a racemic mixture of the different compounds has been produced by the reaction which is catalyzed by penicillin acylase. Penicillinase can also be used for sterility validation tests of β-lactam antibiotics which are basically injectable forms.[228,229]

5.10 FUTURE ASPECTS

Whether it be practical to use as a support for chemical synthesis or in the conversion of natural products, microbial biotransformation has achieved large-scale importance. The cardinal tools for biotransformation are mammalian, plant, or microbial cells and their cell-free enzymes.[230] In the present day, multifarious microbial or enzymatically reactions are being developed for industrial processes yet the efficiency of microbial biotransformation presents a wide-ranging sphere of limitless potentiality for the future. Obtaining new practically usable procedures, demands awareness regarding alternative and broader activities, for instance, the exploration of various substrate structures for intended enzymatic modification, research for unique accessible enzymatic reaction types, and the screening of microflora for suitable microorganisms. Innumerable approaches for chemical synthesis have been developed for the formulation of numerous usable molecules that are pharmaceutically significant with high therapeutic value. However, due to asymmetric centers in some molecules for example testosterone (steroid), manufacturing them synthetically turns into a problematic and non-specific task.[231] Henceforth, in present days, pharmaceutical industries opt for trending microbial transformation than the chemical route, as it provides high specificity, efficient and impressive yield, assists in lowering time consumption, unlike chemical reactions, avoids laborious /non-specific/multistep reactions, production of unwanted side-by-products, strident reaction conditions and an upsurging move toward green synthesis.[3] With the aid of metabolic engineering, various microorganisms, and their enzymes can be engineered successfully to enhance the efficacy of the biosynthesis of essential molecules. Therefore, to produce molecules on a large scale, techniques like metagenomic engineering and high-throughput screening (HTS) must be performed before the fermentation process to obtain an efficient strain. Thus, involving these techniques can help to make industrial processing lead toward a replacement of chemical synthesis.[232] Recent developments in microbial biotransformation and biodegradation can be employed to eliminate toxic environmental pollutants such as polychlorinated dibenzo-p-dioxins (PCDDs) and polychlorinated dibenzofurans (PCDFs), commonly known as dioxins formed from various sources as elimination of these pollutants from the environment is a difficult task due to their persistent nature. Removal of dioxins by biological degradation (biodegradation) is considered a feasible method as an alternative to other expensive physicochemical approaches.[233] Using biocatalysis/ microbial biotransformation procedures crucial and minor reformation can be done in existing molecules to give rise to new molecules with antifungal properties.[234] The biochemical potential of microorganisms can be further taken into advantage to recover and recycle industrial by-products such as polyphenols, carotenoids, sterols, tocopherols, vitamins, or dietary fiber to heighten the sustainability of the industrial and food sectors.[235]

5.11 CONCLUSION

Microorganisms, the undetected minuscule players have undeniably influenced our lives and are highly valued for their multi-faceted performance. Their role in the biotransformation of various compounds and chemicals, food, pharmaceuticals, and agrochemicals has been demonstrated to be an indispensable tool and capture acceptance and demand. Owing to their tremendous rate of cell multiplication, microorganisms are proficient in producing a vast range of enzymes in a limited time frame. Microbial biotransformation has been investigated for generations and since then a plethora of information has been generated. This mechanism allows the modification of compounds using an environmentally friendly passage. However, this phenomenon can occasionally be challenging. The chemical instability of monoterpenes, the toxicity of the substrate and product, and the existence of various transformation pathways in the microorganisms were, for example, the most significant barriers in the microbial biotransformation of monoterpenes, which have enormous commercial appeal to the food and perfume industries. Also, lignin displays several challenges in enzymatic transformation due to its complex structure, composition, and inherent property of resistance to degradation. Scientific collaborations will heighten the collection of afresh information concerning the purification and description of several enzymes coming from microorganisms and microorganisms-driven alterations of natural products, pharmaceuticals, and other compounds. Furthermore, a few of the microbial enzymes have been cloned and sequenced using the genetic information encoding them. In the end, genetic engineering methods might provide modified microorganism strains that can be used to produce the required and relevant product. A substantial number of attempts in the area of microbial biotransformation of diversified compounds are yet to be accomplished. The insight into known biotransformation pathways and their utilities will vitalize additional in-depth research in the field of microbial transformations of innovative and structurally complex drugs and chemical compounds. Advanced techniques in genetic and metabolic engineering can be employed in the development of biological platforms for the dynamic utilization of miscellaneous natural and synthetic compounds.

ACKNOWLEDGMENT

Figure 5.1 to 5.4 and 5.6 were created using PerplexityAI.

KEYWORDS

- **microbial biotransformation**
- **enzymes**
- **microorganism**
- **metabolism**
- **biocatalyst**

REFERENCES

1. Kaup, B.; Bringer-Meyer, S.; Sahm, H. Metabolic Engineering of *Escherichia Coli*: Construction of an Efficient Biocatalyst for D-Mannitol Formation in a Whole-Cell Biotransformation. *Appl. Microbiol. Biotechnol.* **2004,** *64* (3), 333–339. https://doi.org/10.1007/s00253-003-1470-9.
2. Kaup, B.; Bringer-Meyer, S.; Sahm, H. Commentary to: Metabolic Engineering of *Escherichia Coli*: Construction of an Efficient Biocatalyst for D-Mannitol Formation in a Whole-Cell Biotransformation. *Appl. Microbiol. Biotechnol.* **2016,** *100* (2), 1039. https://doi.org/10.1007/s00253-015-7029-8.
3. Singh, R. Microbial Biotransformation: A Process for Chemical Alterations. *J. Bacteriol. Mycol. Open Access* **2017,** *4* (2). https://doi.org/10.15406/jbmoa.2017.04.00085.
4. Kaup, B.; Bringer-Meyer, S.; Sahm, H. D-Mannitol Formation from D-Glucose in a Whole-Cell Biotransformation with Recombinant *Escherichia Coli*. *Appl. Microbiol. Biotechnol.* **2005,** *69* (4), 397–403. https://doi.org/10.1007/s00253-005-1996-0.
5. Hegazy, M.-E. F.; Mohamed, T. A.; ElShamy, A. I.; Mohamed, A.-E.-H. H.; Mahalel, U. A.; Reda, E. H.; Shaheen, A. M.; Tawfik, W. A.; Shahat, A. A.; Shams, K. A. et al.Microbial Biotransformation as a Tool for Drug Development Based on Natural Products from Mevalonic Acid Pathway: A Review. *J. Adv. Res.* **2015,** *6* (1), 17–33. https://doi.org/10.1016/j.jare.2014.11.009.
6. Phang-Lyn, S.; Llerena, V. A. Biochemistry, Biotransformation. In: *StatPearls*; StatPearls Publishing: Treasure Island, 2022.
7. Hegazy, M. E. F.; Mohamed, T. A.; ElShamy, A. I.; Mohamed, A. E. H. H.; Mahalel, U. A.; Reda, E. H.; Shaheen, A. M.; Tawfik, W. A.; Shahat, A. A.; Shams, K. A. et al.Microbial Biotransformation as a Tool for Drug Development Based on Natural Products from Mevalonic Acid Pathway: A Review. *J. Adv. Res.* **2015,** *6* (1), 17–33. https://doi.org/10.1016/j.jare.2014.11.009.
8. Lilly, M. D.; Woodley, J. M. A Structured Approach to Design and Operation of Biotransformation Processes. *J. Ind. Microbiol.* **1996,** *17* (1), 24–29. https://doi.org/10.1007/BF01570144.
9. Cano-Flores, A.; Gómez, J.; S. Escalona-Torres, I.; Velasco-Bejarano, B. Microorganisms as Biocatalysts and Enzyme Sources. In: *Microorganisms*; Blumenberg, M., Shaaban, M., Elgaml, A., Eds.; IntechOpen, 2020. https://doi.org/10.5772/intechopen.90338.
10. Kauffman, G. B. The Language of Biotechnology: A Dictionary of Terms (Cox, Michael; Walker, John M.). *J. Chem. Educ.* **1990,** *67* (5), A150. https://doi.org/10.1021/ed067pa150.
11. Klingenberg, P. J. M. Walker Und M. Cox: The Language of Biotechnology. A Dictionary of Terms. 254 Seiten. American Chemical Society, Washington DC, 1988. *Food Nahr.* **1990,** *34* (2), 170–170. https://doi.org/10.1002/food.19900340220.
12. Shanu-Wilson, J.; Evans, L.; Wrigley, S.; Steele, J.; Atherton, J.; Boer, J. Biotransformation: Impact and Application of Metabolism in Drug Discovery. *ACS Med. Chem. Lett.* **2020,** *11* (11), 2087–2107. https://doi.org/10.1021/acsmedchemlett.0c00202.
13. Winkler, C. K.; Schrittwieser, J. H.; Kroutil, W. Power of Biocatalysis for Organic Synthesis. *ACS Cent. Sci.* **2021,** *7* (1), 55–71. https://doi.org/10.1021/acscentsci.0c01496.
14. Mamoli, L.; Tercellone, A. Über Die Biochemische Hydrierung Des Dehydro-Androsteron. *Hoppe. Seylers Z. Physiol. Chem.* **1936,** *245* (3–4), 93–95. https://doi.org/10.1515/bchm2.1936.245.3-4.93.

15. Krámli, A.; Horváth, J. Microbiological Oxidation of Sterols [19]. *Nature* **1949,** *163* (4136), 219. https://doi.org/10.1038/163219a0.
16. HENCH, P. S. The Reversibility of Certain Rheumatic and Nonrheumatic Conditions by the Use of Cortisone or of the Pituitary Adrenocotropic Hormone. *Ann. Intern. Med.* **1952,** *36* (1), 1–38. https://doi.org/10.7326/0003-4819-36-1-1.
17. Prelich, G. Gene Overexpression: Uses, Mechanisms, and Interpretation. *Genetics* **2012,** *190* (3), 841–854. https://doi.org/10.1534/genetics.111.136911.
18. Schweizer, H. P. Bacterial Genetics: Past Achievements, Present State of the Field, and Future Challenges. *BioTechniques* **2008,** *44* (5), 633–641. https://doi.org/10.2144/000112807.
19. Burmeister, A. R. Horizontal Gene Transfer. *Evol. Med. Public Health* **2015,** *2015* (1), 193–194. https://doi.org/10.1093/emph/eov018.
20. Holmes, R. K.; Jobling, M. G. Genetics. In: *Medical Microbiology*; Baron, S., Ed.; University of Texas Medical Branch at Galveston: Galveston, 1996.
21. Engström, W.; Darbre, P.; Eriksson, S.; Gulliver, L.; Hultman, T.; Karamouzis, M. V.; Klaunig, J. E.; Mehta, R.; Moorwood, K.; Sanderson, T. et al.The Potential for Chemical Mixtures from the Environment to Enable the Cancer Hallmark of Sustained Proliferative Signalling. *Carcinogenesis* **2015,** *36* (Suppl 1), S38–S60. https://doi.org/10.1093/carcin/bgv030.
22. Johnston, C.; Martin, B.; Fichant, G.; Polard, P.; Claverys, J.-P. Bacterial Transformation: Distribution, Shared Mechanisms and Divergent Control. *Nat. Rev. Microbiol.* **2014,** *12* (3), 181–196. https://doi.org/10.1038/nrmicro3199.
23. Bruslind, L.; Bruslind, L. *General Microbiology*, 1st edn.; 2020.
24. Asif, A.; Mohsin, H.; Tanvir, R.; Rehman, Y. Revisiting the Mechanisms Involved in Calcium Chloride Induced Bacterial Transformation. *Front. Microbiol.* **2017,** *8*, 2169. https://doi.org/10.3389/fmicb.2017.02169.
25. Sayavedra-Soto, L. A.; Stein, L. Y. Genetic Transformation of Ammonia-Oxidizing Bacteria; 2011; pp 389–402. https://doi.org/10.1016/B978-0-12-381294-0.00017-1.
26. Transduction (prokaryotes) | Learn Science at Scitable https://www.nature.com/scitable/definition/transduction-prokaryotes-292/ (accessed 15 Feb 2023).
27. Bacteriophages: Viruses That Infect Bacteria · Frontiers for Young Minds https://kids.frontiersin.org/articles/10.3389/frym.2019.00146 (accessed 15 Feb 2023).
28. Chiang, Y. N.; Penadés, J. R.; Chen, J. Genetic Transduction by Phages and Chromosomal Islands: The New and Noncanonical. *PLOS Pathog.* **2019,** *15* (8), e1007878. https://doi.org/10.1371/journal.ppat.1007878.
29. Fillol-Salom, A.; Alsaadi, A.; Sousa, J. A. M. de; Zhong, L.; Foster, K. R.; Rocha, E. P. C.; Penadés, J. R.; Ingmer, H.; Haaber, J. Bacteriophages Benefit from Generalized Transduction. *PLOS Pathog.* **2019,** *15* (7), e1007888. https://doi.org/10.1371/journal.ppat.1007888.
30. Chiang, Y. N.; Penadés, J. R.; Chen, J. Genetic Transduction by Phages and Chromosomal Islands: The New and Noncanonical. *PLoS Pathog.* **2019,** *15* (8), e1007878. https://doi.org/10.1371/journal.ppat.1007878.
31. Howard-Varona, C.; Hargreaves, K. R.; Abedon, S. T.; Sullivan, M. B. Lysogeny in Nature: Mechanisms, Impact and Ecology of Temperate Phages. *ISME J.* **2017,** *11* (7), 1511–1520. https://doi.org/10.1038/ismej.2017.16.
32. Howard-Varona, C.; Hargreaves, K. R.; Abedon, S. T.; Sullivan, M. B. Lysogeny in Nature: Mechanisms, Impact and Ecology of Temperate Phages. *ISME J.* **2017,** *11* (7), 1511–1520. https://doi.org/10.1038/ismej.2017.16.

33. Transduction, Plasmids, and the Foundation of Biotechnology | Joshua Lederberg - Profiles in Science https://profiles.nlm.nih.gov/spotlight/bb/feature/bacgen2 (accessed 15 Feb 2023).
34. Qi, H.; Lu, H.; Qiu, H.-J.; Petrenko, V.; Liu, A. Phagemid Vectors for Phage Display: Properties, Characteristics and Construction. *J. Mol. Biol.* **2012,** *417* (3), 129–143. https://doi.org/10.1016/j.jmb.2012.01.038.
35. Qi, H.; Lu, H.; Qiu, H.-J.; Petrenko, V.; Liu, A. Phagemid Vectors for Phage Display: Properties, Characteristics and Construction. *J. Mol. Biol.* **2012,** *417* (3), 129–143. https://doi.org/10.1016/j.jmb.2012.01.038.
36. Bulcha, J. T.; Wang, Y.; Ma, H.; Tai, P. W. L.; Gao, G. Viral Vector Platforms within the Gene Therapy Landscape. *Signal Transduct. Target. Ther. 2021 61* **2021,** *6* (1), 1–24. https://doi.org/10.1038/s41392-021-00487-6.
37. Bacterial Conjugation - an overview | ScienceDirect Topics https://www.sciencedirect.com/topics/biochemistry-genetics-and-molecular-biology/bacterial-conjugation (accessed 15 Feb 2023).
38. Plasmid Transfer by Conjugation in Gram-Negative Bacteria: From the Cellular to the Community Level - PMC https://www.ncbi.nlm.nih.gov/pmc/articles/PMC7690428/ (accessed 15 Feb 2023).
39. Raleigh, E. A.; Low, K. B. Conjugation. In: *Brenner's Encyclopedia of Genetics*; Elsevier, 2013; pp 144–151. https://doi.org/10.1016/B978-0-12-374984-0.00321-1.
40. Grohmann, E.; Muth, G.; Espinosa, M. Conjugative Plasmid Transfer in Gram-Positive Bacteria. *Microbiol. Mol. Biol. Rev.* **2003,** *67* (2), 277–301. https://doi.org/10.1128/MMBR.67.2.277-301.2003.
41. Llosa, M.; de la Cruz, F. Bacterial Conjugation: A Potential Tool for Genomic Engineering. *Res. Microbiol.* **2005,** *156* (1), 1–6. https://doi.org/10.1016/j.resmic.2004.07.008.
42. Cabezón, E.; de la Cruz, F.; Arechaga, I. Conjugation Inhibitors and Their Potential Use to Prevent Dissemination of Antibiotic Resistance Genes in Bacteria. *Front. Microbiol.* **2017,** *8*. https://doi.org/10.3389/fmicb.2017.02329.
43. plasmid / plasmids | Learn Science at Scitable https://www.nature.com/scitable/definition/plasmid-plasmids-28/ (accessed 15 Feb 2023).
44. Clark, D. P.; Pazdernik, N. J.; McGehee, M. R. Plasmids. In: *Molecular Biology*; Elsevier, 2019; pp 712–748. https://doi.org/10.1016/B978-0-12-813288-3.00023-9.
45. Shintani, M.; Sanchez, Z. K.; Kimbara, K. Genomics of Microbial Plasmids: Classification and Identification Based on Replication and Transfer Systems and Host Taxonomy. *Front. Microbiol.* **2015,** *6*. https://doi.org/10.3389/fmicb.2015.00242.
46. Carroll, A. C.; Wong, A. Plasmid Persistence: Costs, Benefits, and the Plasmid Paradox. *Can. J. Microbiol.* **2018,** *64* (5), 293–304. https://doi.org/10.1139/cjm-2017-0609.
47. Wegrzyn, G.; Wegrzyn, A. Stress Responses and Replication of Plasmids in Bacterial Cells. *Microb. Cell Factories* **2002,** *1*, 2.
48. del Solar, G.; Giraldo, R.; Ruiz-Echevarría, M. J.; Espinosa, M.; Díaz-Orejas, R. Replication and Control of Circular Bacterial Plasmids. *Microbiol. Mol. Biol. Rev.* **1998,** *62* (2), 434–464. https://doi.org/10.1128/MMBR.62.2.434-464.1998.
49. Virolle, C.; Goldlust, K.; Djermoun, S.; Bigot, S.; Lesterlin, C. Plasmid Transfer by Conjugation in Gram-Negative Bacteria: From the Cellular to the Community Level. *Genes 2020 Vol 11 Page* **1239 2020,** *11* (11), 1239. https://doi.org/10.3390/GENES11111239.

50. Nonaka, L.; Yamamoto, T.; Maruyama, F.; Hirose, Y.; Onishi, Y.; Kobayashi, T.; Suzuki, S.; Nomura, N.; Masuda, M.; Yano, H. Interplay of a Non-Conjugative Integrative Element and a Conjugative Plasmid in the Spread of Antibiotic Resistance via Suicidal Plasmid Transfer from an Aquaculture Vibrio Isolate. *PLOS ONE* **2018,** *13* (6), e0198613. https://doi.org/10.1371/journal.pone.0198613.
51. Bacteria - Growth of bacterial populations | Britannica https://www.britannica.com/science/bacteria/Exchange-of-genetic-information (accessed 15 Feb 2023).
52. van Hoek, A. H. A. M.; Mevius, D.; Guerra, B.; Mullany, P.; Roberts, A. P.; Aarts, H. J. M. Acquired Antibiotic Resistance Genes: An Overview. *Front. Microbiol.* **2011,** *2*. https://doi.org/10.3389/fmicb.2011.00203.
53. Babakhani, S.; Oloomi, M. Transposons: The Agents of Antibiotic Resistance in Bacteria. *J. Basic Microbiol.* **2018,** *58* (11), 905–917. https://doi.org/10.1002/jobm.201800204.
54. Active Transposition in Genomes - PMC https://www.ncbi.nlm.nih.gov/pmc/articles/PMC3612533/ (accessed 15 Feb 2023).
55. Iyer, A.; Barbour, E.; Azhar, E.; Salabi, A. A. El; Hassan, H. M. A.; Qadri, I.; Chaudhary, A.; Abuzenadah, A.; Kumosani, T.; Damanhouri, G. et al.Transposable Elements in *Escherichia Coli* & Antimicrobial Resistance. *Adv. Biosci. Biotechnol.* **2013,** *04* (03), 415–423. https://doi.org/10.4236/abb.2013.43A055.
56. Babakhani, S.; Oloomi, M. Transposons: The Agents of Antibiotic Resistance in Bacteria. *J. Basic Microbiol.* **2018,** *58* (11), 905–917. https://doi.org/10.1002/jobm.201800204.
57. Ivics, Z.; Izsvak, Z. Transposons for Gene Therapy! *Curr. Gene Ther.* **2006,** *6* (5), 593–607. https://doi.org/10.2174/156652306778520647.
58. Fabian, B. K.; Tetu, S. G.; Paulsen, I. T. Application of Transposon Insertion Sequencing to Agricultural Science. *Front. Plant Sci.* **2020,** *11* (March). https://doi.org/10.3389/fpls.2020.00291.
59. Lawrence, J. G. Gene Organization: Selection, Selfishness, and Serendipity. *Annu. Rev. Microbiol.* **2003,** *57*, 419–440. https://doi.org/10.1146/annurev.micro.57.030502.090816.
60. Actis, L. A.; Tolmasky, M. E.; Crosa, J. H. Bacterial Plasmids: Replication of Extrachromosomal Genetic Elements Encoding Resistance to Antimicrobial Compounds. *Front. Biosci. J. Virtual Libr.* **1999,** *4*, D43-62. https://doi.org/10.2741/actis.
61. Genomics of microbial plasmids: classification and identification based on replication and transfer systems and host taxonomy - PMC https://www.ncbi.nlm.nih.gov/pmc/articles/PMC4379921/ (accessed 16 Feb 2023).
62. Genome Packaging in Prokaryotes | Learn Science at Scitable http://www.nature.com/scitable/topicpage/genome-packaging-in-prokaryotes-the-circular-chromosome-9113 (accessed 16 Feb 2023).
63. DNA Transposition at Work - PMC https://www.ncbi.nlm.nih.gov/pmc/articles/PMC6380494/ (accessed 16 Feb 2023).
64. Brown T. A. Genome Anatomies. In: *Genomes*; Wiley-Liss, 2002.
65. Vellai, T.; Vida, G. The Origin of Eukaryotes: The Difference between Prokaryotic and Eukaryotic Cells. *Proc. R. Soc. B Biol. Sci.* **1999,** *266* (1428), 1571–1577. https://doi.org/10.1098/rspb.1999.0817.
66. Alberts, B.; Johnson, A.; Lewis, J.; Raff, M.; Roberts, K.; Walter, P. Chromosomal DNA and Its Packaging in the Chromatin Fiber. *Mol. Biol. Cell 4th Ed.*, **2002**.
67. Mattick, J. S. Non-Coding RNAs: The Architects of Eukaryotic Complexity. *EMBO Rep.* **2001,** *2* (11), 986–991. https://doi.org/10.1093/embo-reports/kve230.

68. Sonea, S.; Mathieu, L. G. Evolution of the Genomic Systems of Prokaryotes and Its Momentous Consequences. *Int. Microbiol.* **2001,** *4* (2), 67–71. https://doi.org/10.1007/S101230100015.
69. Lawrence, J. G. Shared Strategies in Gene Organization among Prokaryotes and Eukaryotes. *Cell* **2002,** *110* (4), 407–413. https://doi.org/10.1016/S0092-8674(02)00900-5.
70. Watt, W. B.; Dean, A. M. Molecular-Functional Studies of Adaptive Genetic Variation in Prokaryotes and Eukaryotes. *Annu. Rev. Genet.* **2000,** *34* (February), 593–622. https://doi.org/10.1146/annurev.genet.34.1.593.
71. Pervaiz, I.; Ahmad, S.; Madni, M. A.; Ahmad, H.; Khaliq, F. H. Microbial Biotransformation: A Tool for Drug Designing. *Appl. Biochem. Microbiol.* **2013,** *49* (5), 437–450. https://doi.org/10.1134/S0003683813050098.
72. Heidari, A. Biotranslational Medical and Biospectroscopic Studies of Cadmium Oxide(CdO) Nanoparticles-DNA/RNA Straight and Cycle Chain Complexes as Potent Anti-Viral, Anti-Tumor and Anti-Microbial Drugs: A Clinical Approach. *Transl. Biomed.* **2016,** *7* (2). https://doi.org/10.21767/2172-0479.100076.
73. Singh, S. Preclinical Pharmacokinetics: An Approach Towards Safer and Efficacious Drugs. *Curr. Drug Metab.* **2006,** *7* (2), 165–182. https://doi.org/10.2174/138920006775541552.
74. Burford, E. P.; Fomina, M.; Gadd, G. M. Fungal Involvement in Bioweathering and Biotransformation of Rocks and Minerals. *Mineral. Mag.* **2003,** *67* (6), 1127–1155. https://doi.org/10.1180/0026461036760154.
75. Collins, A. M.; Kennedy, M. J. Biotransformations and Bioconversions in New Zealand: Past Endeavours and Future Potential. *Australas. Biotechnol.* **1999,** *9* (2), 86–94.
76. Strobel, G.; Daisy, B. Bioprospecting for Microbial Endophytes and Their Natural Products. *Microbiol. Mol. Biol. Rev.* **2003,** *67* (4), 491–502. https://doi.org/10.1128/MMBR.67.4.491-502.2003.
77. Thermophilic Fungi: Their Physiology and Enzymes - PMC https://www.ncbi.nlm.nih.gov/pmc/articles/PMC99000/ (accessed 16 Feb 2023).
78. Hotta, Y.; Ezaki, S.; Atomi, H.; Imanaka, T. Extremely Stable and Versatile Carboxylesterase from a Hyperthermophilic Archaeon. *Appl. Environ. Microbiol.* **2002,** *68* (8), 3925–3931. https://doi.org/10.1128/AEM.68.8.3925-3931.2002.
79. Gershwin, L. A. Nematocysts of the Cubozoa. *Zootaxa* **2006,** *1232* (1232), 1–57. https://doi.org/10.11646/zootaxa.1232.1.1.
80. Leresche, J. E.; Meyer, H. P. Chemocatalysis and Biocatalysis (Biotransformation): Some Thoughts of a Chemist and of a Biotechnologist. *Org. Process Res. Dev.* **2006,** *10* (3), 572–580. https://doi.org/10.1021/op0600308.
81. Bonnet, M.; Lagier, J. C.; Raoult, D.; Khelaifia, S. Bacterial Culture through Selective and Non-Selective Conditions: The Evolution of Culture Media in Clinical Microbiology. *New Microbes New Infect.* **2019,** *34*, 100622. https://doi.org/10.1016/j.nmni.2019.100622.
82. Chemocatalysis and Biocatalysis (Biotransformation): Some Thoughts of a Chemist and of a Biotechnologist | Organic Process Research & Development https://pubs.acs.org/doi/abs/10.1021/op0600308 (accessed 16 Feb 2023).
83. Phang-Lyn, S.; Llerena, V. A. *Biochemistry, Biotransformation*; 2022.
84. Almazroo, O. A.; Miah, M. K.; Venkataramanan, R. Drug Metabolism in the Liver. *Clin. Liver Dis.* **2017,** *21* (1), 1–20. https://doi.org/10.1016/j.cld.2016.08.001.

85. Meyer, U. A. Overview of Enzymes of Drug Metabolism. *J. Pharmacokinet. Biopharm.* **1996,** *24* (5), 449–459. https://doi.org/10.1007/BF02353473.
86. Duran, N.; De Conti, R.; Rodrigues, J. A. R. Biotransformations by Microorganisms, Organisms and Enzymes: State of Art. *Boletin Soc. Chil. Quimica* **2000,** *45* (1), 109–121. https://doi.org/10.4067/S0366-16442000000100015.
87. Stewart, J. D. Cyclohexanone Monooxygenase: A Useful Asymmetric Baeyer-Villiger Reactions. *Curr Org Chem* **1998,** *2*, 195–216.
88. Stewart, J. D.; Reed, K. W.; Martinez, C. A.; Zhu, J.; Chen, G.; Kayser, M. M. Recombinant Baker's Yeast as a Whole-Cell Catalyst for Asymmetric Baeyer- Villiger Oxidations. *J. Am. Chem. Soc.* **1998,** *120* (15), 3541–3548.
89. Alphand, V.; Mazzini, C.; Lebreton, J.; Furstoss, R. A New Microorganism for Highly Stereospecific Baeyer--Villiger Oxidation of Prochiral Cyclobutanones. *J. Mol. Catal. B Enzym.* **1998,** *5* (1–4), 219–221.
90. Hu, S.; Tian, X.; Han, G. Novel Microbial Hydroxylation of 13-Ethyl-17β-Hydroxy-18,19-Dinor-17α-Pregn-4-En-20-Yn-3-One. *Steroids* **1998,** *63* (2), 88–92. https://doi.org/10.1016/S0039-128X(97)00139-6.
91. Olivo, H. F.; Hemenway, M. S.; Gezginci, M. H. Synthesis and Microbial Hydroxylation of Some Azabicycloalkanes. *Tetrahedron Lett.* **1998,** *39* (11), 1309–1312. https://doi.org/10.1016/S0040-4039(97)10876-0.
92. Gotor, V.; Quirós, M.; Liz, R.; Frigola, J.; Fernández, R. Fungal and Bacterial Regioselective Hydroxylation of Pyrimidine Heterocycles. *Tetrahedron* **1997,** *53* (18), 6421–6432. https://doi.org/10.1016/S0040-4020(97)00301-3.
93. Butora, G.; Gum, A. G.; Hudlicky, T.; Abboud, K. A. Advanced Intramolecular Diels-Alder Study Toward the Synthesis of (-)-Morphine: Structure Correction of a Previously Reported Diels-Alder Product. *Synthesis* **1998,** *1998* (03), 275–278. https://doi.org/10.1055/s-1998-4529.
94. Füchtenbusch, B.; Wältermann, M.; Steinbüchel, A. Biotransformation of 2, 2-Dimethyl-1, 3-Propanediol to 3-Hydroxypivalic Acid by Acetobacter Acetii DSMZ3508 and Related Bacteria. *Biotechnol. Lett.* **1998,** *20* (5), 507–510.
95. Shafiee, A.; Motamedi, H.; King, A. Purification, Characterization and Immobilization of an NADPH-Dependent Enzyme Involved in the Chiral Specific Reduction of the Keto Ester M, an Intermediate in the Synthesis of an Anti-Asthma Drug, Montelukast, from Microbacterium Campoquemadoensis (MB56. *Appl. Microbiol. Biotechnol.* **1998,** *49* (6), 709–717.
96. Prichanont, S.; Leak, D. J.; Stuckey, D. C. Alkene Monooxygenase-Catalyzed Whole Cell Epoxidation in a Two-Liquid Phase System. *Enzyme Microb. Technol.* **1998,** *22* (6), 471–479.
97. Roberts, S. M. Preparative Biotransformations: The Employment of Enzymes and Whole-Cells in Synthetic Organic Chemistry. *J. Chem. Soc. Perkin 1* **1998,** No. 1, 157–170.
98. Botes, A. L.; Weijers, C.; Van Dyk, M. S. Biocatalytic Resolution of 1, 2-Epoxyoctane Using Resting Cells of Different Yeast Strains with Novel Epoxide Hydrolase Activities. *Biotechnol. Lett.* **1998,** *20* (4), 421–426.
99. Spelberg, J. H. L.; Rink, R.; Kellogg, R. M.; Janssen, D. B. Enantioselectivity of a Recombinant Epoxide Hydrolase from Agrobacterium Radiobacter. *Tetrahedron Asymmetry* **1998,** *9* (3), 459–466.

100. Weijers, C.; Botes, A. L.; van Dyk, M. S.; de Bont, J. A. M. Enantioselective Hydrolysis of Unbranched Aliphatic 1, 2-Epoxides by Rhodotorula Glutinis. *Tetrahedron Asymmetry* **1998,** *9* (3), 467–473.
101. Kroutil, W.; Mischitz, M.; Faber, K. Deracemization of (±)-2,3-Disubstituted Oxiranes via Biocatalytic Hydrolysis Using Bacterial Epoxide Hydrolases: Kinetics of an Enantioconvergent Process. *J. Chem. Soc. - Perkin Trans. 1* **1997,** No. 24, 3629–3636. https://doi.org/10.1039/a704812b.
102. Matoishi, K.; Sano, A.; Imai, N.; Yamazaki, T.; Yokoyama, M.; Sugai, T.; Ohta, H. Rhodococcus Rhodochrous IFO 15564-Mediated Hydrolysis of Alicyclic Nitriles and Amides: Stereoselectivity and Use for Kinetic Resolution and Asymmetrization. *Tetrahedron Asymmetry* **1998,** *9* (7), 1097–1102.
103. Mitra, S.; Murthy, G. S. Bioreactor Control Systems in the Biopharmaceutical Industry: A Critical Perspective. *Syst. Microbiol. Biomanufacturing* **2022,** *2* (1), 91–112. https://doi.org/10.1007/s43393-021-00048-6.
104. Chen, F.-M.; Liu, X. Advancing Biomaterials of Human Origin for Tissue Engineering. *Prog. Polym. Sci.* **2016,** *53*, 86–168. https://doi.org/10.1016/j.progpolymsci.2015.02.004.
105. Wang, S.; Meng, X.; Zhou, H.; Liu, Y.; Secundo, F.; Liu, Y. Enzyme Stability and Activity in Non-Aqueous Reaction Systems: A Mini Review. *Catalysts* **2016,** *6* (2), 32. https://doi.org/10.3390/catal6020032.
106. Hegazy, M.-E. F.; Mohamed, T. A.; ElShamy, A. I.; Mohamed, A.-E.-H. H.; Mahalel, U. A.; Reda, E. H.; Shaheen, A. M.; Tawfik, W. A.; Shahat, A. A.; Shams, K. A. et al.Microbial Biotransformation as a Tool for Drug Development Based on Natural Products from Mevalonic Acid Pathway: A Review. *J. Adv. Res.* **2015,** *6* (1), 17–33. https://doi.org/10.1016/j.jare.2014.11.009.
107. Mustacchi, R.; Knowles, C. J.; Li, H.; Dalrymple, I.; Sunderland, G.; Skibar, W.; Jackman, S. A. Enhanced Biotransformations and Product Recovery in a Membrane Bioreactor through Application of a Direct Electric Current. *Biotechnol. Bioeng.* **2005,** *89* (1), 18–23. https://doi.org/10.1002/bit.20299.
108. Fernandes, P. Miniaturization in Biocatalysis. *Int. J. Mol. Sci.* **2010,** *11* (3), 858–879. https://doi.org/10.3390/ijms11030858.
109. Ghisalba, O.; Meyer, H.-P.; Wohlgemuth, R. Industrial Biotransformation. In: *Encyclopedia of Industrial Biotechnology*; John Wiley & Sons, Inc.: Hoboken, 2010. https://doi.org/10.1002/9780470054581.eib174.
110. Godjevargova, T.; Dayal, R.; Turmanova, S. Gluconic Acid Production in Bioreactor with Immobilized Glucose Oxidase plus Catalase on Polymer Membrane Adjacent to Anion-Exchange Membrane. *Macromol. Biosci.* **2004,** *4* (10), 950–956. https://doi.org/10.1002/mabi.200400058.
111. Glass autoclavable bioreactors https://reseabiotec.ch/our-products/glass-autoclabable-bioreactors/ (accessed 16 Feb 2023).
112. 3 Flow-through Sensors Based on Integrated Reaction and Detection. In: *Techniques and Instrumentation in Analytical Chemistry*; Valcárcel, M., de Castro, L., Eds.; Elsevier, 1994; Vol. 16, pp 81–200. https://doi.org/10.1016/S0167-9244(08)70006-4.
113. *Environmental Microbiology and Biotechnology*; Singh, A., Srivastava, S., Rathore, D., Pant, D., Eds.; Springer Singapore: Singapore, 2021. https://doi.org/10.1007/978-981-15-7493-1.
114. Goldman, E.; Green, L. H. *Practical Handbook of Microbiology: Third Edition*; CRC Press, 2015. https://doi.org/10.1201/b17871.

115. Christopher P. Corbo. Enrichment Cultures, Selective and Differential Media | Microbiology | JoVE.
116. MISAWA, N.; KAWASHIMA, K.; KONDO, F.; KAWAMOTO, H. Development of a Combined Filtration-Enrichment Culture Followed by a One-Step Duplex PCR Technique for the Rapid Detection of Campylobacter Jejuni and C. Coli in Human Faecal Samples. *J. Med. Microbiol.* **2002,** *51* (1), 86–89. https://doi.org/10.1099/0022-1317-51-1-86.
117. Lebeaux, D.; Ghigo, J.-M.; Beloin, C. Biofilm-Related Infections: Bridging the Gap between Clinical Management and Fundamental Aspects of Recalcitrance toward Antibiotics. *Microbiol. Mol. Biol. Rev. MMBR* **2014,** *78* (3), 510–543. https://doi.org/10.1128/MMBR.00013-14.
118. Fitzgerald, G.; Williams, L. S. Modified Penicillin Enrichment Procedure for the Selection of Bacterial Mutants. *J. Bacteriol.* **1975,** *122* (1), 345–346. https://doi.org/10.1128/jb.122.1.345-346.1975.
119. Bacterial stress responses as determinants of antimicrobial resistance | Journal of Antimicrobial Chemotherapy | Oxford Academic https://academic.oup.com/jac/article/67/9/2069/878480 (accessed 16 Feb 2023).
120. Santos, A. G.; de Albuquerque, T. L.; Ribeiro, B. D.; Coelho, M. A. Z. In Situ Product Recovery Techniques Aiming to Obtain Biotechnological Products: A Glance to Current Knowledge. *Biotechnol. Appl. Biochem.* **2021,** *68* (5), 1044–1057. https://doi.org/10.1002/bab.2024.
121. Cooney, C. L. Bioreactors: Design and Operation. *Science* **1983,** *219* (4585), 728–733. https://doi.org/10.1126/science.219.4585.728.
122. Moresi, M.; Sappino, F. Electrodialytic Recovery of Some Fermentation Products from Model Solutions: Techno-Economic Feasibility Study. *J. Membr. Sci.* **2000,** *164* (1–2), 129–140. https://doi.org/10.1016/S0376-7388(99)00186-6.
123. Li, H.; Mustacchi, R.; Knowles, C. J.; Skibar, W.; Sunderland, G.; Dalrymple, I.; Jackman, S. A. An Electrokinetic Bioreactor: Using Direct Electric Current for Enhanced Lactic Acid Fermentation and Product Recovery. *Tetrahedron* **2004,** *60* (3), 655–661. https://doi.org/10.1016/j.tet.2003.10.110.
124. Gehl, J. Electroporation: Theory and Methods, Perspectives for Drug Delivery, Gene Therapy and Research. *Acta Physiologica Scandinavica.* John Wiley & Sons, Ltd April **2003,** pp 437–447. https://doi.org/10.1046/j.1365-201X.2003.01093.x.
125. Dave, V., Khirwadkar, P., & Dashora, K. A Review on Biotransformation. *Indian J. Res. Pharm. Biotechnol.* **2014,** *2* (2), 1136–1140.
126. Renton, K. W. Factors Affecting Drug Biotransformation. *Clin. Biochem.* **1986,** *19* (2), 72–75. https://doi.org/10.1016/S0009-9120(86)80051-0.
127. Esteves, F.; Rueff, J.; Kranendonk, M. The Central Role of Cytochrome P450 in Xenobiotic Metabolism—A Brief Review on a Fascinating Enzyme Family. *J. Xenobiotics* **2021,** *11* (3), 94–114. https://doi.org/10.3390/jox11030007.
128. Fernandez, E.; Perez, R.; Hernandez, A.; Tejada, P.; Arteta, M.; Ramos, J. T. Factors and Mechanisms for Pharmacokinetic Differences between Pediatric Population and Adults. *Pharmaceutics* **2011,** *3* (1), 53–72. https://doi.org/10.3390/pharmaceutics3010053.
129. Dave, V.; Khirwadkar, P.; Dashora, K. A Review on Biotransformation.
130. Štefanac, T.; Grgas, D.; Landeka Dragičević, T. Xenobiotics—Division and Methods of Detection: A Review. *J. Xenobiotics* **2021,** *11* (4), 130–141. https://doi.org/10.3390/jox11040009.

131. Louie, M. T. MICROBIAL TRANSFORMATIONS OF NATURAL PRODUCTS SA NE M SC PL O E – C EO AP LS TE S PL O E –.
132. Ishihara, K.; Hamada, H.; Hirata, T.; Nakajima, N. Biotransformation Using Plant Cultured Cells. *J. Mol. Catal. B Enzym.* **2003,** *23* (2–6), 145–170. https://doi.org/10.1016/S1381-1177(03)00080-8.
133. Godoy, C. A.; Pardo-Tamayo, J. S.; Barbosa, O. Microbial Lipases and Their Potential in the Production of Pharmaceutical Building Blocks. *Int. J. Mol. Sci.* **2022,** *23* (17), 9933. https://doi.org/10.3390/ijms23179933.
134. Bučko, M.; Gemeiner, P.; Krajčovič, T.; Hakarová, M.; Chorvát, D.; Chorvátová, A. M.; Lacík, I.; Rudroff, F.; Mihovilovič, M. D. Immobilized Cell Physiology Imaging and Stabilization of Enzyme Cascade Reaction Using Recombinant Cells *Escherichia Coli* Entrapped in Polyelectrolyte Complex Beads by Jet Break-Up Encapsulator. *Catalysts* **2020,** *10* (11), 1288. https://doi.org/10.3390/catal10111288.
135. Gao, F.; Zhang, J.-M.; Wang, Z.-G.; Peng, W.; Hu, H.-L.; Fu, C.-M. Biotransformation, a Promising Technology for Anti-Cancer Drug Development. *Asian Pac. J. Cancer Prev.* **2013,** *14* (10), 5599–5608. https://doi.org/10.7314/APJCP.2013.14.10.5599.
136. Efferth, T. Biotechnology Applications of Plant Callus Cultures. *Engineering* **2019,** *5* (1), 50–59. https://doi.org/10.1016/j.eng.2018.11.006.
137. Chigorimbo-Murefu, N. T. L.; Njoroge, M.; Nzila, A.; Louw, S.; Masimirembwa, C.; Chibale, K. Biotransformation and Biocatalysis: Roles and Applications in the Discovery of Antimalarials. *Future Med. Chem.* **2012,** *4* (18), 2325–2336. https://doi.org/10.4155/fmc.12.173.
138. Kieran, P. M.; MacLoughlin, P. F.; Malone, D. M. Plant Cell Suspension Cultures: Some Engineering Considerations. *J. Biotechnol.* **1997,** *59* (1), 39–52. https://doi.org/10.1016/S0168-1656(97)00163-6.
139. Khanpour-Ardestani, N.; Sharifi, M.; Behmanesh, M. Establishment of Callus and Cell Suspension Culture of Scrophularia Striata Boiss.: An in Vitro Approach for Acteoside Production. *Cytotechnology* **2015,** *67* (3), 475–485. https://doi.org/10.1007/s10616-014-9705-4.
140. Roy, A. Hairy Root Culture an Alternative for Bioactive Compound Production from Medicinal Plants. *Curr. Pharm. Biotechnol.* **2021,** *22* (1), 136–149. https://doi.org/10.2174/1389201021666201229110625.
141. Basso, A.; Serban, S. Industrial Applications of Immobilized Enzymes—A Review. *Mol. Catal.* **2019,** *479*, 110607. https://doi.org/10.1016/j.mcat.2019.110607.
142. Garay, L. A.; Boundy-Mills, K. L.; German, J. B. Accumulation of High-Value Lipids in Single-Cell Microorganisms: A Mechanistic Approach and Future Perspectives. *J. Agric. Food Chem.* **2014,** *62* (13), 2709–2727. https://doi.org/10.1021/jf4042134.
143. Rolfe, M. D.; Rice, C. J.; Lucchini, S.; Pin, C.; Thompson, A.; Cameron, A. D. S.; Alston, M.; Stringer, M. F.; Betts, R. P.; Baranyi, J. et al.Lag Phase Is a Distinct Growth Phase That Prepares Bacteria for Exponential Growth and Involves Transient Metal Accumulation. *J. Bacteriol.* **2012,** *194* (3), 686–701. https://doi.org/10.1128/JB.06112-11.
144. Lilly, M. D. The Design and Operation of Biotransformation Processes. In: *Recent Advances in Biotechnology*; Springer Netherlands: Dordrecht, 1992; pp 47–68. https://doi.org/10.1007/978-94-011-2468-3_3.
145. Naik, P.; Al-Khayri, J. Abiotic and Biotic Elicitors–Role in Secondary Metabolites Production through In Vitro Culture of Medicinal Plants; 2016; pp 247–277. https://doi.org/10.5772/61442.

146. Ramirez-Estrada, K.; Vidal-Limon, H.; Hidalgo, D.; Moyano, E.; Golenioswki, M.; Cusidó, R.; Palazon, J. Elicitation, an Effective Strategy for the Biotechnological Production of Bioactive High-Added Value Compounds in Plant Cell Factories. *Molecules* **2016,** *21* (2), 182. https://doi.org/10.3390/molecules21020182.
147. Resistance to antibiotics targeted to the bacterial cell wall - PMC https://www.ncbi.nlm.nih.gov/pmc/articles/PMC3945833/ (accessed 17 Feb 2023).
148. Kozłowska, E.; Hoc, N.; Sycz, J.; Urbaniak, M.; Dymarska, M.; Grzeszczuk, J.; Kostrzewa-Susłow, E.; Stępień, Ł.; Pląskowska, E.; Janeczko, T. Biotransformation of Steroids by Entomopathogenic Strains of Isaria Farinosa. *Microb. Cell Factories* **2018,** *17* (1), 71. https://doi.org/10.1186/s12934-018-0920-0.
149. Wudy, S. A.; Schuler, G.; Sánchez-Guijo, A.; Hartmann, M. F. The Art of Measuring Steroids: Principles and Practice of Current Hormonal Steroid Analysis. *J. Steroid Biochem. Mol. Biol.* **2018,** *179*, 88–103. https://doi.org/10.1016/j.jsbmb.2017.09.003.
150. Bhatti, H. N.; Khera, R. A. Biological Transformations of Steroidal Compounds: A Review. *Steroids* **2012,** *77* (12), 1267–1290. https://doi.org/10.1016/j.steroids.2012.07.018.
151. Mensah-Nyagan, A. G.; Meyer, L.; Schaeffer, V.; Kibaly, C.; Patte-Mensah, C. Evidence for a Key Role of Steroids in the Modulation of Pain. *Psychoneuroendocrinology* **2009,** *34* (SUPPL. 1), S169–S177. https://doi.org/10.1016/j.psyneuen.2009.06.004.
152. Tuba, Z.; Bardin, C. W.; Dancsi, A.; Francsics-Czinege, E.; Molnár, C.; Csörgei, J.; Falkay, G.; Koide, S. S.; Kumar, N.; Sundaram, K. et al.Synthesis and Biological Activity of a New Progestogen, 16-Methylene-17α-Hydroxy-18-Methyl-19-Norpregn-4-Ene-3,20-Dione Acetate. *Steroids* **2000,** *65* (5), 266–274. https://doi.org/10.1016/S0039-128X(99)00109-9.
153. Liu, W.-H.; Lo, C.-K. Production of Testosterone from Cholesterol Using a Single-Step Microbial Transformation of Mycobacterium Sp. *J. Ind. Microbiol. Biotechnol.* **1997,** *19* (4), 269–272. https://doi.org/10.1038/sj.jim.2900456.
154. Wang, Z.; Zhao, F.; Hao, X.; Chen, D.; Li, D. Microbial Transformation of Hydrophobic Compound in Cloud Point System. *J. Mol. Catal. B Enzym.* **2004,** *27* (4–6), 147–153. https://doi.org/10.1016/j.molcatb.2003.11.002.
155. Yan, J. L.; Lee, S. S.; Wang, K. C. Microbial Transformation of 3-Hydroxy-5,6-Cyclopropanocholestanes - An Alternative Route to 6-Methylsteroids. *Steroids* **2000,** *65* (12), 863–870. https://doi.org/10.1016/S0039-128X(00)00170-7.
156. Ambrus, G.; Ilkőy, É.; Jekkel, A.; Horváth, G.; Böcskei, Z. Microbial Transformation of β-Sitosterol and Stigmasterol into 26-Oxygenated Derivatives. *Steroids* **1995,** *60* (9), 621–625. https://doi.org/10.1016/0039-128X(95)00077-4.
157. Malaviya, A.; Gomes, J. Androstenedione Production by Biotransformation of Phytosterols. *Bioresour. Technol.* **2008,** *99* (15), 6725–6737. https://doi.org/10.1016/j.biortech.2008.01.039.
158. Manosroi, J.; Saowakhon, S.; Manosroi, A. Enhancement of 17α-Hydroxyprogesterone Production from Progesterone by Biotransformation Using Hydroxypropyl-β-Cyclodextrin Complexation Technique. *J. Steroid Biochem. Mol. Biol.* **2008,** *112* (4–5), 201–204. https://doi.org/10.1016/j.jsbmb.2008.10.003.
159. Madyastha, K. M. Novel Microbial Transformations of Steroids. *Adv. Exp. Med. Biol.* **1996,** *405*, 259–270. https://doi.org/10.1007/978-1-4613-0413-5_23/COVER/.
160. Farooq, A.; Tahara, S. Biotransformation of Testosterone and Pregnenolone Catalyzed by the Fungus Botrytis c Inerea. *J. Nat. Prod.* **2000,** *63* (4), 489–491. https://doi.org/10.1021/np990520b.

161. Ahmed, F.; Williams, R. A. D.; Smith, K. E. Microbial Transformations of Steroids. X. Cytochromes P-450 11α-Hydroxylase and C17-C20 Lyase and a 1-Ene Dehydrogenase Transform Steroids in Nectria Haematocca. *J. Steroid Biochem. Mol. Biol.* **1996,** *58* (3), 337–349. https://doi.org/10.1016/0960-0760(96)00032-5.
162. Choudhary, M. I.; Azizuddin; Atta-Ur-Rahman. Microbial Transformation of Danazol. *Nat. Prod. Lett.* **2002,** *16* (2), 101–106. https://doi.org/10.1080/10575630290019994.
163. Hu, S.; Genain, G.; Azerad, R. Microbial Transformation of Steroids: Contribution to 14α-Hydroxylations. *Steroids* **1995,** *60* (4), 337–352. https://doi.org/10.1016/0039-128X(95)00006-C.
164. Al-Awadi, S.; Afzal, M.; Oommen, S. Studies on Bacillus Stearothermophilus. Part 1. Transformation of Progesterone to a New Metabolite 9,10-Seco-4-Pregnene-3,9,20-Trione. *J. Steroid Biochem. Mol. Biol.* **2001,** *78* (5), 493–498. https://doi.org/10.1016/S0960-0760(01)00115-7.
165. Gupta, A.; Singh, V. K.; Qazi, G. N.; Kumar, A. Gluconobacter Oxydans: Its Biotechnological Applications. *Journal of Molecular Microbiology and Biotechnology*. **2001,** pp 445–456.
166. Tsitsigiannis, D. I.; Bok, J. W.; Andes, D.; Nielsen, K. F.; Frisvad, J. C.; Keller, N. P. Aspergillus Cyclooxygenase-like Enzymes Are Associated with Prostaglandin Production and Virulence. *Infect. Immun.* **2005,** *73* (8), 4548–4559. https://doi.org/10.1128/IAI.73.8.4548-4559.2005.
167. Khurana, S.; Powers, D. B.; Anderson, S.; Blaber, M. Crystal Structure of 2,5-Diketo-D-Gluconic Acid Reductase A Complexed with NADPH at 2.1-Å Resolution. *Proc. Natl. Acad. Sci. U. S. A.* **1998,** *95* (12), 6768–6773. https://doi.org/10.1073/pnas.95.12.6768.
168. Bollag, J. M.; Liu, S. Y. Fungal Degradation of 1-Naphthol. *Can. J. Microbiol.* **1972,** *18* (7), 1113–1117. https://doi.org/10.1139/m72-172.
169. Matsumura, F.; Boush, G. M.; Tai, A. Breakdown of Dieldrin in the Soil by a Micro-Organism. *Nature*. Nature Publishing Group **1968,** pp 965–967. https://doi.org/10.1038/219965a0.
170. Gunner, H. B.; Zuckerman, B. M. Degradation of "Diazinon" by Synergistic Microbial Action. *Nature*. Nature Publishing Group **1968,** pp 1183–1184. https://doi.org/10.1038/2171183a0.
171. Bollag, J.-M. Microbial Transformation of Pesticides; 1974; pp 75–130. https://doi.org/10.1016/S0065-2164(08)70570-7.
172. Ko, W. H.; Lockwood, J. L. Accumulation and Concentration of Chlorinated Hydrocarbon Pesticides by Microorganisms in Soil. *Can. J. Microbiol.* **1968,** *14* (10), 1075–1078. https://doi.org/10.1139/m68-181.
173. Liu, S.-Y.; Bollag, J.-M. Carbaryl Decomposition to 1-Naphthyl Carbamate by Aspergillus Terreus. *Pestic. Biochem. Physiol.* **1971,** *1* (3–4), 366–372. https://doi.org/10.1016/0048-3575(71)90168-4.
174. Kesner, C. D.; Ries, S. K. Diphenamid Metabolism in Plants. *Science* **1967,** *155* (3759), 210–211. https://doi.org/10.1126/science.155.3759.210.
175. Shailubhai, K. Degradation of Chlorinated Derivatives of Phenoxyacetic Acid and Benzoic Acid by Aspergillus Niger. *FEMS Microbiol. Lett.* **1983,** *18* (3), 279–282. https://doi.org/10.1016/0378-1097(83)90341-5.
176. Macrae, I. C.; Alexander, M. METABOLISM OF PHENOXYALKYL CARBOXYLIC ACIDS BY A FLAVOBACTERIUM SPECIES. *J. Bacteriol.* **1963,** *86* (6), 1231–1235. https://doi.org/10.1128/jb.86.6.1231-1235.1963.

177. EL-Kersh, T. A.; Plourde, J. R. Biotransformation of Antibiotics. I. Acylation of Chloramphenicol by Spores of Streptomyces Griseus Isolated from the Egyptian Soil. *J. Antibiot. (Tokyo)* **1976,** *29* (3), 292–302. https://doi.org/10.7164/antibiotics.29.292.
178. Fang, S.-G.; Qiang, T.; Liu, R.-J.; Xu, X.-M.; Zhang, Y.-W. ENHANCED PRODUCTION OF 6-AMINOPENICILLANIC ACID IN AQUEOUS METHYL ISOBUTYL KETONE SYSTEM WITH IMMOBILIZED PENICILLIN G ACYLASE. *Prep. Biochem. Biotechnol.* **2009,** *40* (1), 38–45. https://doi.org/10.1080/10826060903389489.
179. Čvančarová, M.; Moeder, M.; Filipová, A.; Reemtsma, T.; Cajthaml, T. Biotransformation of the Antibiotic Agent Flumequine by Ligninolytic Fungi and Residual Antibacterial Activity of the Transformation Mixtures. *Environ. Sci. Technol.* **2013,** *47* (24), 14128–14136. https://doi.org/10.1021/es403470s.
180. Pajouhesh, H.; Lenz, G. R. Medicinal Chemical Properties of Successful Central Nervous System Drugs. *NeuroRX* **2005,** *2* (4), 541–553. https://doi.org/10.1602/neurorx.2.4.541.
181. Shelly, S.; Pritam, C.; Ieshita, P.; S; Ipan, C.; Pompee, C.; a; Debashis, B.; yopadhyay; Kamala, D. et al.Microbial Transformation of Xenobiotics for Environmental Bioremediation. *Afr. J. Biotechnol.* **2009,** *8* (22), 6016–6027. https://doi.org/10.5897/AJB09.740.
182. Chowdhury, A.; Pradhan, S.; Saha, M.; Sanyal, N. Impact of Pesticides on Soil Microbiological Parameters and Possible Bioremediation Strategies. *Indian J. Microbiol.* **2008,** *48* (1), 114–127. https://doi.org/10.1007/s12088-008-0011-8.
183. Di Gioia, D.; Michelles, A.; Pierini, M.; Bogialli, S.; Fava, F.; Barberio, C. Selection and Characterization of Aerobic Bacteria Capable of Degrading Commercial Mixtures of Low-Ethoxylated Nonylphenols. *J. Appl. Microbiol.* **2007,** *104* (1), 231–242. https://doi.org/10.1111/j.1365-2672.2007.03541.x.
184. Qiu, Y.-L.; Sekiguchi, Y.; Imachi, H.; Kamagata, Y.; Tseng, I.-C.; Cheng, S.-S.; Ohashi, A.; Harada, H. Identification and Isolation of Anaerobic, Syntrophic Phthalate Isomer-Degrading Microbes from Methanogenic Sludges Treating Wastewater from Terephthalate Manufacturing. *Appl. Environ. Microbiol.* **2004,** *70* (3), 1617–1626. https://doi.org/10.1128/AEM.70.3.1617-1626.2004.
185. Zhang, C.; Bennett, G. N. Biodegradation of Xenobiotics by Anaerobic Bacteria. *Appl. Microbiol. Biotechnol.* **2005,** *67* (5), 600–618. https://doi.org/10.1007/s00253-004-1864-3.
186. Kavanagh, E.; Winn, M.; Gabhann, C. N.; O'Connor, N. K.; Beier, P.; Murphy, C. D. Microbial Biotransformation of Aryl Sulfanylpentafluorides. *Environ. Sci. Pollut. Res.* **2014,** *21* (1), 753–758. https://doi.org/10.1007/s11356-013-1985-2.
187. Jones, D. M.; Douglas, A. G.; Parkes, R. J.; Taylor, J.; Giger, W.; Schaffner, C. The Recognition of Biodegraded Petroleum-Derived Aromatic Hydrocarbons in Recent Marine Sediments. *Mar. Pollut. Bull.* **1983,** *14* (3), 103–108. https://doi.org/10.1016/0025-326X(83)90310-7.
188. Daugulis, A. J.; McCracken, C. M. Microbial Degradation of High and Low Molecular Weight Polyaromatic Hydrocarbons in a Two-Phase Partitioning Bioreactor by Two Strains of Sphingomonas Sp. *Biotechnol. Lett.* **2003,** *25* (17), 1441–1444. https://doi.org/10.1023/A:1025007729355.
189. Shimizu, S. Vitamins and Related Compounds: Microbial Production. In: *Biotechnology*; Wiley-VCH Verlag GmbH: Weinheim, 2008; pp 318–340. https://doi.org/10.1002/9783527620937.ch11.

190. Gurung, N.; Ray, S.; Bose, S.; Rai, V. A Broader View: Microbial Enzymes and Their Relevance in Industries, Medicine, and Beyond. *BioMed Res. Int.* **2013,** *2013*, 1–18. https://doi.org/10.1155/2013/329121.
191. Mahmood, Z.; Mahmood, S. Microbial Healthcare Products. In: *Microbial Biotechnology*; 2014; pp 239–278. https://doi.org/10.1201/b17587-12.
192. Sun, X.; Shen, X.; Jain, R.; Lin, Y.; Wang, J.; Sun, J.; Wang, J.; Yan, Y.; Yuan, Q. Synthesis of Chemicals by Metabolic Engineering of Microbes. *Chem. Soc. Rev.* **2015,** *44* (11), 3760–3785. https://doi.org/10.1039/C5CS00159E.
193. van der Maarel, M. J. E. C.; van der Veen, B.; Uitdehaag, J. C. M.; Leemhuis, H.; Dijkhuizen, L. Properties and Applications of Starch-Converting Enzymes of the α-Amylase Family. *J. Biotechnol.* **2002,** *94* (2), 137–155. https://doi.org/10.1016/S0168-1656(01)00407-2.
194. V. Aiyer, P. Amylases and Their Applications. *Afr. J. Biotechnol.* **2005,** *4* (13), 1525–1529.
195. Blanco, C. A.; Caballero, I.; Barrios, R.; Rojas, A. Innovations in the Brewing Industry: Light Beer. *Int. J. Food Sci. Nutr.* **2014,** *65* (6), 655–660. https://doi.org/10.3109/09637486.2014.893285.
196. James, J.; Simpson, B. K.; Marshall, M. R. Application of Enzymes in Food Processing. *Crit. Rev. Food Sci. Nutr.* **1996,** *36* (5), 437–463. https://doi.org/10.1080/10408399609527735.
197. Raveendran, S.; Parameswaran, B.; Ummalyma, S. B.; Abraham, A.; Mathew, A. K.; Madhavan, A.; Rebello, S.; Pandey, A. Applications of Microbial Enzymes in Food Industry. *Food Technol. Biotechnol.* **2018,** *56* (1), 16–30. https://doi.org/10.17113/ftb.56.01.18.5491.
198. Mlichová, Z.; Rosenberg, M. Current Trends of β-Galactosidase Application in Food Technology. *J. Food Nutr. Res.* **2006,** *45* (2), 47–54.
199. Abbas, A.; Irfan, M.; Khan, S.; Hassan, A.; Khan, S.; Javed, R.; Ali, S. Microbes: Role in Industries, Medical Field and Impact on Health. *Saudi J Med Pharm Sci* **2021,** *7* (6), 278–282. https://doi.org/10.36348/sjmps.2021.v07i06.010.
200. Vartoukian, S. R.; Palmer, R. M.; Wade, W. G. Strategies for Culture of 'Unculturable' Bacteria. *FEMS Microbiol. Lett.* **2010,** no-no. https://doi.org/10.1111/j.1574-6968.2010.02000.x.
201. Thallinger, B.; Prasetyo, E. N.; Nyanhongo, G. S.; Guebitz, G. M. Antimicrobial Enzymes: An Emerging Strategy to Fight Microbes and Microbial Biofilms. *Biotechnol. J.* **2013,** *8* (1), 97–109. https://doi.org/10.1002/biot.201200313.
202. Bose, S.; Ghosh, A. K. Biofilms: A Challenge to Medical Science. *J. Clin. Diagn. Res.* **2011,** *5* (1), 127–130.
203. Jozala, A. F.; Geraldes, D. C.; Tundisi, L. L.; Feitosa, V. de A.; Breyer, C. A.; Cardoso, S. L.; Mazzola, P. G.; de Oliveira-Nascimento, L.; Rangel-Yagui, C. de O.; Magalhães, P. de O. et al.Biopharmaceuticals from Microorganisms: From Production to Purification. *Braz. J. Microbiol.* **2016,** *47*, 51–63. https://doi.org/10.1016/j.bjm.2016.10.007.
204. Zhao, K.; Ping, W.; Li, Q.; Hao, S.; Zhao, L.; Gao, T.; Zhou, D. Aspergillus Niger Var. Taxi, a New Species Variant of Taxol-Producing Fungus Isolated from Taxus Cuspidata in China. *J. Appl. Microbiol.* **2009,** *107* (4), 1202–1207. https://doi.org/10.1111/j.1365-2672.2009.04305.x.
205. Svahn, K. S.; Chryssanthou, E.; Olsen, B.; Bohlin, L.; Göransson, U. Penicillium Nalgiovense Laxa Isolated from Antarctica Is a New Source of the Antifungal Metabolite

Amphotericin B. *Fungal Biol. Biotechnol.* **2015,** *2* (1), 1. https://doi.org/10.1186/s40694-014-0011-x.

206. Chanalia, P.; Gandhi, D.; Jodha, D.; Singh, J. Applications of Microbial Proteases in Pharmaceutical Industry. *Rev. Med. Microbiol.* **2011,** *22* (4), 96–101. https://doi.org/10.1097/MRM.0b013e3283494749.
207. Leader, B.; Baca, Q. J.; Golan, D. E. Protein Therapeutics: A Summary and Pharmacological Classification. *Nat. Rev. Drug Discov.* **2008,** *7* (1), 21–39. https://doi.org/10.1038/nrd2399.
208. Gupta, R.; Gigras, P.; Mohapatra, H.; Goswami, V. K.; Chauhan, B. Microbial α-Amylases: A Biotechnological Perspective. *Process Biochem.* **2003,** *38* (11), 1599–1616. https://doi.org/10.1016/S0032-9592(03)00053-0.
209. de Souza, P. M.; e Magalhães, P. de O. Application of Microbial α-Amylase in Industry - a Review. *Braz. J. Microbiol.* **2010,** *41* (4), 850–861. https://doi.org/10.1590/s1517-83822010000400004.
210. Tanyildizi, M. S.; Özer, D.; Elibol, M. Optimization of α-Amylase Production by Bacillus Sp. Using Response Surface Methodology. *Process Biochem.* **2005,** *40* (7), 2291–2296. https://doi.org/10.1016/j.procbio.2004.06.018.
211. Saxena, R. K.; Dutt, K.; Agarwal, L.; Nayyar, P. A Highly Thermostable and Alkaline Amylase from a Bacillus Sp. PN5. *Bioresour. Technol.* **2007,** *98* (2), 260–265. https://doi.org/10.1016/j.biortech.2006.01.016.
212. Nielsen, J. E.; Borchert, T. V. Protein Engineering of Bacterial α-Amylases. *Biochim. Biophys. Acta BBA - Protein Struct. Mol. Enzymol.* **2000,** *1543* (2), 253–274. https://doi.org/10.1016/S0167-4838(00)00240-5.
213. Mitidieri, S.; Souza Martinelli, A. H.; Schrank, A.; Vainstein, M. H. Enzymatic Detergent Formulation Containing Amylase from Aspergillus Niger: A Comparative Study with Commercial Detergent Formulations. *Bioresour. Technol.* **2006,** *97* (10), 1217–1224. https://doi.org/10.1016/j.biortech.2005.05.022.
214. Ahlawat, S.; Dhiman, S. S.; Battan, B.; Mandhan, R. P.; Sharma, J. Pectinase Production by Bacillus Subtilis and Its Potential Application in Biopreparation of Cotton and Micropoly Fabric. *Process Biochem.* **2009,** *44* (5), 521–526. https://doi.org/10.1016/j.procbio.2009.01.003.
215. Chi, Z.; Chi, Z.; Liu, G.; Wang, F.; Ju, L.; Zhang, T. Saccharomycopsis Fibuligera and Its Applications in Biotechnology. *Biotechnol. Adv.* **2009,** *27* (4), 423–431. https://doi.org/10.1016/j.biotechadv.2009.03.003.
216. Toksoy Öner, E. Optimization of Ethanol Production from Starch by an Amylolytic Nuclear PetiteSaccharomyces Cerevisiae Strain. *Yeast* **2006,** *23* (12), 849–856. https://doi.org/10.1002/yea.1399.
217. Abbott, D. A.; Suir, E.; Duong, G.-H.; de Hulster, E.; Pronk, J. T.; van Maris, A. J. A. Catalase Overexpression Reduces Lactic Acid-Induced Oxidative Stress in Saccharomyces Cerevisiae. *Appl. Environ. Microbiol.* **2009,** *75* (8), 2320–2325. https://doi.org/10.1128/AEM.00009-09.
218. Barynin, V. V.; Whittaker, M. M.; Antonyuk, S. V.; Lamzin, V. S.; Harrison, P. M.; Artymiuk, P. J.; Whittaker, J. W. Crystal Structure of Manganese Catalase from Lactobacillus Plantarum. *Structure* **2001,** *9* (8), 725–738. https://doi.org/10.1016/S0969-2126(01)00628-1.

219. Kaushal, J.; Mehandia, S.; Singh, G.; Raina, A.; Arya, S. K. Catalase Enzyme: Application in Bioremediation and Food Industry. *Biocatal. Agric. Biotechnol.* **2018,** *16*, 192–199. https://doi.org/10.1016/j.bcab.2018.07.035.
220. Grigoras, A. G. Catalase Immobilization—A Review. *Biochem. Eng. J.* **2017,** *117*, 1–20. https://doi.org/10.1016/j.bej.2016.10.021.
221. Banci, L. Structural Properties of Peroxidases. *J. Biotechnol.* **1997,** *53* (2–3), 253–263. https://doi.org/10.1016/S0168-1656(97)01677-5.
222. Guzik, U.; Hupert-Kocurek, K.; Wojcieszyńska, D. Immobilization as a Strategy for Improving Enzyme Properties-Application to Oxidoreductases. *Molecules* **2014,** *19* (7), 8995–9018. https://doi.org/10.3390/molecules19078995.
223. Regalado, C.; García-Almendárez, B. E.; Duarte-Vázquez, M. A. Biotechnological Applications of Peroxidases. *Phytochem. Rev.* **2004,** *3* (1–2), 243–256. https://doi.org/10.1023/B:PHYT.0000047797.81958.69.
224. Houde, A.; Kademi, A.; Leblanc, D. Lipases and Their Industrial Applications: An Overview. *Appl. Biochem. Biotechnol. - Part Enzyme Eng. Biotechnol.* **2004,** *118* (1–3), 155–170. https://doi.org/10.1385/ABAB:118:1-3:155.
225. Razzaq, A.; Shamsi, S.; Ali, A.; Ali, Q.; Sajjad, M.; Malik, A.; Ashraf, M. Microbial Proteases Applications. *Front. Bioeng. Biotechnol.* **2019,** *7* (JUN), 1–20. https://doi.org/10.3389/fbioe.2019.00110.
226. Rawlings, N. D. MEROPS: The Peptidase Database. *Nucleic Acids Res.* **2004,** *32* (90001), 160D – 164. https://doi.org/10.1093/nar/gkh071.
227. Brandelli, A.; Daroit, D. J.; Riffel, A. Biochemical Features of Microbial Keratinases and Their Production and Applications. *Appl. Microbiol. Biotechnol.* **2010,** *85* (6), 1735–1750. https://doi.org/10.1007/s00253-009-2398-5.
228. Arroyo, M.; De La Mata, I.; Acebal, C.; Castillón, M. P. Biotechnological Applications of Penicillin Acylases: State-of-the-Art. *Appl. Microbiol. Biotechnol.* **2003,** *60* (5), 507–514. https://doi.org/10.1007/s00253-002-1113-6.
229. Soufer, D. Protecting Penicillin from Penicillinase. *Microbios* **1986,** *45* (184–185), 193–198.
230. Butler, K. T.; Davies, D. W.; Cartwright, H.; Isayev, O.; Walsh, A. Machine Learning for Molecular and Materials Science. *Nature* **2018,** *559* (7715), 547–555. https://doi.org/10.1038/s41586-018-0337-2.
231. Sambyal, K.; Singh, R. V. Production Aspects of Testosterone by Microbial Biotransformation and Future Prospects. *Steroids* **2020,** *159* (December 2019), 108651. https://doi.org/10.1016/j.steroids.2020.108651.
232. Fernández-Cabezón, L.; Galán, B.; García, J. L. New Insights on Steroid Biotechnology. *Front. Microbiol.* **2018,** *9* (MAY). https://doi.org/10.3389/fmicb.2018.00958.
233. Chang, Y. S. Recent Developments in Microbial Biotransformation and Biodegradation of Dioxins. *J. Mol. Microbiol. Biotechnol.* **2008,** *15* (2–3), 152–171. https://doi.org/10.1159/000121327.
234. Bianchini, L. F.; Arruda, M. F. C.; Vieira, S. R.; Campelo, P. M. S.; Grégio, A. M. T.; Rosa, E. A. R. Microbial Biotransformation to Obtain New Antifungals. *Front. Microbiol.* **2015,** *6* (DEC), 1–12. https://doi.org/10.3389/fmicb.2015.01433.
235. Călinoiu, L. F.; Cătoi, A. F.; Vodnar, D. C. Solid-State Yeast Fermented Wheat and Oat Bran as a Route for Delivery of Antioxidants. *Antioxidants* **2019,** *8* (9), 1–2. https://doi.org/10.3390/antiox8090372.

Glossary

A

Aberrant: A growth that deviates from the normal or usual type; as exceptional growths (aberrations) which occur in some tissue cultures.

Abort: To stop or fail to reach full development (maturity) or to develop imperfectly, as some hybrids are prone to embryo abortion or as some attempts to culture a plant may fail (abortive attempt).

Accession: Plant or seed sample, strain, or population held in a genebank or breeding program for conservation and use.

Acetosyringone: One of the phenolic compounds released by wounded plant cells which can stimulate the transcription of the vir region of the Ti plasmid in the initial stages of transfer of DNA from *Agrobacterium* to the plant cell.

Activator: A substance or physical agent that stimulates transcription of a specific gene or operon.

Active Genepool Collection: A collection of germplasm used for regeneration, multiplication, distribution, characterization, and evaluation. Ideally, germplasm in the active collection should be maintained in sufficient quantity to be available on request. Active collection germplasm is commonly duplicated in a base collection and is often stored under medium- to long-term storage.

Adenosine Triphosphate (ATP): It is an organic compound that provides energy to carry out processes in living cells.

Adsorb: The adhesion of atoms, ions, or molecules from a liquid, gas, or dissolved to a solid surface. This process creates a film of the adsorbate (the adsorbed material) on the surface of the adsorbent (the adsorbing agent, such as activated charcoal). The process is adsorption.

Aerate: To supply or mix with air or gas. The process is aeration. Aerobic or acrobiotic refers to an organism that lives in or a process occurring in the presence of molecular oxygen.

Agar or agar-agar: A gelatinous polysaccharide obtained from the red alga *Gelidium corneum* and from several other red algae. It is a solidifying agent that when mixed with nutrient media (0.6–1%), forms a gel for growing tissue-cultured plants and for other purposes. It ranges in quality from relatively inert to very impure (complex, undefined). Its firmness as a gelling agent is affected by medium pH (it is softer when the medium is more dilute). Agar gels melt at about 100°C but solidify at about 44°C.

Age: The period in the life cycle of an organism: the process of growing older, mature. Or the state of being old or senescent.

Agenesis: The absence of development.

Agglutinate: To unite or combine into a group of mass. as protoplasts and bacteria, in the presence of specific antibodies, tend to stick to one another. The process is agglutination.

Agglutinin: An antibody capable of clumping bacteria or other cells.

Agrenet: The computer communications network established by the Agriculture and Food Research Council (AFRC) for the agriculture and botanical research community in the United Kingdom.

Agrobacterium tumefaciens: The bacterium causing crown gall disease in plants, inducing tumors to form. Tumors may be cultured in vitro and are useful in the study of this disease. The Ti plasmid of *A. tumefaciens* is known to cause the disease, and a small portion of it is used as a vector in the genetic modification of higher plant cells. Novel DNA sequences are spliced into the plasmid DNA segment, the DNA is circularized and introduced into the cultured plant cells using *Agrobacterium*.

Agroinfection: A method for initiating viral infection by inoculating host plants containing viral-specific DNA within the T-DNA.

Airlift Fermenter: A cylindrical fermentation vessel in which the cells are mixed by air, which is introduced at the base of the vessel and rises through the column of culture medium. The cell suspension circulates around of the column as a consequence of the gradient of air bubbles in different parts of the reactor.

Aliquot or aliquot part: It contains an exact number of times in something else. An evenly divided unit, portion, or sample (fraction) of the whole.

Alphanumeric: The set of symbols used to represent letters of the alphabet and the integers 0–9.

Ambiguity Code: A character other than the standard codes for representing the DNA and RNA bases (ACGT and U) used to indicate a base with an ambiguous identity.

Amphidiploid: Hybrids between two species or genera containing the sum of the complete chromosome complements, which can be of different ploidy levels, of each parent.

Amplification Fragment Length Polymorphism (AFLP): A variant DNA amplification product produced by DAF, PCR, or RAPD of different sizes.

Anther Culture: A type of tissue culture technique of single pollen grains or the anther, containing the male gametophytes or micropores, with the objective of producing monoploid plants when cultured in nutrients.

Antibiosis: The prevention of the growth or development of an organism by a substance or another organism.

Antigen: A compound that induces the production of antibodies.

Anti-Idiotype Antibody: An antibody that has the properties of an antigen.

Antisense Construct: A plasmid that contains the coding regions of a gene placed adjacent to a promotor in such an orientation as to cause the gene to be transcribed from the DNA strand complementary to that from which RNA is normally transcribed.

Antisense DNA: (1) The sequence of chromosomal DNA that is transcribed. (2) A DNA sequence serves as the template for producing messenger RNA (mRNA).

Antisense RNA: A RNA sequence that is complementary to all or part functional RNA (mRNA).

Artificial Seed: These are artificial encapsulated somatic embryo or other vegetative parts which can serve as a seed and converted into a plant under in vitro or in vivo conditions.

Asepsis: Without infection or contaminating microorganisms.

Aseptic Technique: Procedures used to prevent the introduction of fungi, bacteria, viruses, mycoplasma, or other microorganisms into cell, tissue, and organ cultures.

Aseptic: Free of pathogens, contaminants, algae, bacteria, fungi, viruses, and other. Absence of all microorganisms. Asepsis is the fundamental requirement for plant tissue culture (aseptic culture).

Asparagine (Asp, $C_4H_8N_{203}$ mw 132.12): An amino acid occasionally included in plant tissue culture media, as a source of reduced nitrogen.

Aspirate: To draw something in or out, up or through using suction or a vacuum; as aspiration (vacuum) may be used in the disinfection process to draw disinfectant into the surface layers of plant tissue.

Assay: The substance to be analyzed or the process of examining or testing it (chemically or by other means).

Assembler: It is a program that takes basic computer instructions and converts them into a pattern of bits that the computer's processor can use to perform its basic operations.

Asymmetric Somatic Hybridization/Partial Genome Transfer: The production of viable and fertile hybrid plants by irradiating protoplasts from one parental line (the donor) in order to pulverize the chromosomes and facilitate the elimination of substantial amount of the donor genome. Thereafter, the irradiated donor genome may be transferred into the genome of the recipient by protoplast fusion which can be carried out using three different methods: chemicals, electrical field, or UV laser microbeam. A partial genome transfer system has the potential use for the transfer of epigenetic characters.

Attachment Efficiency: The percentage of cells plated (seeded, inoculated) which attach to the surface of the culture vessel within a specified period. The conditions under which such a determination is made should always be stated.

Autoclave: A closed chamber in which to heat substances under pressure to above their boiling points, to sterilize utensils, liquids, glassware, etc., using steam. The routine method uses steam pressure of 103.4 X 10, Pa at 121°C for 15 minutes, or longer for large volumes to reach temperature. Heat labile components of tissue culture media cannot be autoclaved. These are commonly filter-sterilized. To carry out the process of autoclaving.

Autonomous Replicating Sequence (ARS): Any cloned DNA sequence that initiates and supports extrachromosomal replication of a DNA molecule in a host cell: often used in yeast cells. Also called autonomous(ly) replicating segment, autonomously replicating sequence.

Autoradiography: A technique that captures the image formed in a photographic emulsion as a result of the emission of either light or radioactivity from a labeled component that is placed next to unexposed film.

Auxin: It is a group of naturally occurring and artificially synthesized plant hormones. They play an important role in the regulation of plant growth. It causes cell enlargement and elongation but in excised, cultured systems they may promote cell division.

Auxotroph: A microorganism or plant cell line possessing atypical nutritional requirements for some item(s) (growth factors) in addition to those required by the control or wild-type which can synthesize them; as auxotrophic cells requiring certain amino acids.

Axenic Culture: A culture without foreign or undesired life forms. An axenic culture may include the purposeful co-cultivation of different types of cells, tissues or organisms.

Axenie: A pure culture of one species. This implies that cultures are free of microorganisms (aseptic or germ-free).

Axis: The main plant stem.

B

B-Replicase: It helps in the replication of viral RNA.

B Vitamin: One of a complex of vitamins made in plants and thought to be essential for healthy growth. Some are routinely added to plant tissue culture media to promote growth. The latter include thiamine (B), niacin (B3), adenine (B), and pyridoxine (B).

Bactericide: A substance or agent that kills bacteria (usually rapidly) is bactericidal. Batch culture: A cell suspension grown in a liquid medium of a set volume.

Beteriostat: A chemical or other agent that does not kill but prevents bacterial growth and multiplication: is Bacteriostatic.

Bergmann, L. (1960): He was the first to obtain callus by plating cells from suspension cultures onto the solid medium.

Bioassay: A biological assay performed on living cells or a living organism; sometimes used to detect minute amounts of substances which influence or are essential to growth.

Biotechnology: The industrial use of biological processes such as yeast fermentation for alcohol production or plant cell culture for extraction of secondary products. The scope of biotechnology has been dramatically increased through genetic engineering.

Biotin ($C_{10}H_{1603}N_{25}$ mw 244.31): Part of the vitamin B complex. It is also called vitamin H. It is present in all living cells, bound to polypeptides or proteins, and is of important in fat, protein, and carbohydrate metabolism. It is a common addition to plant tissue culture media.

Bleach: A fluid, powder, or other whitening (bleaching) or cleaning agent. Bleach contains calcium hypochlorite or sodium hypochlorite and is a common disinfectant used for cleaning working surfaces, tools, and plant materials in plant tissue culture.

Buffer: A substance in solution, or system that resists pH change, (despite the addition of some small amount of acid or base). The degree to which pH change is resisted is a measure of the solution's buffering capacity.

Bung: A stopper or closure.

Burgeon: To flourish, bud, sprout, or grow.

Calcium Hypochlorite (Ca(OCI)2.4H20 mw 215): An algicide, bactericide, and fungicide capable of disinfecting and bleaching. It is used in dilute solution (5–10% w/v) as a plant tissue disinfectant often when agitating or using a vacuum. Tissue damage occurs if contact is prolonged.

Callus, pl. calli or Calluses: It is defined as a group of cells derived from component source tissue that cultures under in vitro conditions to form an undifferentiated mass of cells and vary widely in texture, appearance, and rate of growth; a function of the tissue type (species and explant) and the composition of the medium, the process of callus formation is callogenesis.

Cannula: A small tube for insertion into tissue; as a cork borer.

C

Carbon Source: A source of the non-metallic element carbon (C); as organic substances like sugars taken up and metabolized by plant tissue cultures.

Casamino Acid: An amino acid obtained through the digestion (acid hydrolysis) of the milk protein casein. Protein digest yields casamino acids and other substances. These are undefined constituents of some plant tissue culture media.

Casein Hydrolysate (edamin): A milk protein digest product composed of amino acids (casamino acids) and other substances. This complex (undefined) product is sometimes used as an addition (0.02–2.10%) in nutrient solutions used in plant tissue culture as a non-specific source of organic nitrogen.

Casein: It is the chief protein in milk and the essential ingredient of cheese, in pure form, it is an amorphous white solid.

Caulogenesis: Shoot formation; as de novo shoot development from callus.

Cell Count: The number of cells per unit suspension volume or callus weight. Tissue is treated with chromic acid (5–8%) or pectinase (0.25%) for up to 15 minutes followed by mechanical dispersion, then cell numbers are estimated with a hemocytometer (hemocytometer).

Cell Culture: The culture of single or groups of cells on solid or dispersed in liquid nutrient media (cell suspension cultures) usually following a set of defined growing conditions (protocols).

Cell Hybridization: The technique of combining two cells from different tissue or species in a cell culture.

Cell Line: Developmental history or descent, through cell division from a single original cell; as callus may constitute a group of cells, all descendants are from a single cell plating. Numerous cell lines may be present in a culture. Any deviation in culture technique may favor one cell line over another.

Cell Number: The absolute number or approximation of the number of cells per unit area of a culture or medium volume.

Cell Selection: Selection within a group of genetically different cells; involves competition between cells often under some stress. Selection criteria may involve cell viability a unique phenotype or biochemical activity; or another basis for choice. Select cells or cell lines are generally relocated to fresh medium for continued selection and often are exposed to an increased level of the stress agent. The final objective is usually to regenerate plants from those select cells, in the hope that the plants will exhibit the traits selected for at the cellular level.

Cell Suspension: Cells and small aggregates of cells suspended in a liquid medium; as in cell suspension cultures. Explants, or callus derived from them are transferred to a liquid medium and the cultures are then agitated on a mechanical shaker. The ensuing single cells and small cell clusters are used for a number of purposes in plant tissue culture; as in single-cell cloning.

Centrifuge: An apparatus used for separating particles from suspension using centrifugal force. Balanced tubes containing the suspension are rapidly rotated causing sedimentation of particles to the bottom of the tubes (pelleting) based on their weight and the density of the suspending medium.

Chelate: A chemical compound with which metal atoms may be combined in order to prevent them from precipitating out of solution and so being unavailable to plants. For example as ethylenediamine tetra-acetic acid, disodium salt (NazEDTA) is complexed to iron (and other metal ions) in nutrient solutions used for plant tissue culture. The process is chelation.

Chemically Defined Medium: A nutrient medium for plant tissue culture in which all of the chemical components are fully known and stated. Quality (high purity) chemicals are utilized, and no undefined constituents are included.

Chemotherapeutant: A chemical used to pre-treat diseased source plants prior to excision or incorporated into media to support some therapeutic objective; as malachite green or Virazole (ribavirin) are used to eliminate the virus in meristem tip culture. The process is chemotherapy.

Chimera or Chimaera: A plant or tissue composed of more than one kind of genetic tissue. In a periclinal chimera, one genotype occurs in a superficial layer, covering a genetically different core. In a medicinal chimera, one genotype occurs in a localized part of the plant and the rest is occupied by another genotype. In a sectorial chimera, two genotypes share distinct sectors of the plant. Chimeras can occur naturally: be established artificially by grafting, or developed through somatic mutation by chemicals such as colchicine, or by other means in plant tissue cultures.

Choline Chloride (CsH14CINO mw 139.63): An alkaloid, (B complex vitamin), occasionally added to plant tissue culture media. Choline, found in many plant organs, is the basic constituent of lecithin.

Clonal Multiplication or Clonal Propagation: Vegetative (asexual) propagation from a single cell or plant.

Clone: (1) A basic category of cultivar. The original cultivar or variety. Not to be confused with the source clone, indicating different originals within the clone. Members have a common origin, are the extension of a single cell (via mitosis) or plant have been produced by vegetative means only. Genetic uniformity is accepted. A clone may have one or more sources. (2) While the expectation is that members of a clone are both phenotypically and genotypically identical, this assumption is not always true and members may not be genetically or phenotypically equivalent (variants). The objective of much tissue culture propagation is to propagate very large numbers of selected plants with the same genotype. The process is cloning.

Closed Continuous Culture: A cell suspension culture with a continuous influx of fresh medium, maintained at constant volume by the efflux of spent medium. All cells are retained within the unit. See continuous culture.

Cocking, E.C. (1960): Obtained the first higher plant protoplasts using root cells subjected to fungal cellulase and demonstrated new cell wall regeneration on protoplasts from tomato fruit locule tissue.

Coconut Milk or Coconut Water: it is the liquid endosperm from the center of the coconut seed. A complex, undefined addendum of variable quality and effects in some nutrient solutions (2–15%, v/v) for plant tissue culture. It has growth-promoting effects and cell division factors. It is replaceable in some cases by cytokinins and/or sugar. It was first used by van Overbeek et al. (1941) to stimulate Datura embryo cultures.

Coculture: It is the act or process of culturing two types of cells. it is done in various dual culture systems or the nurse culture technique.

Colony: (1) A group of interdependent cells or organisms. 2. An aggregate of cells developed from a single cell; as in single cell platings, a clone composed of one cell line.

Complex Substance: A complicated and undefined substance; as in some addenda to nutrient solutions used in plant tissue culture. Examples include protein hydrolysate, yeast or malt extracts, endosperm from corn or coconuts, juices like oranges or tomato juice and others.

Contaminant: An undesirable bacterial, fungal, or algal microorganism accidentally introduced to a culture or culture medium. It may overgrow the plant cells or inhibit their growth through the release of toxic contaminate: To accidentally introduce a substance or organism (contaminant) into a medium or culture of metabolites. Rigorous exclusion of potential contaminants must be practiced in plant tissue culture. The process is contamination.

Continuous Culture: A ceil suspension culture with a continuous influx of fresh medium, maintained at constant volume by the efflux of spent medium (closed continuous) or with the efflux of cells and spent medium (open continuous).

Control: 1. The untreated plant or unchanged (standard) protocol or treatment for comparison with the experimental treatment. 2. To direct or regulate; as in induction of organogenesis in cultures through hormone regulation.

Controlled Environment: A chamber, room, or situation in which the environmental parameters of light, temperature, and relative humidity are controlled. The partial pressure of gases may also be controlled.

Corn Milk: Corn endosperm tissue. An undefined complex addendum to some plant tissue culture media

Critical concentration: The chemical concentration above or below which reaction or developmental process will not proceed.

Crown Gall: A disease of plants in which tumors form. The causal agent is the bacterium, *Agrobacterium tumefaciens*. Gall tissue has been grown in vitro. A tumor inducing portion of the bacterial genome (theTi plasmid) my be used experimentally as a genetic tool to incorporate (vector) genetic information into plant cells.

Cryoprotectant: A substance used to protect biological tissue from freezing and thawing damage to cells as they are frozen or defrosted. They are classified either as permeating (glycerol and dimethyl sulfoxide) or non-permeating (sugars, dextran, ethylene glycol, polyvinyl pyrrolidone and hydroxyethyl starch) agents.

Culture Alteration: A persistent change in a cultured cell or tissue anatomy or physiology, including a change in one or more nutritional requirements or a change in its proliferative capacity.

Culture Medium: A prepared nutrient solution (substrate) that may be chemically defined for growing plant tissues or other organisms in vitro.

Culture Room: A controlled-environment room (light, temperature, relative humidity, often air conditioned, etc.) for incubation of plant cultures.

Culture Tube: A test tube used to contain medium and an explant or culture derived from it, usually incubated in a controlled environment. Culture vessel: A container used to hold medium and an explant or culture derived from it, usually incubated in a controlled environment.

Culture: In vitro aseptic cultivation of microorganisms or of cells, tissues, or organs of plants, animals in or on prepared nutrient media under controlled, aseptic conditions.

Cybrid: A cell or plant cytoplasmic hybrid (heteroplast) with the nucleus of one and cytoplasmic organelles of another, or of both cells or plants.

Cysteine Hydrochloride (C3H7NOS. HCI mw 157.63): An occasion, amino acid (cysteine) salt additive in plant tissue culture media. It is included primarily for its antioxidant properties.

Cytochimera or Chromosomal Chimera: Cells within a tissue possessing different chromosomal numbers as when a callus consists of diploid and polyploid regions of cells.

Cytodifferentiation: A functional and/or morphological differentiation occurring in cells during ontogenesis, affecting their phenotype.

Cytolysis: Cell dissolution or disintegration.

Cytoplasmic Hybrid or Cybrid: A cell or plant hybrid (heteroplast) with the nucleus of one and the cytoplasmic organcils of another or of both cells or plants.

Cytoplasmic Variant: A maternally inherited change in a cellular trait.

Cytotoxic: A chemical or other agent toxic to cells.

D

De novo: (Latin: from the beginning, anew). Arising, sometimes spontaneously, from unknown or very simple precursors.

Decereration Phase: It is the declining growth rate phase following the linear phase

Decontaminate: To free from contamination or surface sterilize.

Dedifferentiation: It is a cellular process by which cells grow in reverse, from a partially differentiated stage to a less differentiated stage within their own lineage. Cell division leading to the formation of small, microvacuolate, isodiametric cells with prominent nuclei, that are capable of organized development; as when adventitious buds or roots develop on mature tissue.

Dehumidifier: An apparatus that removes moisture from the air.

Deionized Water: Water that has been passed through an ion exchange device to remove soluble minerals and some organic salts. The process is deionization.

Deletion: The lost portion of a chromosome or a nucleotide sequence in nucleic acid.

Derepression: The mechanism by which repression is alleviated; as when a repressing metabolite is removed, resulting in the increased level of a protein or enzyme.

Detergent: A cleaning agent; one of numerous synthetic cleaning preparations chemically different from soap. Detergents are commonly used to prepare work areas for aseptic plant tissue culture. manipulations and to remove dirt and microorganisms from plant material prior to explantation.

Dewar Flask: A double-walled glass flask used to keep liquids at other than ambient temperatures; as a thermos does.

Direct Embryogenesis: Embryoid formation directly on the surface of zygotic or somatic embryos or on seeding plant tissues in culture, without an intervening callus phase.

Direct Organogenesis: Organ formation directly on the surface of relatively large intact explants, without an intervening callus phase.

Dispense: To give, deal or portion out; as nutrient medium is portioned into glassware for plant tissue culture.

DNA Ligases: Enzymes which bind DNA fragments together Donor plant (mother plant): The source that the plant used for propagation, whether an explant, graft, or cutting. Source plants used for micropropagation are usually pathogen-tested (disease-free).

Dry Ice: Frozen (solid) carbon dioxide (CO_2). It is commonly used as a refrigerant.

Dual Culture: A culture system that includes plant tissue and one organism (such as a nematode species) or microorganism (such as a fungus). Dual cultures are used to study host–parasite interactions or in the production of axenic cultures for a variety of purposes. Usually, the microorganism selected is an obligate parasite.

E

Embryo Culture: (1) The growth of any species in its embryogenic stage in an artificial medium tto enhance its survival is called embryo culture. (2) Cultures in which embryos are induced to form (embryogenesis) whether in suspension or on a variety of explants or cultures on solidified media. More correctly, these non-zygotic or somatic embryos are termed embryoids.

Enation: Outgrowth on a plant surface.

Enhancer Element: These are short regions of DNA which have the effect of increasing the levels of transcription, but unlike promotors, they affect genes from chromosomal loci at a long distance.

Environmental Chamber: A controlled environment cabinet (incubator) in which temperature, light quality, intensity, and duration; and preferably also the relative humidity and airflow are controlled.

Enzyme: Any one of a number of specialized proteins produced in living cells, that speed the rate of specific chemical reactions, even at very low concentrations (organic catalyst), but is not used up in the reaction.

Epigenetic Gene Control: This is a new concept. It describes that the control of gene expression in higher eukaryotic organisms is related to the methylation of cytosine residues in the gene-coding sequence.

Equimolar: The same amount of solute per liter of solution, the same molar concentration.

Established Culture: (1) Achievement of an aseptic viable explant, synonymous with Stage I culture (culture establishment) or transplant, synonymous with Stage IV culture (establishment in soil). (2) A suspension culture subjected to several passages, adjusted so that cell number per unit of time is constant from subculture to subculture.

Ethylenediaminetetraacetic Acid, Disodium Salt (NazEDTA, mw 372.25): A chelating agent which reversibly binds positive ions, such as iron, magnesium, etc. It is commonly added to plant tissue culture media to keep iron and other salts available by releasing them slowly into the medium as required.

Ex vitro: (Latin; from glass). Organisms are removed from culture and transplanted, generally to soil or potting mixture.

Excise: To cut, sever, or otherwise remove or extract an organ or a segment of tissue from a plant or plant part: as the surgical removal of shoot tips. The process is excision.

Explant Donor: The source plant or mother plant from which the explant used to initiate a culture is taken.

Explant: The excised plant portion is used to initiate a tissue culture. The process of dissection and removal to culture of these small organs or tissue sections is explantation. Explant choice, the timing of excision, and pre-treatment are important determinants of culture success.

F

Factor: A condition, influence or constituent to be taken into account.

Filter Sterilize: To sterilize by passing a solution through a porous material capable of separating out suspended microbes or their spores; as heat-labile components of nutrient media are sterilized.

Fixative: A compound that stabilizes, sets, or fixes other compounds or structures securely so that their structural integrity is retained; as the process of fixation utilizes chemical agents to permanently prepare cells or tissues for microscopy.

Floccule: An aggregation (coalescence) of microorganisms or colloidal particles floating in or on a liquid. Flocculation is seen in some contaminated liquid media appearing as a cloud.

Folic Acid (C19 H19N706 mw 441.40): A member of the vitamin B group (Be). It is also known as vitamin M. It is present in green leaves and has some coenzyme activities. It is occasionally added to plant tissue culture media.

Freeze-Dry or Lyophilize: The process of preserving tissue by freezing it very quickly and then subjecting it to a vacuum which removes ice.

Fusogen: A fusion-inducing agent used for protoplast agglutination in somatic hybridization studies; as is polyethylene glycol.

G

Gametoclone: A plant regenerated from a tissue culture originating from gametic tissue.

Gelatin: A glutinous, translucent, colorless, flavorless food ingredient, commonly derived from collagen taken from animal connective tissues. It is sometimes used to gel or solidify nutrient solutions for plant tissue culture. Other gelling agents, specifically agars, are usually preferred.

Geirite: The brand name of a synthetic (*Pseudomonas*-derived) refined polysaccharide used as a gelling agent and agar substitute.

Generation Time: The time between successive generations of cells or organisms within a population.

Genetic Engineering or Recombinant DNA Technology: Technology involving man-made changes in the genetic constitution of cells (apart from selective breeding). This technology usually employs a vector (such as the Ti plasmid of *Agrobacterium tumefaciens*) for transferring useful genetic information from a donor organism into a cell or organism that does not possess it. Genetic engineering has many potential uses.

Genetic Selection: Selection of genes, cells (cell selection), clones, etc. by man within populations of between populations or species. The usual purpose is to alter a specific phenotypic character. Such selection usually results in differential success rates of the various genotypes, reflecting many variables including selection pressure and genetic variability in populations Genetic transformation: The transfer of extracellular DNA (genetic information) among and between species, as, for example, with the use of bacterial or viral vectors.

Genetic Variation: Differences in individuals derived from the same genotype in distinction to differences caused by the environment. Genetic variation denotes the proportion of phenotypic variance caused by differences in the genetic makeup of an individual.

Genome: The basic haploid chromosome set of an individual; the sum of its genes.

Genotoxic: Carcinogenic; toxic to the chromosomes.

Germplasm: The genetic material of an individual which may be transferred from one generation to another by somatic or sexual means is known as germplasm.

Gibberellic Acid (gibberellin A3 or GA3, C19H2206 mw 346.37): One of the gibberellins, a group of growth hormones promoting cell division and elongation. The first of the group to be isolated and the most widely used gibberellin in plant tissue culture. Isolated from the fungal pathogen *Gibberella fujikuroi*. It dissolves in base (ca, IM KOH or NaOH).

Gyrase: It is Type-II topoisomerase that catalyzes the conversion of relaxed DNA to a superhelical form and requires ATP.

Hemicellulase: An enzyme from *Aspergillus niger*, available as a commercial preparation, that degrades hemicellulose to galactose.

HEPA Filter: An acronym for high-efficiency particulate air filter. A filter capable of screening out particles larger than 0.3 m. They are used in laminar air flow cabinets (hoods) for sterile transfer work. Heterograft or xerograft: An interspecific graft. The likelihood of success of this graft is proportional to the degree of relatedness between the donor and recipient.

Heteroplasmon or Heteroplast: A cell with a mixture of two types of cytoplasm (cytoplasmic hybrid); a cell with foreign organelles.

H

Hood: Laminar air flow cabinet.

Hydrate: To add or incorporate water. The process is hydration.

Hypertrophy: An abnormal increase in cell size causing irregular swelling or growth. Hypertrophy is usually a disease or other stress-induced response.

Hypochlorite: A salt of hypochlorous acid (sodium hypochlorite, potassium hypochlorite, or calcium hypochlorite). All are oxidizing agents used for disinfecting and for bleaching.

I

In Situ: In the natural, original place or position; as in the location of the explant on the mother plant prior to excision.

In Vitro: (Latin in glass). Experimentation on organisms or portions performed or taking place in a test tube, culture dish under artificial conditions as in tissue culture. The antonym of in vivo.

In Vivo: (Latin in life) Experimentation on organisms under natural conditions within intact living organisms; the antonym of in vitro.

Incubate: To maintain under conditions favorable for development, often in an incubator. The process is incubation.

Incubator: An apparatus providing controlled environmental conditions (light, photoperiod, temperature, humidity, etc.) suitable for (incubating) plants or plant cultures. This term is sometimes used as a synonym for cabinet (growth cabinet) but usually implies a greater degree of environmental control.

Indirect Embryogenesis: Embryoid formation on callus tissues derived from zygotic or somatic embryos, Indirect organogenesis: Organ formation on callus tissues derived from explants. the antonym is a direct seedling plant or other tissues in culture. The antonym is direct embryogenesis.

Induction Media: (1). Media that can induce organs or other structures to form, (2) Media that will cause organogenesis. variation or mutation in the tissues exposed to it.

Infestation: Occupation by threatening numbers of insects, mites, or potential disease agents such as mite infestations in incubators. An infesting organism can cause contamination of tissue cultures. This should not be confused with infected, as infection is not necessarily implied.

Infrared Gas Analyzer (IRGA): An instrument for measuring the proportion of a particular gas in a mixture such as changes in CO_2 concentration can be monitored to evaluate a plant's photosynthetic or respiratory activity.

Innoculum: Material introduced (inoculated) onto or into a host or medium or an inoculum can be defined as the population of microorganisms or cells that is introduced in the fermentation medium or any other suitable medium.

Inoculate: (1) To deliberately introduce something into; as inoculum is placed into (or onto) medium to initiate a culture. This process is inoculation, but should not be confused with contamination.

Inositol; myo, meso, or i-inositol or hexahydroxycyclohexane (C6H1206 mw 180.16): A widely distributed sterioisomeric alcohol in plants and essential in animals. Inositol is included by many sugars in the B complex vitamin group. It is involved in the synthesis of phospholipids, cell wall pectins, and membrane systems in cell cytoplasm. It is commonly added to the nutrient medium used for plant tissue culture (ca. 100 mg/1) for its growth-promoting effects.

Introns: Sequences in DNAs which are in fact transcribed into mRNAs but are subsequently split out and therefore form no permanent constituents of the mRNAs (only in eucaryotes).

Irradiance: The total radiation on an exposed surface.

Irradiate: (1) To illuminate. (2) To emit or expose to waves of light, heat, or nuclear emissions. (3) To treat by exposure to radiation. The process is irradiation. Isograft or syngraft: A graft or transplant among isogenic (genetically identical) individuals, as on the same organism.

Isolation Medium: A medium suitable for explant survival and development. It may be synonymous with Stage I medium, or contain antioxidant(s), reduced hormone concentration, bacterial indicators, or other addenda, and precedes Stage I culture.

J

Jiffy Pots: A brand name for peat pots that are sometimes used for ex vitro transplantation.

K

Kinases: They catalyze the transfer of the a-phosphate of a nucleoside 5'-triphosphate to the 5'-OH terminus of a DNA or RNA molecule.

Kinin: The original class name for substances promoting cell division to which the prefix cyto has been added (cytokinins) to distinguish them from kinins in animal systems.

L

Laminar Air Flow Cabinet or Hood: A structure in which a uniform flow of filtered air prevents the settling of particles in the work area. The air is filtered through prefilter (furnace-filter quality), then a high-efficiency particulate air (HEPA) filter that strains out particles greater than 0.3 m. These hoods are commonly employed for aseptic plant tissue culture manipulations and are designed to accommodate one to several persons, depending on the model.

Layering: (1) Covering stems, runners, or stolons with soil causing adventitious roots to form at the nodes enabling propagation by rooted cuttings. This procedure is used commercially to propagate many plants. such as the brambls. (2) In vitro layering involves the horizontal placement on agar of cultured shoots (with or without leaves) or nodal segments to promote axillary bud proliferation.

Lecithin: A naturally occurring, choline-containing phospholipid present in animal and plant tissue. Chemically lecithins are similar to fat, but they also contain phosphorus and nitrogen.

Ligases: They catalyze the formation of phosphodiester linkages between DNA chain segments.

Ligation: Direct linkage of nucleic acid fragments.

Limiting Requirement or Factor: An environmental variable whose absolute levels at a given time limit the growth or other activity of an organism.

Linear Phase: The constant increase in cell number following the exponential growth phase and proceeding to the deceleration phases in most batch suspension cultures.

Liquid Culture: The culture of plant cells on a liquid medium, in suspension, or on supports, cultures are either held steady (stationary culture) or are shaken (agitated culture or shake culture).

Liquid Medium: Medium not solidified with a gelling agent. Liquid media are used for suspension cultures and for a wide range of research purposes. They are also useful for Stage I culture in some micropropagation protocols. Usually, a support structure or wick is used to hold the tissue above the nutrient medium.

Lysis: Cell rupture or destruction, as through enzymatic action.

M

Macerase: A brand name for pectinase. Pectinase is useful in the isolation of intact cells and protoplasts of higher plants.

Macerate: To disintegrate or separate tissues through cutting, soaking, enzymatic or other action, resulting in cell dissociation.

Magnetic Stirrer: An apparatus used for stirring solutions. A magnetic stir bar is placed at the bottom of the container and is twirled in the solution that requires mixing by an electrically driven, rotating magnet below the platform of the apparatus.

Malachite green (C22H25N2Cl mw 364.66): An organic compound that is used as a dyestuff and controversially as an antimicrobial in aquaculture.

Mannose (C6H1206 mw 180.16): A hexose component of many polysaccharides and mannitol. It is occasionally employed as a carbohydrate source in plant tissue culture media.

Maternal Inheritance: Inheritance controlled by extra-chromosomal (cytoplasmic) hereditary determinants.

MDA: Microdroplet array technique.

Medium Formulation: One of many available, usually derived formulas that are used in plant tissue culture. The formulas commonly contain the microelements and microelements (high and low salt formulas are available), some vitamins (B vitamins, inositol), hormones (auxin, cytokinin, and sometimes gibberellin). A carbohydrate source (usually sucrose or glucose) and often other substances such as amino acids (the most common addition is glycine) or complex growth factors. The media may be liquid or solidified with agar, the pH is adjusted (ca. 5-6) and the solution is sterilized (usually by filtration or autoclaving).

Medium: The substrate for plant growth; as a nutrient solution, soil, sand, etc. This is a general term for the liquid or solidified formulation upon which plant cells tissues or organs develop in plant tissue culture.

Mericlinal: A meristem tip source clone. This term can be synonymous with calli clone or may not involve callus.

Meristem Culture: The culture of meristems; meristematic dome tissue without adjacent leaf primordia or stem tissue. In practice, this usually means meristem tip culture. This may more generally imply the culture of meristemoidal regions of plants or of meristematic growth (associated with or sharing the characteristics of the meristem) in culture.

Meristem Tip Culture: Cultures derived from meristem tip explants. They are excised for virus elimination or axillary shoot proliferation purposes, less commonly for callus production.

Meristem Tip: A common explant comprising the meristem (meristematic dome) and usually one pair of leaf primordia. Also refers to apical (apical meristem tip) or lateral (lateral or axillary meristem tip) origin. The term meristem tip is often confused with the term shoot tip, which is much larger and usually has more immature leaves and some stem tissue.

Meristeming: The utilization of or process of utilizing meristem tips for explants. This term is often used to refer to micropropagation via axillary shoot proliferation.

Meristemoid: A cluster of small, isodiametric meristematic cells with a meristem, or cultured tissue, with the potential for developmental (totipotential) growth. These cells have a dense microvacuolated cytoplasm and a high nucleo-cytoplasmic ratio. They may give rise to plant organs (shoots, roots) or entire plantlets in culture.

Methionine (Met, C5H11NO2S mw 149.21): An amino acid precursor in ethylene synthesis and occasionally added to plant tissue culture media.

Methyl Methanesulfonate (MMS, C2H6SO3 mw 110.13): A frequently used, very potent chemical mutagen which acts by adding methyl groups to guanine and subsequently causes base-pairing errors as binds to adenine.

Methylases: They catalyze the transfer of methyl groups from donor S-adenosyl-L-methionine (AdoMet) to specific bases of acceptor DNA molecules.

Microclimate or Microhabitat: The climate in the immediate vicinity or surrounding parts.

Microcutting: A tiny cutting; as an explant such as a meristem tip removed for culture with the use of a dissecting microscope.

Microdroplet array or multiple drop array (MDA) or hanging droplet technique: Introduced by Potrykus, Harms, and Lorz (1979), this technique is used to evaluate large numbers of media modifications, employing small quantities of medium into which are placed small numbers of cells. Droplets of liquid culture (medium and suspended cells or protoplasts) are arranged on the lid of a petri dish, inverted over the bottom half of the dish containing a solution with a lower osmotic pressure, and the dish is sealed. The cells or protoplasts form a monolayer at the droplet meniscus and can easily be examined.

Micrograft: An in vitro graft. This procedure involves placing a meristem tip or shoot tip explant onto a decapitated rootstock, that has been grown aseptically from seed or micropropagated. The purpose may vary, from virus elimination, using meristem tips and virus-free rootstocks, to viral assays or to expedite grafting that is usually done in the greenhouse. The explanted scion is placed directly on the cambium of the severed stock or inserted into a T or inverted T-shaped epidermal incision on the stock This has been used to eliminate viruses, viroids, and mycoplasmas from citrus species.

Micromole (M): One millionth (0.000001) of mole (10^{-6} M). The unit of concentration is used for microelements, vitamins, hormones, and some other organic addenda used in plant tissue culture media.

Micropropagation: Refers to propagation in culture by axillary or adventitious means. It is a general term for vegetative (asexual) in vitro propagation. It sometimes refers specifically to axillary bud proliferation.

Minimum Inoculum Size: The smallest inoculum that can be successfully used for subculture.

Mitotic index (MI): The ratio of nuclei undergoing mitosis (including prophase) to total nuclei. MI = (Number of nuclei in mitosis/Total number of nuclei examined in the sample) X 100. A MI of 0.3 means that 30% of cells in the population are observed in mitosis.

Mitotic Index (MI): The ratio of nuclei undergoing mitosis (including prophase) to total nuclei. MI = (number of nuclei in mitosis/Total number of nuclei examined in the sample) X 100. A MI of 0.3 means that 30% of cells in the population are observed in mitosis.

Mitotic Nondisjunction: Occurs when sister chromatids fail to migrate to opposite poles of the cell during mitotic anaphase. The result is daughter cells with hyperploid and hypoploid chromosome counts.

Monoculture: A one-crop agriculture system. An intensified monoculture could refer to a clonal (one genotype) agricultural system.

Mononucleotide: A nucleic acid basic building block. Composed of a pentose, a phosphoric acid plus a purine or pyrimidine base.

MS: An abbreviation for the Murashige, T. and F. Skoog (1962) medium formulation. Mutability: the propensity of an individual's genes or genotype to undergo heritable mutation.

Mutagen: A chemical or physical treatment or agent capable of producing genetic mutation. Causes insertion, deletion, or alteration of a base or part of the nucleic acid chain (DNA). Artificial mutations can be induced in cells using chemicals or ionizing radiation. The process is mutagenesis electromagnetic and particle radiation, such as UV, X-rays, and β particles all increase mutation frequency. So do chemical mutagens, such as ethyl methanesulphonate, acridine, nitrous oxide, and many others.

N

Natural Complex: A complicated, non-synthetic, often undefined addendum in plant tissue culture media. such as orange or tomato juice or coconut milk.

Neoplasm: Localized cell multiplication or tumor, a collection of cells which have undergone a genetic transformation. These cells differ in structure and function from the original cell type.

Nobecourt, J.R. (1937): was among the first to describe the prolonged culture of carrot callus.

Nucellar Embryony: The process by which individual cells of the nucellus give rise to somatic embryos (embryoids). Plants on which this occurs naturally are considered good candidates for tissue culture of somatic embryoids.

Nucleases: An enzyme capable of cleaving the phosphodiester bonds between nucleotides of nucleic acids.

Nucleo-Cytoplasmic Ratio: The ratio of cell nuclear to cytoplasmic volume. This ratio is elevated in meristematic cells and low in differentiated cells.

Nutrient Film Technique (NFT): Hydroponic plant growth whereby plant roots are suspended in shallow, slowly circulating nutrient solutions delivered as a continuous film of liquid rather than in bursts of liquid.

Nutrient Gradient: A diffusion gradient of nutrients and gases is set up in tissues where only a portion of the tissue is in contact with the medium. This is common in callus culture and causes differential growth rates and senescent regions of the callus. Similarly, diffusion gradients of nutrients are set up in the medium adjacent to the tissue. Gradients are less likely to form in liquid media.

O

Ontogeny: The developmental (ontogenetic) life history of an organism, its development from fertilized egg (zygote) to seed-forming adult (reproductive phase), with emphasis on embryonic development.

Open Continuous Culture: A cell suspension culture with a continuous influx of fresh medium, maintained at constant volume by the efflux of cells and spent medium.

Operator: Regulatory section in a nucleic acid with binding capacity for a repressor protein for blocking the synthesis of mRNAs.

Organ Culture: The growth in aseptic culture of plant organs such as roots or shoots, beginning with organ primordia or segments and maintaining the characteristics of the organ.

Organized Growth: The in vitro development of organized explants such as meristem tips or shoot tips, floral buds or organ primordia, or their de novo formation from unorganized tissues.

Organogenesis: The initiation (de novo) and growth of organs (roots and shoots, usually) from cells or tissues; as in organ culture. Organs may form on the surface of explants (direct organogenesis) or upon an intervening callus phase (indirect organogenesis).

Organoid: An anomalous organ-like structure formed in culture; as on leaves, roots, or callus.

Ortet: The original mother plant or donor plant from which vegetatively propagated plants are derived. Or in another term, an organoid is a miniaturized and simplified version of an organ produced in vitro in three dimensions that show realistic microanatomy.

Osmium Tetroxide or Osmic Acid (Os04 mw 254.20): A fixing agent used to prepare tissues for electron microscopy.

Osmolarity: The total molar concentration of the solutes affecting the osmotic potential of a solution or nutrient medium. (in vitro)

Ovale Culture: A culture derived from an explanted ovule; which may be fertilized in culture fertilization). This technique is used to study the development of zygotes and young embryos and is sometimes used to rescue embryos susceptible to abortion when embryo culture is not possible.

Oxygenase: An enzyme enabling an organism or system to utilize atmospheric oxygen.

P

Packed Cell Volume (PCV): A quantitative method of estimating cell growth. It is based on the total cell volume in an aliquot of suspension culture. The aliquot is centrifuged for 5 minutes at 200 g after which the packed cell volume is expressed as a percentage of the aliquot volume.

Panicle Culture: Aseptic culture of grain panicle segments, usually in an effort to induce microspore pantothenic acid

Papain: A water-soluble proteolytic enzyme (protease) extracted from papaya fruit and used especially as a meat tenderizer.

Paper Raft Technique: A technique developed by W.H. Muir (1953) to promote the development of single cells taken from suspension cultures in which cells are placed onto filter paper squares set on actively growing callus (nurse tissue). Growth factors and nutrients from the callus tissue diffuse through the filter paper, promoting cell growth and development.

Para-amino-benzoic acid (PABA, CH, NO2 mw 137.12): An occasional addition (vitamin Bx) to plant tissue culture media.

Parahormone: A substance with hormone-like properties that is not a secretory product; such as ethylene or carbon dioxide.

Parasexual Hybridization: Refers to genetic recombination by means other than through fertilization of germ cells (parasexual) leading to hybrid cells or individuals; as in hybrid cells or plants derived from somatic cell fusion.

Particle Radiation: Refers to particles (positively charged) and B-particles (negatively charged), electrons. Protons, and neutrons. These particles are used to produce mutant cells or organisms in plant tissue Parts per million (ppm): This term has largely been replaced by the equivalent mg/L of solution or mu1/1 for culture liquids or gases.

Passage number: The number of subculture intervals. Culture age is a function of passage number and the culture dilution ratio.

Passage time: The interval between subcultures or the culture period.

Pasteur pipette: A causal agent of a disease; as a pathogenic fungus, bacterium or virus.

PCV: Packed cell volume.

Pectinase: An enzyme that degrades pectin, the adhesive material that cement cells together. It is used alone or with other enzymes to digest the polygalacturonic acid of plant cell walls (to sugar and galacturonic acid) in protoplast production.

Peptide: A compound consisting of two or more amino acids covalently linked. Peptides with three or more amino acids are polypeptides. Periclinal chimera: Refers to a condition in which genotypically or cytoplasmically different tissues are arranged in concentric layers

Periclinal Chimera: Refers to a condition in which genotypically or cytoplasmically different tissues are arranged in concentric layers.

pH: A measurement of the degree of acidity or alkalinity of a solution. The negative log of the hydrogen ion concentration, in moles per liter (M/1). On the pH scale from 0 to 14, there is a 10-fold difference for each unit of change of pH. The lower the value, the more alkaline (more hydroxyl ions it contains). Most culture media are adjusted in the range pH 4–6 using IM NaOH or HCl.

Phenocopy: A non-hereditary phenotypic change that is environmentally induced, during a limited developmental phase of an organism, that mimics the effect of a known genetic mutation.

Phenolic oxidation: The process by which many plant species which contain phenolics blacken through oxidation when they are wounded, as in vitro, soon explantation or subculture. This may lead to growth inhibition, or in severe cases, to tissue necrosis and death. Antioxidants are incorporated into the sterilizing solution or isolation medium to prevent or reduce oxidative browning.

Phenols or Phenolics: A class of organic compounds with hydroxyl group(s) attached to the benzene ring forming esters, ethers, and salts, phenolic substances may bleed from newly explanted tissues, oxidizing to form colored compounds visible in nutrient media or on the filter wicks supporting the tissue.

Phloretic acid: A phenolic acid formed by the oxidation of a hydroxyl (OH) of 1,3,5-trihydroxy benzene. An occasional constituent of some plant tissue culture media. Photoheterotroph: A photosynthetic organism that requires an organic hydrogen source. The condition is photoheterotrophy.

Photometer: An instrument for measuring the luminous intensity of light sources.

Photosynthetically active radiation (PAR): The photon flux expressed as moles or micro insteins per meter squared per second (mol/m^2 or Em^2/s) or as watts per meter squared (Wm^2) over the wavelength range of 400–700 nm; the part of the light spectrum which is primarily absorbed by plants and used in photosynthesis.

Phytokinin: An absolute term for cytokinin.

Phytotron: A controlled environment building or chamber for studying plant growth under defined conditions.

Plant Tissue Culture: A general term encompassing the in vitro culture of plant cells, tissues, whole plantlets.

Plantlet: A shoot, sometimes exclusively a rooted shoot, growing in culture or derived from culture.

Plasmid: Annular nucleic acid in cells outside the central chromosome with a lower molecular weight than the central chromosome; bearer of nonessential additional information, for example, sex factors (F factors) and resistance factors (R ractors), plasmids and viral nucleic acids with known functions such as resistance or sequences are denoted by trivial abbreviation derived from their function or origin, for example, SV 40 for simian virus, pBR322 for a plasmid imparting resistance to antibiotics.

Plasmotype: A cell type displaying features which are expressions of cytoplasmic, rather than nuclear inheritance.

Pleiotropy: The condition in which several characteristics are affected by a single gene.

Pollen Culture: The culture of pollen grains, which germinate in vitro. Such cultures may eventually form a monoploid callus, from which shoots or embryoids develop into monoploid plants.

Poly(A) adding enzymes: They attach AMP residues, from ATP, to the 3'. 5' phosphodiester bond.

Polyethylene Glycol (PEG) or carbowax: A fusion-inducing agent (fusogen) for agglutinating protoplasts. which is used in somatic hybridization studies. This compound is also sometimes used in media as a non-metabolite osmoticum and is available in various molecular weights, ranging from ca. 200 to 6,000.

Polygalacturonic Acid: A long-chain sugar acid polymer composed of galacturonic acid and hexose subunits.

Polygenes: Systems of genes associated with quantitative character variation in which each gene individually affects the phenotype in a minor way.

Polymerases: They catalyze addition of nucleoside triphosphates to 3-OH terminals of single strands of preexisting polynucleotides (primers) with the release of pyrophosphate.

Polypropylene: A strong, flexible, transparent thermoplastic formed by the polymerization of propylene. It is used in many labware products.

Population density: Cell number per unit medium area or medium volume.

Potassium Hypochlorite (KCIO mw 90.6): A water-soluble bleaching or disinfecting agent used in plant tissue culture practices.

Preadaptation: In possession of advantageous characters enabling an organism to be well suited (adapted) to environmental conditions previously unknown to it.

Precipitin test or microprecipitin test: A serological assay in which visible particulates (precipitates) form when soluble antigen and antibody react. This precipitin reaction detects and identifies antigens.

Prefilter: A coarse filter (furnace filter) such as those used in a laminar air flow cabinet to screen out large particles before the air is forced through a filter (HEPA filter).

Premix: To mix before use; as nutrient mixtures are commercially available as dry powders of pre-weighed ingredients, which are put into solution when required.

Primary Culture: Primary cell culture is the *ex vivo* culture of cells freshly obtained from a multicellular organism, as opposed to the culture of immortalized cell lines.

Probe DNA: A radioactively lebeled (usually 32P) DNA molecule used to detect complementary-sequence nucleic acid molecules by molecular hybridization. To localize the probe DNA sequence and reveal the complementary hybridization sequence auto-radiography is often used.

Promeristem: The embryonic meristem-containing organ initials or foundation cells. Propagule: The form or portion of an organism used for reproduction or propagation; as new shoots or callus derived from explants are subdivided into propagules and re-cultured for further multiplication.

Protein Digest: The enzymatic hydrolysis of proteins to yield their building block components, amino acids, and short-chain peptides (chains consisting of two to several amino acids normally without enzymatic function).

Protein Stability: Protein, in eukaryotic cells, have very diverse half-lives ranging from a few minutes to several hours. The degradation of unnecessary proteins is dependent upon the small protein ubiquitin, which when conjugated to proteins leads to their rapid degradation.

Protoclone: Distinct phenotypic regenerates from a plant protoplast. A clone initiated from a protoplast or protoplast-fusion product.

Protocol: A sequence of activities, techniques or procedures linked to achieve a specific goal; as a schedule for protoplast preparation.

Protoplast Culture: The isolation and culture of plant protoplasts by mechanical means or by enzymatic digestion of plant tissues or organs, or cultures derived from these. Protoplasts are utilized for selection or hybridization at the cellular level and for a variety of other purposes.

Protoplast Fusion: The coalescence of the plasmalemma and cytoplasm of two or more protoplasts in contact with one another. Initial adhesion is a random process but coalescence may be promoted in various ways (induced fusion). When adhesion occurs between adjacent protoplasts during enzymatic wall degradation or between freshly isolated protoplasts in the absence of a fusion agent, it is termed spontaneous fusion.

Pure Line: All cells or individual members are homozygous for one or more characters or genes and will give rise to more cells or organisms with the character(s) under consideration.

Q

Quiescent: Quiet, at rest, not necessarily dormant, and having the potential for resumed activity; can apply to cells unlikely to divide, the non-meristematic cells.

R

Radiation: Rays of heat, light, or particles in waveform.

Ramet: The vegetatively propagated offspring of an ortet. An individual of a clone.

Reaction Norm: The range of phenotypic responses of a particular genotype in response to the environmental influences.

Reciprocating Shaker: A platform shaker with a back and forth action used for agitating culture flasks at variable speeds.

Recombinant DNA: Genetic material with novel gene sequences produced by cross-overs, chromosome reassortment, by other natural means, or through genetic engineering.

Reconstructed Cell or Recon: A viable cell hybrid, cybrid, or a transformed cell resulting from genetic engineering.

Reculture or Subculture: The aseptic transfer of a culture or a portion (division when necessary) of that culture (inoculum) to a fresh nutrient medium.

Redifferentiation: Cell or tissue reversal in differentiation from one type to another type of cell or tissue.

Rejuvenation: Treatment that leads to culture invigoration (such as subculture) or revival (dormancy breaking).

Repelcote: The brand name of a material (dimethyl dichloro-silane) used to coat glassware (soda glass); tissue culture containers and for other purposes.

Repression: Altered gene expression resulting in the failure of a specific protein synthesis.

Repressor: Regulatory protein with binding capacity for operators.

Resistance transfer factor: A plasmid, present in certain types of bacteria such as *E. coli*, that can impart resistance to antibiotics in animals exposed to them.

Resistivity: The degree of resistance (in ohm-cm or ohm-m) or interference to the flow of an electrical current or the movement of particles from place to place. A measure of water purity; the purer the water, the greater its resistivity. The reciprocal of conductivity.

Restriction Endonucleases: Enzymes, which specifically cleave DNAs at a definite information-bearing sequence of nucleotides and thereby permit insertion of foreign sequences, Restriction endonucleases are named in genetic engineering by trivial cell, for example, EcoRI from *Escherichia coli* RY 13, Hind from Haemophilus influenza H pall from *Haemophilus parainfluenza*

Reverse transcriptase: It catalyzes transcription of RNA nucleotide sequence into complementary DNA.

Rhizogenesis: Root formation and growth; as in root development de novo from callus. Rhodamine isothiocyanate or tetramethyl rhodamine isothiocyanate.

Root Culture: Isolated root tips of apical produce in vitro root systems with indeterminate growth habits. These were among the first kinds of plant tissue cultures (White, 1934) and remain important research tools in the study of developmental phenomena; mycorrhizal: symbiotic, and plant–parasitic relationships.

Rotary Shaker: A platform shaker with a circular motion used for agitating culture flasks at variable speeds.

Rotating Culture or Rotator: A wheel-like device for slowly (ca. rpm) rotating and gently cultures, usually in a vertical plane.

S

S phase: The cell cycle phase during which DNA synthesis occurs.

Schiff's Reagent: A mixture of pararosaniline. HCI (an analine dye) and sodium bisulfite ($NaHSO_3$) are used in staining chromosomes and nuclear material. Scion or cion: The portion of a bud or shoot used for grafting onto another plant (the stock).

Screen House: A metal or plastic mesh enclosure similar in function to a greenhouse but with more exposure to the elements.

Scutellum: That part of the cotyledon in Gramineae (grasses) that absorbs food from the endosperm at germination. It is very thin with a high surface area. An explant source for plant tissue culture.

Selection Culture: Utilizes difference(s) in environmental conditions or more usually in culture medium composition, such that preferred variant cells or cell lines (presumptive or putative mutants) are favoured over other variants or the wild-type.

Selection Pressure: The measure of the effectiveness of natural or experimental selection in altering the genetic composition of a population.

Selection Unit: Single cells or small clusters, units of optimum size for isolating and regenerating variants or mutants; the minimum number of cells effective in the screening process.

Semicontinuous Culture: The maintenance of cells in a culture vessel in an actively dividing state by periodically draining the medium and adding fresh medium.

Serial Float Culture: Sunderland's (1977–1979) technique of floating anthers on liquid medium and subculturing them to a new medium at several day intervals an anther dehiscence, pollen release, and development occur, increasing anther productivity.

Shake Culture: An agitated suspension culture. Usually, a flask (commonly an Erlenmeyer flask) containing the culture is attached to a horizontal or platform shaker, or agitated with a magnetic stirrer to provide adequate aeration for cells in the liquid medium.

Shakeror Platform Shaker: A platform fitted with clips for grasping flasks (usually Erlenmeyer flasks) or with surfaces suitable for attaching flasks, with set or variable speed control. It is desirable to adjust the shaking speed for gentle, even agitation of suspension cultures.

Shoot Tip Graft or Micrograft: The grafting of a very small shoot tip or meristem tip onto a prepared seedling or micropropagated rootstock in culture. Meristem tip grafting is used for in vitro virus elimination with Citrus and for other plants as an alternative to grafting in the greenhouse.

Sodium Hypochlorite (NaOH mw 74.44): A frequently used plant tissue sterilant at 0.5–2% w/v. The usual source is dilute (10–21% v/v) commercial laundry bleach, sometimes with surfactant is common, Tissue damage may occur if contact with tissue is prolonged. Thorough washing with sterile water generally follows treatment.

Somacione: A plant regenerated from a tissue culture originating from somatic tissue.

Somatic Cell Embryogenesis: The production of embryos from somatic cells of explants (direct embryogenesis) or by induction on callus formed by explants (indirect embryogenesis). These two processes may not be materially different in results.

Somatic Cell Variant or Embryoid: An organized embryonic structure morphologically similar to a zygotic embryo but initiated from somatic (non-zygotic) cells. These develop into plantlets in vitro through developmental processes that are similar to those of zygotic embryos.

Somatic Hybrid: A technique that allows the fusion of two different plants to obtain a new hybrid plant with characteristics from both plants.

Somatic Mutation: Mutation occurring in vegetative cells or tissues.

Somatic Organogenesis: The production of shoots, roots, or other organs on somatic tissues of explants (direct organogenesis) or by induction on callus formed by explant (indirect organogenesis).

Spontaneous Fusion: Uninduced protoplast fusion which may occur between freshly isolated protoplasts or following adhesion of adjacent cells during enzymatic cell wall degradation. Spontaneous variation: The variation in plant populations derived from tissue cultures not exposed to mutagens but occurring as a result of the culture conditions.

Stages of Culture (I–IV): Stage I: Aseptic explanation or establishment of the explant in culture. Stage II: Multiplication of the propagules. Stage III: Rooting of the propagules and preparation for transplant to soil. Soil IV: Establishment of Stage II or III propagules ex vitro in soil or potting mix.

Stationary Culture: A non-agitated culture. the antonym is shake culture.

Stock Plant: The source plant from which cuttings or explants are made. These are usually maintained carefully in an optimum state for (sometimes prolonged) explant use. Preferably they are certified, pathogen-free plants.

Stock Solution: A solution, usually concentrated (10–100 times the final medium concentration), of select medium constituents that re-grouped for compatibility to avoid precipitation and prepared beforehand to save time during medium preparation. Usually, they are frozen or stored in the refrigerator and portions are utilized as media are prepared.

Subculture or Passage: A culture derived from another culture or the aseptic division and transfer of a culture or a portion of that culture (inoculum) to a fresh nutrient medium. Sub-culturing is usually done at set time intervals, the length of which is called the subculture interval or passage time.

Subline: A cell line regenerated from a unique cell line of a hybrid callus colony. Supra optimum: An amount (level) greater than required; as an inhibitory concentration of an exogeneous growth factor.

Surface Sterilization: The removal of plant surface microflora prior to aseptic excision of explants. Surface sterilization is accomplished by immersion of tissue in one of many sterilants, such as calcium hypochlorite or sodium hypochlorite, hydrogen peroxide, mercuric chloride, silver nitrate, or bromine water for an empirically determined period of time.

Suspension Culture: Cells and groups of cells (aggregates) dispersed in an aerated, usually agitated, liquid culture medium. These are obtained by adding friable callus to the medium. The plant species, explant size, and nature of the cell aggregates, which may appear different during each culture growth phase. Cultures are used to study cell division, differentiation, and metabolism in secondary product synthesis or form the basis for single cell lines, callus cultures, somatic cell embryogenesis, and many other culture purposes.

Synchronized Cells: Synchronized mitosis in a group of cells in culture by natural or artificial means.

Synchronous Culture: A plant cell or microbial culture treated in such a way as to have all (or most) cells or individuals in the same stage of development or mitosis. This can be achieved in various ways including via temperature variation and nutrient limitation.

T

Ti Plasmid: A portion of the genome of the bacterium *Agrobacterium tumefaciens*, the agent involved in crown gall disease. This plasmid is a useful (experimental) vector for the transfer of genetic information into plant cells.

Totipotency: The potential or inherent capacity of a plant cell or tissue to develop into (recreate) an entire plant if suitably stimulated. Totipotency implies that all the information necessary for the growth and reproduction of the organism is contained in the cell. Although theoretically all plant cells are totipotent the meristematic cells are best able to express it.

Toxic: Poisonous; as are some chemicals (toxicants) or any substances present to excess and detrimental to normal plant function or growth.

Tracer: A substance that can be followed within a reaction or an organism; as radioactive isotopes and certain dyes.

Transcription: Conversion of a DNA (gene) information sequence into a mRNA sequence

Transduction: Introduction of foreign genetic material into a cell by means of a virus or, in the case of bacteria, by means of phages.

Transfer Room: A small room, sometimes sterilized internally by a bactericidal lamp (u.v. irradiation) and provided with clean (filtered) air, employed for sterile transfer work.

Translation Control: In certain cases, after a gene has been transcribed, the mRNA can fold back upon itself to prevent its translation. This process is known as attenuation.

Tumor-Inducing Principle (TIP): The plasmid carried by *Agrobacterium tumefaciens*, the crown gall organism. Through incorporation into the host genome, the host tissue is transformed into tumor tissue.

Tyrosine: An amino acid, sometimes used in plant tissue culture media as a source of nitrogen.

U

Ultrasonic cleaner: A device to include high-frequency vibration of materials, removing adhering substances from surfaces by mechanical action. This device is useful for cleaning glassware and for disinfecting plant material.

Ultraviolet Light (UV): Radiation with wavelength (100–400 nm) at the violet end of the visible spectrum.

Unorganized Growth: In vitro formation of tissues with few differentiated cell types and lacking recognizable structure.

Unwinding Enzymes: They catalyze the separation of complementary strands of DNA; require ATP. Valine (Val, $CHNO_2$ mw 117.1: An amino acid found in seeds and proteins). Occasionally added to plant tissue culture media. Helicase unwinds the DNA.

V

Vector: Plasmid or virus that after ligation with a foreign DNA is used to introduce new information into the cell.

Virus Elimination: Thermotherapy, chemotherapy, and meristem or meristem tip culture, used alone or in combination have been employed for the elimination of systemic viruses from plants.

Virus-Tested or Virus-Free: A plant that appears healthy and repeatedly tests negatively for the presence of one or more identifiable viruses. Such a plant may then be used as a stock or donor plant (explant source) for propagation purposes, and may be certified as virus tested (certified virus tested). The term virus-free is incorrect in most cases, as such a plant may contain one or more viruses which have not been assayed.

Vitamin B Complex: Includes a large group of water-soluble vitamins that function as coenzymes; thiamine (B_{2}) riboflavin or vitamin G(B_{2}) niacin or nicotinic acid (B_{3}) pantothenic acid (B_{5}) pyridoxine (B_{12}) biotin or vitamin H, folic acid or vitamin M (BC), inositol, choline, and others.

Vitamin B_{x} : para-amino-benzoic acid.

Vitamin C: Ascorbic acid. A water-soluble vitamin.

Vitamin D: A fat-soluble vitamin composed of a group of related steroids. Present in all plants.

Vitamin K: A fat-soluble quinone involved in photosynthetic electron transfer in plants and blood clotting in animals.

W

Water of Hydration: The amount of water chemically bound to a substance. This amount may be variable and must be taken into account when solutions of salts are prepared; as in medium preparation.

X

Xenograft: A graft of tissue taken from a donor of one species and grafted into a recipient of another species.

X-Ray or Roentgen Ray (r): Electromagnetic radiation of short wavelength produced by high-speed electrons impacting on a metal object under vacuum.

Xylose ($C_5H_{10}O_5$ mw 150.1): An aldopentose sugar (wood sugar) present in the woody tissues of many plants as polymeric xylan. It is occasionally used as a carbohydrate source in plant tissue culture media.

Y

Yeast Extract: A complex, undefined, B vitamin-containing addendum to some plant tissue culture media.

Yield Test: Productivity assessment.

Index

B

C

D

H

I

K

L

M

U

V

W

X

Y

Z

www.ingramcontent.com/pod-product-compliance
Lightning Source LLC
LaVergne TN
LVHW010627300926
849162LV00001B/88
* 9 7 8 1 7 7 4 9 1 9 7 8 1 *